ESSENTIAL ELECTRONICS
FOR PC TECHNICIANS

ESSENTIAL ELECTRONICS
FOR PC TECHNICIANS

John W. Farber

CHARLES RIVER MEDIA, INC.
Hingham, Massachusetts

Acquisitions Editor: James Walsh
Cover Design: The Printed Image

CHARLES RIVER MEDIA, INC.
10 Downer Avenue
Hingham, Massachusetts 02043
781-740-0400
781-740-8816 (FAX)
info@charlesriver.com
www.charlesriver.com

This book is printed on acid-free paper.

John W. Farber. *Essential Electronics for PC Technicians.*
ISBN: 1-58450-317-3

All brand names and product names mentioned in this book are trademarks or service marks of their respective companies. Any omission or misuse (of any kind) of service marks or trademarks should not be regarded as intent to infringe on the property of others. The publisher recognizes and respects all marks used by companies, manufacturers, and developers as a means to distinguish their products.

Library of Congress Cataloging-in-Publication Data
Farber, John.
 Essential electronics for PC technicians / John Farber.
   p. cm.
 Includes index.
 ISBN 1-58450-317-3 (pbk. with cd-rom : alk. paper)
 1. Electronics. 2. Computer technicians. I. Title.
 TK7816.F295 2004
 621.381—dc22

                          2004012809

Printed in the United States of America
04 7 6 5 4 3 2 First Edition

CHARLES RIVER MEDIA titles are available for site license or bulk purchase by institutions, user groups, corporations, etc. For additional information, please contact the Special Sales Department at 781-740-0400.

Requests for replacement of a defective CD-ROM must be accompanied by the original disc, your mailing address, telephone number, date of purchase and purchase price. Please state the nature of the problem, and send the information to CHARLES RIVER MEDIA, INC., 10 Downer Avenue, Hingham, Massachusetts 02043. CRM's sole obligation to the purchaser is to replace the disc, based on defective materials or faulty workmanship, but not on the operation or functionality of the product.

I dedicate this book to my older brother, Steve,
who left us way too soon, and my two sons,
Eric and Devon—may they always follow their dreams.

—John W. Farber

Acknowledgments

I'd like to thank several people who have helped spur me on to undertake this book project, or encouraged me to stay on course and see it completed. First, thanks to Mr. Verley O'Neal, my former Dean, and Dr. Charles A. Lindauer, PhD, my present Dean of Computers, Technology, and Information Systems (CTIS) at Foothill College, for encouraging me early on to undertake the book. And to my good friend Walter E. Miller, thanks for giving me much helpful advice and encouragement to keep going. Thanks to Mr. Jim Walsh at Charles River Media for never letting me stop until the work was done. Finally, to my son, Eric, who put up with a dad that had less time to spend with him for the duration of the writing project.

Contents

12 Alternating Current Circuits   309

Preface

WHO SHOULD USE THIS BOOK?

This book is intended for use by those looking to expand their knowledge of the electrical and electronic workings of personal computers. Students enrolled in A+ PC Certification courses will find much useful information, as well as more indepth coverage of basic electric circuits than is usually presented in the typical PC hardware course. Students pursuing the CCNA or MCSE programs will also benefit greatly form the information in this book

My primary reason for writing this book is my perception that the typical PC troubleshooting or A+ textbook either skips vital subject matter on electrical and electronics circuits and devices typically used in or with PCs, or they "dumb it down" so much that it is nearly useless. On the other extreme are those books written for electrical engineers or electrical technicians who are pursuing 2–5 year college degree programs. Textbooks designed for that audience are written way beyond what the typical PC tech or home PC builder needs or wants.

This book assumes the reader has at least a basic knowledge of PCs and is familiar with some of the more common terms used to describe PC subsystems, such as RAM, motherboards, and CPUs. Some readers may be taking other PC courses, such as MCSE, CCNA, A+ training, or PC Troubleshooting and can use this book as an adjunct to their other PC training courses.

Mathematics Level

The level of math required to solve the problems presented in the book is usually described as "college algebra." In an attempt to make it easier for students who are new to using scientific calculators, calculator button sequences are provided for many of the solutions. Often, there are multiple methods that can be used to solve problems using a calculator; the ones shown have proved to work for the author.

Where this Book Fits

This book fills a particular niche: providing a level of subject matter and instruction that properly fits with what most PC techs and those interested in working with LANs, routers, switches really need. It focuses on the electrical and electronics subjects needed by people working with PCs who are neither "dummies" nor electrical engineers.

HOW THIS BOOK IS ORGANIZED

This book is organized into eighteen chapters, starting with Chapter 1, "PC Overview, Safety, and Tools," since the reader needs to get a quick overview of what makes up a PC and how to work safely with PCs while installing, upgrading, or testing them.

Chapter 2 presents the fundamental concepts of conductors, insulators, matter and electrons, semiconductors; all the basic elements and ideas used to describe electricity, electronic devices, and circuits.

Chapter 3 describes the commonly used metric prefixes and scientific and engineering notations. It still amazes me, having taught electronics courses since 1978 and PC courses since 1992, that students have no trouble using terms like megabyte (MB) and Gigahertz (GHz) yet are not familiar with milliamps (mA) of current, megohms (MΩ) of resistance or kilovolts (kV) of voltage. Therefore, Chapter 3 introduces all those and other commonly-used terms, as well as the proper way of representing numbers in with engineering notation system.

Chapter 4 presents the essential elements of electric circuits, the source, the control element, the conductors and the load, and gives common examples of each.

Chapter 5 introduces the digital multimeter (DMM) and explains its use in typical PC troubleshooting applications.

Chapter 6 presents resistors, common electrical devices used for a variety of applications.

Chapters 7–9 present the three basic electric circuit forms: series, parallel, and series-parallel circuits, with lots of solved examples for each type circuit.

Chapter 10 describes many of the ways a PC uses magnetism and electromagnetism in motors, hard and floppy drives, fans, and more.

Chapter 11 presents two more basic electrical circuit devices: inductors and capacitors, which are commonly used in PC power supplies, motherboards, and adapter cards.

Chapter 12 presents the concept of alternating current (AC) and describes the reasons it is the preferred type of electricity for long-distance electric power transportation and distribution. It also shows common AC circuits used in PCs, such as filters.

Chapter 13 presents transformers, another type of extremely useful device used to perform many functions at once.

Chapter 14 describes power supplies in their various forms, and specifically, the types used in PCs.

Chapter 15 gives the reader a brief history lesson in the development of amplifying and switching devices, beginning with vacuum tubes and ending with the current use of microprocessors.

Chapter 16 presents the basic concept of radio or wireless transmission and reception and gives examples of both wired and wireless equipment used with PCs.

Chapter 17 presents the essential elements of building electric power wiring, as well as some of the more common wiring faults and their remedies.

Chapter 18 is intended primarily as a summary of the book as a whole, although some new material is also presented, so don't make the mistake of skipping this chapter. Because students are "saturated" with information by the time final exams rolls around, they certainly could use a good refresher, but enough little tidbits of new material are included to keep the reader from becoming bored.

Key Words

Be sure to study the key words listed at the end of each chapter, and be sure you understand their meaning. If in doubt, go back and re-read the chapter until you feel confident you understand all the terms.

Exercises

Always work the exercises and the end of each chapter. Answers to the odd-numbered problems are in the back of the book. Instructors will have access to all the exercise answers via the publisher, *www.charlesriver.com.*

Appendixes

The Appendixes include vital tables and charts listing color codes and common parts values for copper wire, resistors and capacitors, electronic formulas, and a guide to motherboard manual interpretation. Appendix A describes typical jobs one can achieve in the field of electronics, as well as some of the typical hobbies involving electronics.

The Accompanying Laboratory Manual

There is an accompanying laboratory manual available from the publisher with activities using test equipment and PCs. It is highly advisable to use the lab manual along with this book if it is being used as a classroom text, as "learn by doing" is a great way to master a skill. Self-taught students are also encouraged to work the exercises in the lab manual to hone their skills.

STUDY TIPS

To the classroom student: Courses on computers and electronics require learning many new terms and concepts, which requires careful and repeated study of the text and handout materials and requires your complete attention in the classroom. Be sure to do all the end-of-chapter exercises in the book. The answers to the odd-numbers problems are in the back of the book. Don't allow distractions such as TV or radio when you study. An excellent way to improve your comprehension of the material is to meet with other students before class, in the library or student center, to go over important concepts. Remember, instructors can only guide you and aid in your learning, but you must do the learning.

To the self-study student: Glance through the entire book first, and decide which areas you need to review or study intently, then carefully read those chapters. Be sure to do all the end-of-chapter exercises. The answers to the odd-numbers problems are in the back of the book.

PHOTO CREDITS

Unless otherwise specifically noted below a photograph, all photographs taken by the author.

Photo Equipment Used

The author used the following to take the photographs in this textbook:

Camera: Nikon D-100

Lenses: AF Nikor 105 mm, 1:2.8 D, AF Nikor 24–120 mm, 1:3.5–5.6 D

Flash Units: Nikon SB-50DX, Nikon SB16 with remote trigger, Sunpak Auto 411 with remote trigger, custom small flash unit with remote trigger, several incandescent spot lamps.

Post-processing: Photoshop Elements 2

PC: Author-built AMD-2400, A-bit KD7 motherboard, 512 Mb DDR RAM, RAID 1+0 four 80 MB Maxtor drive system.

1 PC Overview, Safety, and Tools

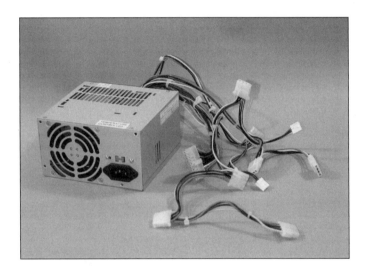

B efore starting the study of electric circuits in detail, it will be helpful to get a good overview of a typical PC and the major subsystems within it.

1.1   PC OVERVIEW

Modern PCs are machines that are capable of performing amazing things. Surfing the Internet, managing stock portfolios, playing games with people across town, across the country, or even in other countries, using e-mail, doing scientific research, exploring our genealogy, printing out maps to nearly any location—all these things are possible using a PC today. They are marvelous devices and are extremely complex. They all run on electricity and use sophisticated electronic devices to perform the magic that is increasingly taken for granted in our daily lives.

1.2   THE SYSTEM UNIT

The large metal box housing the main part of a PC is called the *system unit*, or some (such as folks at IBM®) refer to it as the *CPU* (Central Processing Unit). The author prefers the term system unit, since to most people the term CPU means the microprocessor inside, located on the motherboard. Inside the system unit are most of the subsystems that make up a PC: the power supply, motherboard (main board or system board), expansion cards, memory, various drives, possibly a single small speaker, cooling fans, and so on. Notebook and laptop PCs are all-in-one affairs that include built-in keyboards, pointing units, and even speakers. They can be considered to be system units as well.

This chapter provides a brief overview of the important subsystems inside the system unit as well as the most commonly used peripheral units used with a PC. Figure 1.1 shows a typical PC system unit.

FIGURE 1.1   A typical PC system unit.

1.2.1   The Power Supply

The PC's power supply must accept a relatively high-voltage *alternating current* (AC) ranging in different countries from 100 to 240 V (volts) provided by the electric power companies, and convert it into much lower voltage *direct current* (DC), which operates the electronic devices inside. The power supply delivers several different DC voltage levels and they must all stay very close to the correct values. The supply provides + and –12 VDC, + and –5 VDC, +3.3 VDC and +5 VDC standby. Voltage variation outside the accepted standards (usually +/–1%, or +/–5% depending on the quality of the supply) can cause data errors or system crashes. The power supply must provide these well-regulated DC voltages despite changes in the load currents inside the PC, and it must also hold the output voltages steady even if the AC input voltage from the wall outlet changes. A more detailed explanation of PC power supplies can be found in Chapter 14, "Power Supplies." At this point you should be aware that the power supply runs all subsystems inside the system unit and is vital to the proper operation of any PC. You should also know that different voltage levels, one voltage level for a 1 and another voltage level for a 0, represent the digital 0s and 1s in a PC. Refer to Figure 1.2 for a picture of a typical PC power supply.

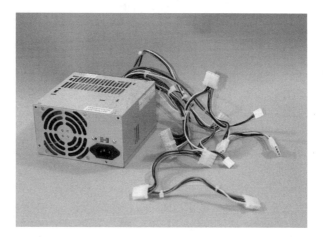

FIGURE 1.2   A PC power supply.

The United States, Canada, and a few other countries use 120 V, while much of the rest of the world uses 220–240 V. More information is available at http://kropla. com/electric2.htm.

1.2.2   The Motherboard

The motherboard (AKA main board or system board) is the largest circuit board in a PC. It contains the microprocessor(s) with heat sink, solid-state random access memory (RAM), support chips, Basic Input/Output System (BIOS), backup battery, plus most of the features that were once added to a PC only by installing expansion cards. Many motherboards today include built-in modems, sound, video, Ethernet, SCSI host adapters, RAID, support for fast I/O such as USB 2.0, Firewire, and other options. These highly integrated motherboards reduce the parts count (fewer individual chips) and increase the reliability of the product. Refer to Figure 1.3 for a picture of a PC motherboard.

FIGURE 1.3   A PC motherboard.

1.2.3   The Microprocessor

Most people are familiar with the company names Intel® and AMD®, the largest producers of microprocessors used in PCs today. There are other players in the game as well, to a lesser extent. IBM®, Motorola®, Transmeta®, VIA Technologies®, and others also make and sell microprocessors. The vast majority of PCs, however, use chips made by Intel and AMD. The bulk of computing power is performed by the main microprocessor, so it has a major effect on the overall speed and efficiency of the PC. Commonly used indirect gauges of microprocessor power are the *clock speed* (in cycles per second) and the *data bus width* (number of bits transferred at once along conductors) of the processor. Figure 1.4 shows a microprocessor.

FIGURE 1.4 A typical PC microprocessor.

Processor Speed

The speed at which microprocessors operate is measured in repetitions of a regulating signal, called the *clock signal*, per second. The clock signal synchronizes operations in the PC. The unit used for repetitions per second is the Hertz (Hz). Microprocessors today operate at millions or billions of repetitions per second. The units used to express these speeds are in a metric form. Millions of repetitions per second is commonly represented as *MHz* (megahertz) and billions of repetitions per second is commonly represented by *GHz* (gigahertz). Refer to Chapter 3, "Electrical Terms, Notations, and Prefixes," for a more thorough presentation of the commonly used metric prefixes.

Processor Width

Microprocessors are built to operate on, or handle a certain number of digital bits at once. This is called the native *word size* of the microprocessor. Special circuits inside the microprocessor called *registers* store, manipulate, and transfer the data and instructions represented by these bits. The registers that are used to do most of the manipulation of digital numbers are called *accumulator* registers. The word size of a microprocessor usually relates directly to the size of the processor's registers. In general, a microprocessor that can handle a larger word size (more data bits) is more capable than one that can handle a smaller word size. But in reality, many factors contribute to the overall processing speed. Microprocessors are therefore referred to as being an "8-bit chip," a "16-bit chip," a "32-bit chip," a "64-bit chip," and so on, based on the chip's native word (and major registers) size.

1.2.4 **The Chipset**

The so-called *chipset* usually consists of one or more *VLSI integrated circuits* (Very Large Scale ICs), which support the main microprocessor. The main microprocessor is what is called a general-purpose processor. The chipset integrated circuits are usually processors designed to do very specific and limited jobs. Typical jobs handled by the chipset include interfacing all the drives and memory with the main microprocessor, and handling communications with the main memory without forcing the memory contents to pass through the main processor. This last process is called *DMA* (Direct Memory Access) and greatly speeds overall processing, since it frees the main microprocessor from having to do this job. Interfacing with the keyboard and performing most timing functions are other jobs performed by the chipset. Chipsets are the bridge between the native microprocessor bus (also called the local bus) and other external busses such as the ISA, PCI, AGP, memory, SCSI bus, and so on. These busses in turn connect to the actual peripherals. Some modern chipsets include sound and video; all these functions previously required the installation of dedicated expansion cards. Figure 1.5 shows a typical chipset.

FIGURE 1.5 A motherboard chipset.

1.2.5 Memory

Memory is the temporary storage area in the PC. There are two main types of memory, *RAM* and *ROM*. These terms are commonly misunderstood, so they should be examined in more detail. When asked, most beginning students taking PC courses will describe their view of RAM as just "random access memory." While true, this simple answer is not complete. *Random access* means being able to access any particular memory location just as easily (quickly) as any other, nothing more, nothing less. But what operations can be performed once a given memory address (location) is found? The contents can be both examined and copied (known as a *read* operation), and the contents can be changed (called a *write* operation). Memory able to allow both these operations is called Read/Write (R/W) memory. So the term RAM, as applied to the type of RAM used in a PC, usually refers to random access, read/write memory. In addition, this type of memory will lose its contents if power is lost, so it is also known as *volatile* memory.

NVRAM stands for *nonvolatile memory*, a type that will not lose its contents when power is removed from the chip. It is used to hold *firmware*—software held in hardware—that is always needed by the PC for booting and other jobs such as input/output (I/O) routines. A common physical form for this type memory is a chip called an *EEPROM* chip, which stands for *Electrically Erasable Programmable Read-Only Memory*. PCs use EEPROMs to hold both the system setup information, commonly called the *CMOS* setup information, as well as the BIOS code. BIOS is the Basic Input/Output System. It is software stored in the EEPROM.

The second type of memory is called ROM. Again, most students taking PC courses, when asked, will state that *ROM* stands for Read-Only Memory, which is again, a correct, but incomplete definition. *Read-only* means once a memory address is accessed; the memory contents of that location can be examined, but not changed, hence the term read-only. But what about the ease (speed) involved with going to a given ROM address? It turns out that ROM is also random access. So ROM is really random access, read-only memory. The author was always surprised at these acronyms for memory types. RAM stands for how easy a given location can be accessed, but does not describe the operations possible once the location is accessed. ROM, on the other hand, describes what operation can be performed (read-only) once an address is accessed, but not the access speed relative to any other ROM memory address. It can be rather confusing. Figure 1.6 shows some typical RAM modules.

FIGURE 1.6   RAM modules.

1.2.6 **Drives**

A typical PC has one or more hard disk drives (hard drives) as well as at least one optical drive. These drives are used for storage of programs and data. Optical drives include CD-R, CD-RW, DVD, and combinations of CD-RW and DVD. Current CD formats can hold approximately 700 MB per disk. Newer DVD drives can also write as well as play DVDs. DVD capacity is around 4.7 GB with newer, higher capacity drives on the horizon. Most new PCs include a single 3½ inch, 1.44 MB floppy drive as well. Figure 1.7 shows a selection of hard drives. Figure 1.8 shows some optical drives.

FIGURE 1.7   Examples of hard drives.

FIGURE 1.8 Some optical drives.

1.3 PERIPHERALS

Peripherals include all the things that plug into a PC, such as mice, keyboards, printers, plotters, modems, hubs, switches, routers, video displays, joysticks, and so on. All of these are electronic devices as well. Many have their own power supplies. Some draw power from the PC, so a PC technician needs to understand how they operate. Figure 1.9 shows a keyboard, mouse, and joystick.

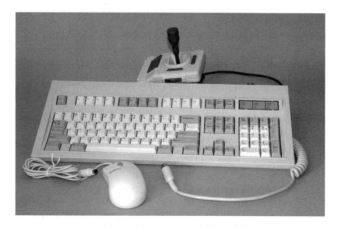

FIGURE 1.9 Keyboard, mouse, and joystick.

1.4   SAFETY

Safety practices should be learned early and always adhered to when working with any electrical device, including personal computers and peripherals. This is for the protection of not only the people working with them, but also for the devices themselves.

1.4.1   Power Supplies

Because the power supplies used in modern electrical and electronic devices are sometimes designed to provide a very large potential difference (high-voltage) to circuits, technicians need to become aware of the general safety rules involved with such equipment. Too many times someone is injured or killed by not paying proper attention to details when working with electronics. Never assume that a circuit is de-energized—always make sure it is not. Never take someone else's word that a circuit has no power applied; check it yourself. Some types of devices used in electronics can store a lethal charge for hours or even days after the circuit has been turned off and unplugged. Learn how to safely discharge capacitors before ever attempting to work around them. Electrolytic capacitors can explode if installed in a circuit backwards. Applying an excessive voltage to an electrolytic capacitor can also cause an explosion. Electrolytic capacitors that have not had a voltage applied to them for a few years time should be *formed*. This means to apply a lower voltage than they are rated for, and to slowly increase the voltage over time, in order to allow the capacitors to form a new layer of oxide between the plates. This will prevent destruction of the capacitor due to a rapid gas pressure build-up as the oxide layer rapidly forms again. The reality is that few, if any, PC techs bother to form caps today. A PC's motherboard contains several dozen or more electrolytic capacitors. Figure 1.10 shows electrolytic capacitors mounted on a PC motherboard.

The presence of the AC line voltage in a plugged-in PC is a very real danger. In addition, modern PCs have voltages present on the motherboard even when the PC is turned "off." The power supply provides standby power for *local area network* (*LAN*) and modem cards so they may "wake" the PC. This capability is termed *WOL* (*Wake On LAN*) and *WOM* (*Wake On Modem*), respectively. The standby voltage is present to allow faster booting as well, since the "on switch" is now a simple low-current pushbutton that temporarily signals the power supply to provide all the other voltages to the PC as well. PCs can even be programmed in the CMOS to allow a single keystroke to turn the machine on. The ATX power standard allows for this.

1.4.2   Battery-powered Circuits Can Be Dangerous!

Many people mistakenly assume that circuits employing batteries cannot hurt a person. Such is not the case. Even a few volts are sufficient to stop a person's heart,

FIGURE 1.10 Electrolytic capacitors on a PC motherboard.

if the current through the body is high enough. Conditions that affect how well the human body will resist current are moisture and whether the contact is made on or under the skin. (See Chapter 3 for more on voltage.) Remember, the human heart is an electrically operated organ; too much electricity will permanently damage or destroy it. Batteries placed in your pocket can short out to keys or coins and rapidly discharge, causing burns. Always be aware of this danger when transporting batteries and wrap or enclose them with a layer of insulating material to prevent inadvertent discharge.

Motherboard Batteries

Motherboards contain small batteries, called *backup* batteries, that keep the configuration settings alive in *CMOS* (a list of important configuration settings). These batteries can last for several years or more, due to the extremely low power draw from the CMOS BIOS chip kept "alive" by it. Never place a motherboard on a sheet of aluminum foil, as this can short out the battery and cause damage. Refer to Figure 1.11 for a picture of a motherboard battery.

Battery Disposal

Electric cells and batteries contain heavy metals such as lead, nickel, cadmium, mercury, and other potentially dangerous chemicals. Never throw batteries away in the trash; take them to a proper recycling center. Some communities have centralized recycling facilities, and many automotive parts stores are now accepting

FIGURE 1.11 A motherboard backup battery.

batteries for recycling for free or a small handling charge. Your local RadioShack® store may also take rechargeable batteries for recycling.

1.4.3 The One Hand Rule

There is an old safety rule from the early days of electrical work that is still very useful today. It is called the "one-hand-in-your-pocket" rule. It states that whenever working on electric circuits; always assume that the circuit is live. Keep one hand in your pocket to prevent electricity from flowing across the chest area of your body should you accidentally come in contact with a live circuit. The idea is this: most electrocutions take place when the person's hands grasp points of potential difference. The resulting current flow through the heart area causes injury or death. Preventing the current from flowing directly across the chest area (the heart area) lessens, but does not eliminate, the danger of electrocution.

1.4.4 Safety in Laboratory Exercises

A good general rule to follow during the construction and use of laboratory exercises involving electricity is this: if in doubt about a hook-up, ask the instructor. Never apply power to a circuit that has not been double-checked for correct parts placement and correct component values. Most lab problems involve incorrect parts values used in their construction, or improper wiring, due to a hurried job of circuit construction. Take the time to do it correctly the first time. This will surely pay off in ease of getting the circuit to work the first time power is applied. It will also reduce the frustration that can easily set in if a student has difficulty making a lab circuit operate correctly. Always try to reason out why a circuit does not work properly. Do not simply wiggle wires and components trying to achieve a "magic" fix for a problem. There is always a logical reason for a circuit malfunction.

The industry term for those with the skills to reason out a circuit problem and quickly repair it is *troubleshooter*. A good troubleshooter uses a combined knowledge of electronics with a logical approach to problem solving as well as common sense to earn very good money in today's job market.

1.4.5 Electrostatic Discharge

Electrostatic discharge (ESD) is the equalizing of electric charges between two bodies, such as a person and a PC. This process causes a spark, which depending on the voltage involved, may or may not be perceived by the person touching the object. It is common for a person walking on a carpet to accumulate a static charge of several thousand volts. This results in a current flowing to equalize the charge differential when another object is touched. The problem this causes when working on PCs is due to their extreme sensitivity to only a few volts (as little as 30 volts can damage many chips used in PCs). Microprocessors, memory, and most of the chips in a PC can be damaged by ESD caused by improper handling. Compared with moist air, dry weather is more of a problem for the buildup of static charges, and will allow a larger charge to accumulate on objects. A person working on a PC or handling circuit boards should always observe proper ESD precautions. Commonly used safety devices designed to minimize or eliminate static damage include ESD wrist straps and conductive desk and floor mats. An *ESD wrist strap* consists of a conducting band that is worn around the wrist, which is connected via a coiled cord to a clip that is connected to the metal PC chassis. The cord connecting the wrist strap and the PC has a high-value resistor of 1 M (1 million ohms) inside to limit the current flow, which helps to minimize the chance of electrocution should the ESD wrist strap wearer come into contact with a live circuit. A PC being worked on should be unplugged from the wall socket. Modern PCs have standby voltages present on the motherboard whenever the PC is plugged in, whether the PC is turned on or not. The utility company-provided line voltage is also present on some connections inside the power supply. The idea of "grounding" in this case merely means connecting the person and the PC together electrically so they will be at the same electrical potential. It has nothing to do whatsoever with a connection to the earth. In fact, if connected to the earth, the PC presents a very real shock hazard if a person touching the grounded PC also comes into contact with a live circuit.

ESD mats are used to protect boards and other devices that have been removed from a PC—place the removed devices on the mat. The mats are usually also connected to the PC chassis. Many commercial manufacturing, repair, and test facilities use grounded floor mats as well, to minimize ESD damage. An inexpensive ESD wrist strap should be part of every PC technician's tool kit. Decent quality ESD wrist straps can be purchased separately for $5 or so, and desk mats can be had for $15. Avoid the "one time use" wrist straps, as they are very flimsy and don't last long. Refer to Figure 1.12 for a picture of an ESD wrist strap and conductive mat.

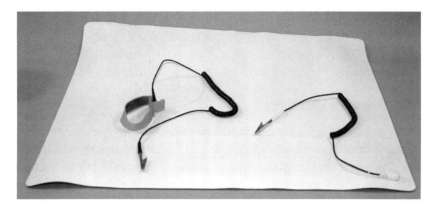

FIGURE 1.12   ESD wrist strap and conductive mat.

1.4.6   AC Outlet Tester

Another very useful safety item is an AC circuit outlet tester, available for under $10. It can be used to determine if an AC wall outlet is properly wired. It can indicate if outlet wires are not connected, reversed, and so on. Improperly wired outlets can cause shocks or electrocution as well as all sorts of random computer lockup problems. Refer to Figure 1.13 for a picture of an AC outlet tester.

FIGURE 1.13   An AC outlet tester.

1.5   PC HAND TOOLS

The most commonly used PC repair tools can be found prepackaged in any good electronics or computer store. Many experienced PC techs prefer to add some tools to the basic kits, however. The following tool lists will aid in assembling your own set of PC troubleshooting tools.

1.5.1 Basic Tools

- A flat head screwdriver about 6 inches long.
- A #2 Phillips' head screwdriver about 6 inches long.
- A small parts grabber tool, which is useful for retrieving dropped screws and other small parts inside a PC. A magnetized parts retrieval tool is also a good choice to include in this tool category.
- A pair of needle nose pliers about 6 inches long.
- A small, inexpensive flashlight. A type that will give very good service and still fit inside a small tool kit is one taking two "AA" cells. A type offering longer battery life is a two-AA cell light using four or more light emitting diodes (LEDs).
- An antistatic electrostatic discharge (ESD) wrist strap. Units costing $5–10 are fine. Avoid the "disposable" types.
- A small digital voltmeter (DVM) or digital multimeter (DMM). Useful units can be had today for $20 or less and are small enough to pack in one's PC repair toolkit.
- An ESD mat. Useful for safe temporary placement of removed components.
- Several antistatic plastic bags for storing removed boards and other small, static-sensitive parts.
- A small plastic or metal box or bin for temporary storage of small parts such as screws.
- A small toolbox or bag to house all the tools.
- A spare-parts kit consisting of mounting screws for drives, cover screws, extra jumpers, a power "Y" cable, and so on.

Figure 1.14 shows a basic PC repair tool kit.

FIGURE 1.14 A basic PC repair tool kit.

1.5.2 Additional recommended tools

Most savvy PC technicians will want to add most of the following tools to the basic kit.

- An AC circuit outlet tester (discussed earlier in this chapter).
- An extendable mirror. Used to view connectors on a PC's rear panel.
- An extendable parts grabber.
- Electric screwdriver and bit set. Used to assemble and disassemble PC cases and install/remove boards.
- A flexible shaft for use with an electric screwdriver. Available at many auto parts stores.
- A sharp knife, such as an X-ACTO® knife set. Useful for trimming plastic bezels, and for other uses.
- A pair of needle nose pliers about 12 inches long.

Refer to Figure 1.15 for a picture of some optional PC repair tools.

FIGURE 1.15   Additional recommended PC repair tools.

1.6   SUMMARY

- The main box housing the majority of PC subsystems is called the system unit.
- The PC's power supply converts 120 VAC or 240 VAC into several DC voltages required by the PC. Older PC power supplies provided +/–12 VDC, and +/–5 VDC. Modern PC power supplies provide these voltages as well as +3.3 VDC and +5 VDC (the 5 VDC is the *standby voltage*—it is always on).
- The motherboard, also known as the main board or system board, is the largest circuit board in the PC and contains the main system processor, RAM, BIOS, backup battery, and the chipset.

- The chipset consists of one or more highly integrated VLSI chips that support the main system processor. They have evolved to replace many smaller, less-integrated chips used in the past.
- Motherboard batteries can be dangerous, so proper safety precautions should be taken when handling them.
- The "one-hand-in-your-pocket" rule is designed to help prevent current flowing through a person's heart in case that person comes into contact with a live circuit.
- When performing in-class laboratory exercises, be sure to follow all safety procedures, and when in doubt, ask the instructor before proceeding. As you gain valuable experience diagnosing and repairing problems, you will become a PC troubleshooter.
- Always consider ESD and take proper precautions when handling PCs or circuit boards, RAM, processors, and other chips. Use of antistatic wrist straps and conductive mats is the standard accepted procedure when assembling, repairing, or upgrading PCs.
- Using an inexpensive AC outlet tester is a good way to test for proper AC outlet wiring.
- A basic set of PC repair hand tools is required to service most problems normally encountered in a PC.
- Additional tools will enhance the technician's ability to quickly repair PCs.

1.7 KEY TERMS

System Unit	Word size
CPU	Registers
Microprocessor	VLSI
Clock Signal	Chipset
Power Supply	DMA
AC	Read
DC	Write
Motherboard	Formed
MHz	WOL
GHz	WOM
RAM	CMOS
BIOS	Troubleshooter
ROM	ESD

1.8   EXERCISES

1. The system unit is . . .
   a. the large box containing most of the PCs subsystems
   b. the microprocessor
   c. the complete PC including all peripheral components
   d. the box the PC arrives in from the shipper
   e. None of the above

2. A CPU is . . .
   a. the large case containing the PC and drives
   b. the processor on the motherboard
   c. what IBM calls the main PC case and everything inside
   d. both b and c
   e. none of the above

3. Clock signals are used to . . .
   a. keep track of how long a PC has been in service
   b. keep track of the number of PC running hours
   c. maintain synchronization in a PC
   d. keep the actual time
   e. both c and d

4. AC stands for . . .
   a. always connected
   b. automatic control
   c. alternating current
   d. asymmetric clock
   e. none of the above

5. The power supply inside a PC must do which of the following?
   a. maintain very constant AC levels inside a PC
   b. maintain very constant DC levels inside a PC
   c. correct for some variation in the AC from the wall socket
   d. correct for varying load currents
   e. b, c, and d

6. Millions of cycles per second is known as . . .
   a. kilohertz
   b. megahertz
   c. gigahertz
   d. millihertz
   e. none of the above

7. The type of RAM used in a PC is . . .
   a. random access
   b. read-only
   c. read/write
   d. both a and c
   e. none of the above

8. The computer word size is the number of bits normally operated on by the processor in its native mode.
   a. true
   b. false

9. A processor's register size usually matches the computer's word size.
   a. true
   b. false

10. Electrolytic capacitors must be treated carefully. Excessive voltage can cause them to explode.
    a. true
    b. false

11. WOL and WOM refer to methods to do what?
    a. Modify the BIOS
    b. Install new processor chips
    c. Increase the airflow inside a PC case
    d. Allow incoming traffic to wake a PC
    e. none of the above

12. CMOS refers to . . .
    a. the chip that stores configuration settings
    b. the same thing as BIOS
    c. the microprocessor chip
    d. the ROM chips
    e. none of the above

13. A person who can efficiently diagnose and repair problems is called a . . .
    a. troubleshooter
    b. fixer
    c. repair whiz
    d. system engineer
    e. system administrator

14. ESD is a bigger problem . . .
    a. during dry weather
    b. during wet weather
    c. when working with ICs using a wrist strap
    d. when working with ICs using a conductive mat
    e. none of the above

15. VLSI stands for . . .
    a. a type of BIOS chip
    b. the PC's clock signal
    c. a Very Large-Scale Integrated circuit
    d. all of the above
    e. a new standard of I/O

2 Basic Concepts of Electricity

2.1   INTRODUCTION

Before beginning the study of electronics, a couple of terms you will be reading many times throughout this text should be defined: *electrical* and *electronic*. *Electrical* usually means the large-scale generation, distribution, and use of electric power. *Electronic* usually means controlling electrical energy on a much smaller scale.

Examples of electrical systems include the large generators installed in a hydroelectric power plant, the very large transformers used to raise the electrical "pressure" (voltage) of the generated electricity, the "high-tension" (high-voltage) power lines that transport the electricity across long distances to the power substation, and the power lines that connect this power to your home or business. These are examples of common electrical systems and subsystems.

The wiring in your home or business is considered an electrical system. The workers who install the wiring in a home, place of business, or in the military are known as *electricians*. They are concerned with installing and maintaining electrical systems.

The realm of electronics includes such things as communication, including radio, TV and telemetry; data systems using computers and their peripheral equipment; and sensors, microchips, and most small components used in any large electrical system. There is a great emphasis toward understanding the operating principles (theory) of existing electronic devices. Research and development (R&D) electronics workers also help develop and invent new applications for existing devices and circuits. The advanced areas of electronics are concerned with the physics of materials used to construct new devices. These devices include the control, data storage, and protective components used in electronic systems. In the past fifty years, electronics has undergone dramatic changes in devices and the materials used to construct them. Materials such as germanium, silicon, selenium, gallium arsenide, and others have greatly supplanted the use of the materials once used to produce early vacuum tubes and other parts. Vastly increased use of semiconductors (also known as *solid-state* components) as data storage devices has completely eliminated the use of vacuum tubes for this application. Except for television picture tubes, older oscilloscope tubes, and CRT-based computer monitors, and tubes used for the high power transmission of radio and TV signals, lighter, more rugged, sometimes cheaper, and generally more reliable semiconductors now perform the vast majority of the jobs formerly performed by vacuum tubes. Some high-end audio amplifiers and guitar amplifiers are still made with vacuum tubes. The "audiophiles" (sound lovers) who purchase these swear they sound "sweeter" and have less distortion than amplifiers made using solid-state components.

As in the study of any subject, it is necessary to develop a solid foundation of the basic terms and principles involved in electricity and electronics. So it is necessary to learn many new terms used to describe these ideas, devices, and circuits.

2.2   WORK AND ENERGY

The concepts of work and energy are closely related. *Energy* is the capacity, or ability to do *work*. Energy can be expressed in many forms, such as mechanical, thermal, nuclear, chemical, electrical, and electromagnetic. Energy is expended at a rate that can be specified in various ways. The standard unit of energy is the *joule*. The rate that energy is expended (work is done) is called *power*. The unit of power is the *watt*, which is a rate of energy expenditure of 1 joule per second. The units specified for the expenditure of energy depend on their application. In electrical and electronics applications, electron volts, joules, watt-hours, and kilowatt-hours are used. Some of these terms will be described further in subsequent chapters.

2.3   CONSERVATION OF ENERGY

The principle of *conservation of energy* is as follows: energy can neither be created nor destroyed, but only changed from one form to another. Devices and systems that operate on this basic principle are common in electrical and electronics work. An ordinary light bulb converts electrical energy into heat and light energy. An automobile battery converts chemical energy into electrical energy, and so on. *Transducer* is the technical term for a device that responds to one form of energy and converts that energy into another form. A more commonly used term for a transducer is *sensor*. Transducers include devices such as loudspeakers (conversion from electrical to mechanical, or sound energy), microphones (mechanical to electrical), strain gauges (mechanical to electrical), thermocouples (thermal to electrical), and many others. The overall idea is that the amount of energy applied to a transducer is exactly equal to the amount of energy (in its new form) produced by the device, plus all unwanted losses such as heat energy. In the case of an incandescent light bulb, heat is the unwanted byproduct of the conversion. Heat is usually the price that must be paid due to the less-than-100-percent conversion efficiency of transducers.

Conversion efficiency (Figure 2.1) indicates the amount of useful energy output compared to the energy input of a device or system. It is equal to dividing the total energy input to a given system or device in question by the intended energy output from the system. This ratio is then multiplied by 100 to give the answer as a percentage. Some electrical devices are quite efficient. Transformers (covered in detail in Chapter 13, "Transformers"), which are devices generally used to increase or decrease alternating electrical voltage, can approach 100 percent efficiency. Many of the semiconductor components used in radios and TVs are roughly 50 percent efficient. Solar cells, semiconductor devices producing electricity from light, are only about 15 percent efficient.

As an example, assume that a certain electrical device consumes 100 units of energy. The energy output in the desired form from this device equals 85 units. What is the conversion efficiency of the device?

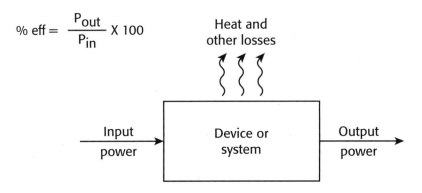

$$\% \text{ eff} = \frac{P_{out}}{P_{in}} \times 100$$

FIGURE 2.1   Conversion efficiency.

Notice there are no units used to express conversion efficiency. The units cancel out in the calculation. Efficiency is nearly always expressed as a simple percentage.

2.4   MATTER AND ELECTRONS

As you probably recall from other science courses, matter is the physical "stuff" the world is made of. Matter consists of all the known materials such as elements and compounds. Matter can exist in four possible states: solid, liquid, gas, and a highly ionized state known as the *plasma* state. An *ion* is an atom with a net electrical charge. Atoms that have either gained or lost their normal number of electrons will have a net electrical charge. All matter is made up of atoms. From chemistry and physics classes you already should know that atoms are made up of smaller constituent parts. The center of the atom is called the *nucleus.* The atomic nucleus consists of at least one *proton*, a very small particle with a *positive* electrical charge. Some atoms also have one or more *neutrons* in their nucleus. A neutron has approximately the same mass as a proton, but with no apparent electrical charge. A relatively very large distance away (on the atomic level) from the atomic nucleus, there exists one or more *electrons*. Electrons are very much smaller than a proton or neutron, but carry an electrical charge that is equal to but opposite from that of a proton. An electron's electric charge is considered *negative*. In ordinary atoms the net electric charge is zero since there are an equal number of protons and electrons. Early models of atoms depicted electrons as being in "orbit" around the atomic nucleus. Figure 2.2 shows the simplified atomic structure.

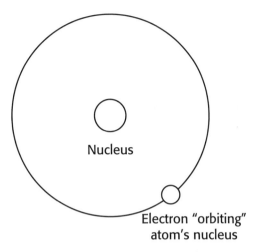

Nucleus

Electron "orbiting"
atom's nucleus

FIGURE 2.2 Simple atomic structure.

Modern atomic models attempt to describe electrons as existing in more of a "cloud" condition, with their average distance from the nucleus expressed as a finite level or *electron shell*. Pictures attempting to show this averaged condition of the electrons look like a cloud. The electron shells have an equivalent energy, depending on their distance away from the nucleus.

Modern quantum physics holds that an electron can exist only in certain discrete energy levels, and must either gain or lose an exact amount of energy in order to move from one shell to another, or from a shell to completely out of the influence of the nucleus. It is these free electrons that constitute electrical *current*, when they have a net direction in a circuit or device. Current will be defined further in Chapter 3, "Electrical Terms, Notations, and Prefixes."

When an electron gains sufficient energy to escape an atom, it leaves behind the empty spot it once occupied. Since the electron in its *valence shell* (the outermost energy position it could occupy and still be attached to the atom) helped to balance the positive electric charge exhibited by the proton (or protons) of the atomic nucleus, the electron's negative influence on the atom is gone once it becomes free. This effectively leaves behind a spot that acts positive. This positive-acting spot once occupied by the electron is called a *hole*. Another way of stating this idea is to say that when sufficient energy is applied to an atom to ionize it, an *electron-hole pair* is created (Figure 2.3). Both the electron and the atom it was freed from now have an electrical charge. Both are now considered to be electrically charged particles.

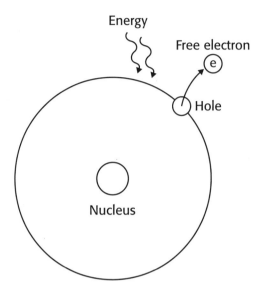

FIGURE 2.3   An electron-hole pair.

2.5   CONDUCTORS

The materials used to construct electrical devices and components are generally classified into three groups according to their overall electrical properties. *Conductors* are those materials that offer very little opposition to electric current. Most metals are good conductors. Some of the more common metals used to make electric conductors, usually in the form of wires, are: copper, silver, aluminum, brass, steel, nichrome (an alloy of nickel and chromium), and tungsten.

Copper and aluminum are most commonly used to manufacture wires. These wires are used for everything from industrial plants and machinery, house wiring, and motor vehicles, to microchips. In the microchip example, tiny hairlike threads of aluminum and copper are used to connect a solid-state microchip to the much-larger pins of the integrated circuit plastic package.

Silver, which has a very high conductivity, is widely used as a thin coating applied to other metals in order to increase their electric conducting properties. High-powered radio and TV transmitters often use silver-coated components in the critical high-power output sections of the systems to reduce circuit losses.

Gold is often used in modern electronics as a protective coating on copper or aluminum traces on circuit boards and their edge connectors, since it is a good conductor and does not corrode as does silver. Despite the high price, it is considered good engineering practice, and cost effective to use gold this way.

Nichrome is used to make the wires for heating elements in toasters, electric heaters, hot plates, and all sorts of heat-producing electric appliances. Nichrome offers considerably more opposition to electric current than does copper for the same diameter and length of wire. So by using a material with higher resistance, the heating elements can be reduced to a more manageable size.

Tungsten is drawn into very fine wire that is tightly coiled in order to fit a great length of it into an incandescent light bulb. When electric current flows through the tungsten wire, the opposition offered by the wire produces heat and light (incandescence). Figure 2.4 shows examples of incandescent lamps.

FIGURE 2.4 Incandescent lamps.

Other materials that readily allow the flow of electricity are ionized liquids and gasses. Ions are atoms that have either one or more extra electrons or they are missing one or more electrons. Examples of ionized liquids are the acid-water solutions used in automobile batteries. This type of battery is composed of several "wet-cells." The wet part of the cell is the electrolyte solution. The electrolyte of a typical "dry-cell" consists of a pastelike compound that chemically acts on the inner and outer electrodes to produce ions. These ions are responsible for the resulting electric current flow when the cell is connected in a circuit, which provides a complete pathway for the current to flow. When a dry cell really dries out, it becomes useless.

The gasses inside a fluorescent lamp tube are ionized by an electric discharge and allow a good pathway for the flow of electrons through the tube. A thin coating of a colored material, usually white or off-white, is deposited on the inside of the tube. The high-energy ultraviolet (UV) photons (packets of light) produced by the electric discharge strike the atoms of this fluorescent coating, causing it to glow with visible light. Recall the principle of energy levels in atoms. In the case of the fluorescent lamp tube, the excited ions within the gas in the tube strike the coating on the inside of the tube and add energy to the atoms of the coating material. The outermost electrons temporarily attain a higher energy level. When the unstable electrons in the coating's atoms fall back to their original levels, they must give off a specific amount of energy; that energy is in the form of the light given off by the lamp.

It is known today that conductors act the way they do because of their atomic makeup. An atom will always have a definite number of electrons existing at the farthest allowable point from the nucleus, and yet those electrons are still controlled, or held in position, by the nucleus. This is because unlike charges attract each other. In chemistry and physics courses, you probably learned there are specific letter and number designations given to the various energy levels or electron shells. The study of electronics is generally not concerned with any electrons except for those existing in the outermost shell. Electrons in this outermost shell are called *valence electrons*. Science has determined that all chemical activity is determined by the valence electrons. It turns out that elements whose atoms have only a few electrons (generally one or two) in their valence shell make good electrical conductors. Recall that electrons can exist only in shells a specific distance from the atomic nucleus; therefore, they can exist only at a discrete energy level. In order for them to move out of the valence shell beyond the influence of the nucleus, they must receive enough additional energy to break free.

Conductors usually have many free electrons at room temperature. The reason is that it takes very little energy applied to the valence shell electrons in order to dislodge them from the influence of the nucleus. Recall from the definition of energy that it can take many forms. One of those forms is thermal (heat) energy. At room temperature, the background or ambient thermal energy is sufficient to create many free electrons in the conductor. When they are thus "freed," they are available to flow through the conductor.

A good conductor has a "sea" of free electrons. The unbound electrons are free to move under the influence of an electric field. This is the principle of conduction, or current flow, in a wire.

2.6   INSULATORS

Insulators are materials that offer a large opposition to the flow of electricity. These materials are used to construct devices used to constrain electric flow to conductors, as in the case of the insulating covering of wires and cables (bundles of wires). Insulators are made from materials such as glass, paper, plastic, mica, rubber, DuPont™ Teflon®, wax, oil, and certain ceramics. Insulators made of nylon, a man-made plastic material, are commonly used to hold a *printed circuit board* (PCB) away from the conducting metal chassis to prevent unwanted electrical contact, which would be destructive to the components mounted on the board. Insulating materials are used to cover most small electrical components including resistors, capacitors, inductors, solid-state relays, and most types of switches. Insulators serve to help prevent unwanted electrical pathways from interfering with the desired pathway. Electric current always tends to take the path of least resistance. In a given circuit, giving the electricity a choice controls the electric pathway. The choice is between an easy-to-follow pathway through a conductor, or a difficult pathway through an insulator. Unless there is an interruption in the conductor pathway preventing the flow of electricity, or a severe breakdown in the insulator material causing its opposition to electricity to become greatly reduced, the electric flow will go where directed. Insulators are made from materials whose atoms have electrons that are tightly bound, and not able to participate in the flow of current. Figure 2.5 shows some typical insulators.

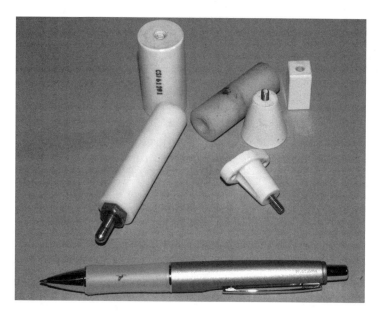

FIGURE 2.5   Electric insulators.

Insulators come in many shapes, sizes, and colors. Some are very small, such as the coating of silicon dioxide (SiO_2), better known as sand, used to insulate consecutive layers of conductors on a microchip. Some very large insulators are used to suspend the wires on a high-voltage power transmission tower.

2.6.1 Insulator Breakdown

The primary reason insulators fail is if excessive voltage is applied to them. When they fail, their resistance can fall to a very low value. This catastrophic decrease in opposition can make the insulator act as a fairly good conductor, allowing dangerously large values of electric current to flow in a circuit. The result is usually damage to the system, with replacement of the insulator required. Insulators fail due to high temperatures, exposure to oil, gas, water, or other chemicals, or as a result of excessive voltage applied to them.

Insulators are rated for their capability to prevent current flow. This capability is greatly dependent on the material used to construct the insulator. Paper, wool, rubber, wax, and mica were all materials used in early electrical devices as insulators. Modern plastics, ceramics, and even rare gasses such as helium are used in today's insulators. Industrial sand is used in many semiconductors and integrated circuits (ICs) as an excellent insulating material.

Insulators are made from elements having atoms with a completely filled or near-completely filled valence shell. Most elements used to make insulators have atoms with filled or nearly filled (seven or eight) valence shells. The result is that it requires a relatively large amount of energy applied to these atoms in order to cause one or more of the valence shell electrons to become dislodged from the atom, or "free." Insulators do break down on occasion, allowing electricity to flow through them. The breakdown is a direct result of sufficient energy applied to the insulator material's atoms, creating free electrons that are available for flow in a circuit.

A common example of insulator breakdown is the spark plug wires in a car or truck. Exposure over time to chemicals and temperature extremes causes the insulating covering on the plug wires to break down, allowing some of the electrical flow intended to go to the spark plugs to instead escape into the air surrounding the wires. The result is called a "weak spark" that causes the engine to misfire or "miss." The engine runs roughly and vibrates due to the uneven power produced. Periodic replacement of the spark plug wires is the preventive measure taken to ensure that all the electricity gets to the intended destination. Similar deterioration happens to virtually any insulator material, although usually over a much longer time period. The insulators in our home computers, TV sets, CD players, or any other piece of electrical or electronic equipment can ultimately fail. Ozone, a three atom molecule of oxygen (O_3) in the atmosphere attacks most plastics, causing breakdown of their insulating qualities. Ozone is produced during lightning discharges in a thunderstorm, or from any other spark. The ultraviolet (UV) rays produced by the sun or even by arc welding can cause surface breakdown in many insulating materials.

In general, keeping insulators dry, shielded from the sun's rays, and at a cool, steady temperature will greatly extend their capability to perform their intended function. This, in turn, will greatly extend the useful life of the components and systems using these insulators.

2.6.2 PC Motherboard Insulators

PC motherboards, up until the ATX style, used small *standoff* insulators made from plastic to suspend the motherboard from the inside of the PC's metal case. The insulators have a pointy end that snaps through the motherboard mounting holes from the bottom, and they have a round flat shape that suspends the insulator and therefore the motherboard a safe distance off the metal case, preventing short-circuiting of the board's circuit traces. Figure 2.6 shows a motherboard standoff insulator.

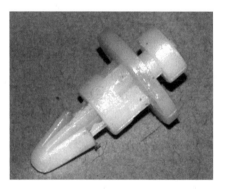

FIGURE 2.6   Motherboard insulator.

ATX motherboards use a set of between four and eight brass standoff supports that are electrical conductors. These multiple points of contact to the PC chassis for the motherboard provide a more stable and less noisy electrical design.

2.6.3 Circuit Boards

The base material used for a nearly all circuit boards is a type of insulator. A common fabrication method is to combine fiberglass and epoxy. This combination offers high strength as well as very good insulating properties and does not deteriorate much over time. A PC's motherboard and all the add-on cards use this type of material as the base material upon which the circuit traces are deposited and the individual integrated circuit chips and sockets are mounted. Circuit boards are present on all hard drives, optical drives, and floppy drives. They are also used in every modem, hub, switch, router, and in virtually all electronic devices made since the late 1960s or so.

2.7   SEMICONDUCTORS

Semiconductors are materials that offer an opposition to electric flow greater than that of conductors, but less than that offered by true insulators. Recall the atomic make-up of conductors. Conductors have only a few electrons in their valence shell. They require only a small amount of energy to create free electrons. For all the commonly used conductor materials, the ambient thermal energy at room temperature is sufficient to cause the creation of free electrons. The valence shell of conductors is not very stable.

The elements of most conductors, insulators, and semiconductors are made up of atoms more complex than the simple atom presented earlier in the text. These elements are made up of atoms having many different electron shells. The atoms of semiconductor material are the most stable, and chemically and electrically balanced when they have eight valence electrons. This idea will become more important later in our presentation of semiconductor manufacturing. By contrast, insulators have a completely or near completely filled valence shell. They require a large amount of energy (at room temperature) in order to create free electrons. The valence shells of insulators are very stable.

Semiconductors are made from elements having four or five valence shell electrons. Elements such as germanium and silicon are frequently used in the manufacture of semiconductor devices such as diodes, transistors, and complex *integrated circuits* (ICs). The stability of the valence shell for these elements is between that of conductor atoms and those of insulators. One of the first semiconductor materials used in electrical work was carbon. Carbon was used in microphones used for early radio work, and later the telephone employed carbon-filled elements to convert sound waves into a time-varying current known as a *signal*. The carbon element contained many tiny particles of carbon sandwiched between two conductors. An electric current supplied by dry cells passed through the microphone element. The electricity then flowed through an amplifier. An *amplifier* is an electronic circuit that increases the intensity of a signal applied to it. As the sound waves vibrate the *diaphragm* (the front conductor of the microphone element), the carbon granules are compressed more tightly. This decreases the opposition the carbon offers to the electric flow, and causes a corresponding increase in the electricity produced by the microphone element. As the diaphragm springs back to its original position, the carbon granules become less compressed, increasing the resistance to electricity and the output from the microphone drops. So the vibrations of sound are converted into an AC signal.

Silicon Valley, located in northern California, got its name because so many businesses manufacturing and using semiconductor devices are located there. Silicon, the most abundant element on the Earth, is still used in great quantities to manufacture semiconductor devices such as diodes, transistors, and integrated circuits. Figure 2.7 shows a semiconductor wafer, containing many individual integrated circuits.

FIGURE 2.7 A semiconductor wafer.

This changing electric current produced by the microphone is only a fairly accurate reproduction—in electric terms—of the sound waves arriving at the microphone. The carbon element is very durable and relatively inexpensive to produce; this fact more than makes up for its low sound fidelity. Carbon microphone elements are seldom used for high-fidelity sound reproduction installations. They are the perfect choice, however, for applications involving wide ranges in temperature and humidity, and subject to high vibration such as a telephone handset.

Carbon is also extensively used to make carbon-composition resistors, the older form of resistors. Resistors are devices designed to offer a specific amount of opposition to current flow. Because of its semiconducting qualities, carbon offers a good choice for the manufacture of resistors. Chapter 5, "Using a Digital Multimeter," presents the construction and use of resistors in detail. Figure 2.8 shows some typical medium-wattage-sized (25-watt) resistors.

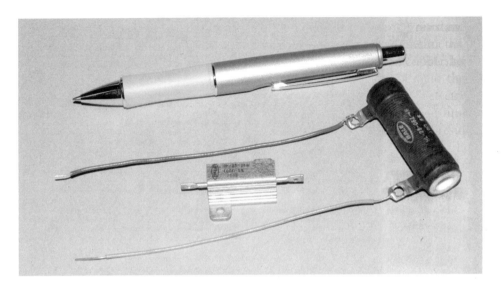

FIGURE 2.8   Medium-wattage-sized (25 W) resistors.

2.8   ELECTRICAL POTENTIAL DIFFERENCE

There are two methods in which electricity can be built up and controlled. *Static* electricity is the buildup of charges in different places. *Current* electricity is the type we normally think of, and consists of the movement of charges in a circuit.

2.8.1   Static Electricity

One force that causes movement of charged particles is the uneven distribution of charges. Whenever there is more charge at one point with respect to another point, there may be the movement of charges between the two points. In a previous section, you learned about the idea of positive and negative charges. Those charges were applied to parts of the atom: the proton and the electron. "Positive" and "negative" are also frequently used in a much broader context, however.

Figure 2.9 shows two equal-sized spheres. The sphere on the right is electrically balanced—it has a net zero electrical charge on it. The sphere on the left has one

One million excess
electrons

Neutral electric
charge

FIGURE 2.9 The left sphere has an excess
of one million electrons, and the right
sphere has a neutral charge.

million extra electrons on it. There is a difference of one million electrons between
the two spheres. Since electrons carry an electric charge, the sphere on the left has
a *potential difference* of one million electrons compared to the sphere on the right.
If brought together, half of the extra electrons on the left sphere would flow over
to the right sphere, causing the spheres to have equal charges. The difference of
potential caused the movement of charges between the two spheres.

Stated another way, since electrons carry a negative electric charge, it can be
stated that before touching the right sphere, the sphere on the left was negative with
respect to the sphere on the right. In fact, it can also be stated that the left sphere was
one million electrons more negative than the right sphere. It could also be stated
another way: that the right sphere was positive with respect to the left sphere by one
million electrons worth of electric charge. Before the two made contact, the sphere on
the left was negative, and the sphere on the right was positive, relative to each other.

Figure 2.10 shows another two equal-sized spheres. In this case the left sphere
has an excess of two million electrons. The sphere on the right has an excess of one
million electrons. Electrons all have an equal, negative charge. But in this case, one
sphere has more of these negatively charged electrons. The sphere on the left has
one million more electrons than the right sphere, so it is negative compared to the
right sphere. The sphere on the right is positive compared to the sphere on the left.
If touched together, there would be a movement of charges. Half of the difference
of one million electrons would flow from the left sphere to the right sphere.

Two million excess
electrons

One million excess
electrons

FIGURE 2.10 The left sphere has two million
excess electrons while the right sphere has
one million excess electrons.

Figures 2.9 and 2.10 show merely that it is the relative difference of electric potential that determines whether there will be a movement of charges from one point to another. It is this same principle of difference of potential between two points that causes lightning.

2.8.2 Electric Current Flow

In an electric circuit, devices such as cells, batteries, and power supplies are used to provide a difference of potential. They provide electric fields that influence the electrons in the conductors and other circuit components. Electrons in a circuit flow from a place of relative abundance to a place of relative scarcity, or from the negative terminal of a source, through the circuit, and back to the positive terminal of the source. In the early days of electrical experimentation, when the real nature of electricity was not well understood, it was assumed that electricity flowed from a positive place to a negative place. Electric circuits will be explained more completely in Chapter 3.

2.9 SAFETY

Because the power supplies used in modern electrical and electronics work are sometimes designed to provide a very large potential difference to circuits, one needs to be aware of the general safety rules involved with such equipment. Too many times someone is injured or killed by not paying proper attention to details in electronics. Never assume or take someone else's word that a circuit is de-energized; always make sure it is not. Some types of devices used in electronics can store a lethal charge for hours or even days after the circuit has been turned off and unplugged. Learn how to safely discharge capacitors before ever attempting to work around them. Electrolytic capacitors can explode if installed in a circuit backwards. Applying an excessive voltage to an electrolytic capacitor can also cause it to explode.

Do not wear watches, rings, or other decorative metal objects, as they can cause a conductive path between your body and an energized circuit.

Many people mistakenly assume that circuits employing batteries cannot hurt a person. Such is not the case. Lawnmowers can provide extremely high potential differences, sufficient to stop a person's heart. The spark plug ignition system can produce several thousand volts of electrical potential difference. (See Chapter 3 for more on voltage.) Remember, the human heart is an electrically operated organ; too much electricity will permanently damage or destroy it.

Remember the "one-hand-in-your-pocket" rule from Chapter 1, "PC Overview, Safety, and Tools."

2.9.1 Turned Off Does Not Always Mean Nonpowered

Modern PCs use a type of power supply that constantly supplies at least a part of its normal output to circuits inside the PC that can be "awakened" by signals from outside the PC. Typically, the network interface card (NIC) and telephone modem can be activated, bringing the PC to a fully booted and functioning state. The + 5 VDC voltage, called a *standby voltage*, is always present on the motherboard as long as the PC is plugged in. In addition, the commercial power at 120 VAC or 240 VAC is also present inside the power supply. So one should never assume it is safe to work on a PC that is turned off if that PC is still plugged into the wall outlet.

2.9.2 ESD Awareness Is Vital

PCs are filled with static-sensitive devices and circuits that can be damaged or destroyed with the static electrical charge on a person's body. Always use proper antistatic equipment, such as an ESD wrist strap and conductive mat, when working on a PC. Don't try to get by simply by touching the computer chassis. If you forget and walk away from the computer, you need to touch the chassis each time you return, or the static charge on your body built up by walking on carpet could damage many of the chips inside the PC should you forget to ground yourself to the chassis upon returning. Get into the habit of always using an ESD wrist strap whenever you are inside the PC case.

2.10 SUMMARY

- Electrical potential difference refers to the relative number of electric charges between two objects or points in a circuit. Potential difference is measured in volts, abbreviated V. Potential difference is required in order to cause current to flow.
- Electric current, the flow of charges past a point, can occur anytime there is sufficient difference in potential and a complete pathway or circuit between the two points.
- The movement of electrons in a circuit from negative to positive is known as electron flow.
- Since the early days of electricity, current direction in a circuit has been assumed to be from positive to negative. This view became the accepted convention, and today is the prevailing way to view current flow in a circuit. It is know as conventional current flow.
- Proper electrical safety requires constant awareness and attention to detail.

- Never work on electrical equipment when you are tired or under the influence of drugs or alcohol.
- Never work on electrical equipment in a wet environment.
- Always wear good quality shoes with insulated soles when working on powered equipment.
- Never work on high-voltage equipment alone.
- Keep one hand in your pocket when working inside electrical equipment.
- Never assume a piece of equipment is unplugged—check it yourself first.
- Avoid wearing rings, watches, or other jewelry when working on electronic equipment.
- Whenever plugged in, modern PCs using ATX-style power supplies are always on, supplying a backup voltage to the system.
- PCs contain many static-sensitive devices that can be easily destroyed. Always use proper ESD protection.
- When working in the lab, always seek help from the instructor first if you have any concerns about safety.
- Troubleshooting involves making a quick and accurate diagnosis of the problem, and then repairing it.

2.11 KEY TERMS

Electrical

Electronic

Energy

Work

Joule

Power

Watt

Conservation of Energy

Positive

Negative

Nucleus

Proton

Neutron

Electron

Electron Shell

Valence Shell

Free Electron

Hole

Conductor

Ambient

Insulator

Breakdown

Transducer

Sensor

Conversion Efficiency

Semiconductor

Integrated Circuits

Potential Difference

Troubleshooter

2.12 EXERCISES

1. The person who comes to your workplace or home to rewire a 120 VAC circuit is most likely a (an)…
   a. engineer
   b. technician
   c. electrician
   d. electronics troubleshooter
   e. programmer

2. The ability to do work is called…
   a. energy
   b. power
   c. motivation
   d. force
   e. drive

3. The standard unit of energy is the…
   a. joule
   b. watt
   c. kelvin
   d. ohm
   e. volt

4. The rate work is done is defined as…
   a. energy
   b. power
   c. voltage
   d. current
   e. resistance

5. Energy can neither be created nor destroyed.
   a. true
   b. false

6. The sticker on the power supply of a PC lists the input requirements as 805 watts. The total output is listed as 262 watts. What is the conversion efficiency of the supply?
   a. 94%
   b. 88%
   c. 64%
   d. 33%
   e. 12 %

7. A single atom consists of at least one electron and one...
   a. proton
   b. neutron
   c. lepton
   d. isotron
   e. duron

8. Electrons carry what type of electric charge?
   a. Positive
   b. Neutral
   c. Negative
   d. Zero

9. When an electron leaves its location in an atom and wanders or is pulled away, the location it left is known as a...
   a. vacuum
   b. hole
   c. proton
   d. nucleus
   e. insulator

10. Conductors have _____ valence shell electrons.
    a. many
    b. few
    c. 10
    d. 12
    e. 20

11. Glass is considered to be a good...
    a. semiconductor
    b. insulator
    c. generator
    d. conductor

12. California's Silicon Valley is so named because so many manufacturers of equipment using these are located there.
    a. Conductors
    b. Insulators
    c. Semiconductors
    d. Resistors

13. An atom with an overall electrical charge is known as a (an)…
    a. unequal charge
    b. ion
    c. proton
    d. electron
    e. neutron

14. In the original view of electric current flow, current was thought to flow…
    a. from positive to negative
    b. from negative to positive
    c. from neutrons to protons

15. When working on electrical equipment, you should avoid wearing what?
    a. Watches, rings or other metal jewelry
    b. Wrist straps
    c. Hats
    d. Suspenders
    e. Well-insulated shoes

3 Electrical Terms, Notations, and Prefixes

3.1 INTRODUCTION

Students with good math skills may want to skip this chapter. If it's been a while since you've used a calculator, you might want to review this chapter. Students with little math background should consider this chapter essential.

3.1.1 Scientific Notation and Engineering Notation

If you are already familiar with the concepts of scientific notation and engineering notation, you might want to skip this section. Consider using it as a review, especially if you have not used scientific or engineering notation for a while.

Scientific notation is a convenient shorthand method for dealing with very large, as well as very small numbers. Very large or very small numbers are quite cumbersome to write in the normal fashion. A number expressed in scientific notation consists of a decimal number between 1 and 10 followed by a power of 10. The decimal number indicates the first few digits of the actual value as it would be written in longhand, or conventional, form. The power of 10 gives the factor by which the decimal number is to be multiplied. The exponent (the power of 10) is always an integer (whole number); it may be positive or negative. The decimal part of a scientific notation expression may have any number of significant figures, depending on the required accuracy.

For electronics and engineering work, a system called *engineering notation* is normally used. It is similar to scientific notation, but uses numbers in the range of 1 to 1,000 times a power of ten in steps of 3. For instance, 10^3 corresponds with the prefix *kilo*, 10^6 with *mega*, 10^9 with *giga*, 10^{-3} with *milli*, 10^{-6} with *micro*, 10^{-9} with *nano*, and 10^{-12} with *pico*. See Table 3.1, later in this chapter, for a full list of prefixes and their corresponding engineering notation numbers.

Example 3.1

The number 2.3×10^6 is an approximation accurate to two significant figures of 2,345,678.

When two numbers are multiplied or divided in scientific notation, the decimal numbers are first added (for multiplication) or subtracted (for division). Finally, the product or quotient is reduced to standard form. That is, the decimal part of the expression should be at multiplied or divided by each other. Then the powers of 10 are added (for multiplication) or subtracted (for division). The final answer should have a number of 1 to 10 times a standard metric prefix.

Example 3.2

$$(3 \times 10^2)(7 \times 10^3)$$

$$= 21 \times 10^5$$

$$= 2.1 \times 10^6$$

When working with scientific notation, it is important to be aware of the value of the significant figures in the expressions. Values should not be expressed more precisely than is warranted by the number of significant digits.

3.1.2 Significant Figures

When mathematical operations are performed between or among numbers, the resultant cannot have more significant digits than the numbers themselves. If the original values have different numbers of significant figures, then the resultant must be *rounded off* to have the smallest number of significant figures consistent with the smallest number of significant digits in the original values. In a decimal number, all nonzero digits are significant. All zeros between significant digits are significant. A zero following a nonzero digit may or may not be significant, depending on whether it is accurate or whether its only function is to help place the decimal point in an approximate number. For example, the number 673,924 has six significant digits, but its approximation of 674,000 has only three. However, if 674,000 were the exact number it would have six significant digits.

Example 3.3

Perform the indicated operations:

$$(1.5535 \times 10^2)(7.4 \times 10^3)$$

Solution

Multiply the integer parts first.

$$1.5535 \times 7.4 = 11.4959$$

Then deal with the powers of 10 like this:

$$10^2 \times 10^3 = 10^5 \text{ (the exponents are added)}$$

The complete product is:

$$(1.5535 \times 10^2)(7.4 \times 10^3)$$

$$= 11.4959 \times 10^5$$

Then use the (3rd) blue calculator button and then the (ENG)—the "+" button—to get…

$$= 1.14959 \times 10^6$$

The answer does not have six significant figures. The original decimal expression, 1.5535 and 7.4, have only five and two significant figures, respectively. After rounding off the answer to two significant figures, the answer is 1.1×10^6.

Example 3.4

Perform the indicated operations:

$$(743 \times 10^6) \div (12.65 \times 10^3)$$

Σx'	ITC	►lb	n	x'	σxn	►in	►gal	►OZ	x'	SLP	n	►lb	►DD
7	4	3	EE	6	÷	1	2	•	6	5	EE	3	=

Solution

First perform the division on the decimal parts of the numbers.

$$743 \div 12.65 = 58.735178$$

Next, perform the subtraction of the power of 10 in the denominator from the power of 10 in the numerator.

$$10^6 \div 10^3 = 10^3 \quad \text{(the exponents are subtracted)}$$

The answer must be truncated (shortened) to the number of significant digits that the original number with the least number of significant digits has. In this case, the number 743 has three significant digits, so the answer must have no more than three significant digits.

Finally, express the answer in the scientific form, 743×10^3 units.

3.1.3 Metric Prefixes

The metric system is a decimal system for measurement of length, mass, and time. It is also used as a worldwide standard for most scientific units. The principle unit of length is the meter. The decimeter is 0.1 meter; the millimeter is 0.001 meter; the kilometer is 1,000 meters. The decimal nature of the metric system makes it much easier to use than the older English system. However, the English system is still used for nonscientific applications in the United States. Engineering notation is commonly used because it uses metric prefix powers of ten. Table 3.1 lists the commonly used metric prefixes for electronics work.

Values are commonly given by using only one of these *metric prefixes*. Normally in electronics work, 10^1, 10^2, 10^4, 10^5, 10^7, 10^8, 10^{10}, and 10^{11} are not used to give a

TABLE 3.1 Most Commonly Used Metric Prefixes in Electronics Work

Prefix Name	Power of 10	Symbol
Giga	10^9	G
Mega	10^6	M
Kilo	10^3	k
Units	10^0	None
Milli	10^{-3}	m
Micro	10^{-6}	μ
Nano	10^{-9}	n
Pico	10^{-12}	p

value in its final form. For instance, milliamps, millivolts, microamps, microvolts, kilohms, kilovolts, and megohms are commonly used.

In addition, the shortened forms for these commonly used metric expressions are used.

Table 3.2 lists examples of electrical metric units.

TABLE 3.2 Examples of Electrical Metric Units

Unit	Abbreviation	Value
gigavolts	GV*	10^9 volts
gigamperes	GA*	10^9 amps
gigohms	GΩ	10^9 ohms
megavolts	MV*	10^6 volts
mega-amperes	MA*	10^6 amps
megohms	MΩ	10^6 ohms
kilovolts	kV	10^3 volts
kiloamperes	kA*	10^3 amps
kilohms	kΩ	10^3 ohms
millivolts	mV	10^{-3} volts
milliamperes	mA	10^{-3} amps

(continued)

TABLE 3.2   Examples of Electrical Metric Units *(continued)*

Unit	Abbreviation	Value
milliohms	mΩ	10^{-3} ohms
microvolts	μV	10^{-6} volts
microamperes	μA	10^{-6} amps
microhms	$\mu\Omega$	10^{-6} ohms
nanovolts	nV	10^{-9} volts
nanoamperes	nA	10^{-9} amps
nanohms	nΩ	10^{-9} ohms
picovolts	pV*	10^{-12} volts
picoamperes	pA*	10^{-12} amps
picohms	pΩ*	10^{-12} ohms

* Infrequently used symbols.

Example 3.5

Perform the indicated operations and express the result in the standard form for this text. Use as many significant digits as are justified, and a standard metric prefix.

$$862.235 \; Mega \div 56.3 \; kilo$$

Solution

First, perform the indicated division on the decimal parts of the numbers.

$$\langle 862.235 \div 56.3 \rangle = 15.15009$$

Next, perform the required operation on the powers of ten.

Mega $= 10^6$ and kilo $= 10^3$. Since this is division, the 3 in the power of ten is subtracted from the 6 in the power of ten leaving 3 or 10^3 (kilo).

Finally, combine the results to put the answer into standard form. Since the answer can be expressed only to three significant digits, and must follow the rules for rounding, the final answer is 15.3 kilo.

Example 3.6

Multiply 34.8 milli by 57.2 mega.

Solution

First, write each quantity in scientific notation.

$$(34.8 \times 10^{-3})(57.2 \times 10^{6})$$

▸ lb	ITC	▸ OZ	Σy	n	P▸R	▸ lb	σyn	SLP	Σx	▸ OZ	▸ gal	n	x'	▸ DD
3	4	•	8	EE	+/−	3	×	5	7	•	2	EE	6	=

Next, perform the operation on the decimal part of the equation.

$$(34.8 \times 57.2) = 1990.56$$

▸ lb	ITC	▸ OZ	Σy	σyn	SLP	Σx	▸ OZ	▸ gal	▸ DD
3	4	•	8	×	5	7	•	2	=

Next, add the exponents.

$$10^{-3} \times 10^{6} = 10^{3} \quad \text{(or kilo)}$$

Combine the results.

$$= 1990.56 \text{ kilo}$$

Finally, put in standard form.

$$= 1.99 \text{ kilo}$$

Notice that the answer is expressed using three significant digits. This was justified, as the least number of significant digits in the original values is three.

3.1.4 Rounding Off

The standard for rounding off numbers used in this text is the same as is used by the scientific community. When rounding off a number to a certain number of significant digits, look at the next least significant digit, that is, the digit to the right of the last significant digit required. If the next least significant digit is 5 or greater, round up the digit to the left. If the next least significant digit is 4 or less, do not round up the next more significant digit.

Example 3.7

Round off 123.578 to four significant digits.

Solution

First, write the first four most significant digits,

123.5

Second, examine the next least significant digit; in this case, it is 7. Since 7 is 5 or greater, round up the digit in the next most significant place. In this case, round up the 5 to a 6. The answer then is 123.6, to four significant digits.

Example 3.8

Use the same original number from example 3.7, and round the number to three significant digits.

Solution

The original number was 123.578.
Write the first three most significant digits of the original number:

123

Then look at the next least significant digit; in this case it is 5.
Finally, since 5 is 5 or greater, round up the next most significant digit; in this case the 3 becomes a 4.
The answer is 124, to three significant digits.

Example 3.9

Express the same original number used in the last two examples in a two-significant-digit form.

Solution

123.578 becomes 12. Then examine the next least significant digit, the 3. Because 3 is less than 5, do not round up the next more significant digit. The answer is 120, to two significant digits.

Standard Form Used in This Text

The standard form used throughout this text is engineering notation with standard metric prefixes and no more than three significant figures.

3.1.5   Calculators

For the type of electronics calculations done in this text, a relatively simple calculator is all that is required. It should be the "scientific type" and can be purchased for around twenty dollars. It should have buttons marked **SCI** and **ENG**, which stand for scientific and engineering notation forms respectively. Figure 3.1 shows the type of calculator recommended for use with this text.

FIGURE 3.1   The recommended calculator, a Texas Instruments® model 36X Solar.

3.2   ELECTRICAL TERMS

It is essential to understand the different measurement units of electricity in order to understand the rest of this book. Following are specific discussions of these measurement units.

3.2.1 Voltage

Voltage is the existence of a potential difference in charge between two points in an electric circuit. As previously defined, electrical potential difference is normally manifested by an excess or abundance of electrons at a point, and/or a deficiency of electrons at another point. Sometimes, other charge carriers, such as holes or protons, can be responsible for a voltage between two objects or points in a circuit. Electric voltage is measured in units called *volts*. A potential difference of one volt will cause a current of one ampere to flow through a resistance (opposition) of one ohm. Voltage can be either alternating or direct. The letter *E* is used to designate voltage rises, and the letter *V* is used for voltage drops. Voltage rises and drops will be further explained in Chapter 4, "The Electric Circuit and Voltage Generation."

3.2.2 Current

Electric *current* is a flow of charge carriers past a point, or from one point to another, in an electric circuit. The charge carriers may be electrons, holes, or ions. In some cases, atomic nuclei may carry a charge. The standard for electric charge is the charge carried by one electron. Electric current is measured in units called *amperes*. A current of one ampere means the movement of one coulomb of charge per second. A *coulomb* is a quantity of electric charges equal to 6.28×10^{18}. Current may be either alternating (AC) or direct (DC). Current is symbolized by the letter *I,* stemming from the French word *intensitie.* The letter "C" was not used for current, because it is used for capacitance, as explained in Chapter 11, "Inductors and Capacitors." As mentioned earlier in the text, there are two views of current flow, as will be explained in the following sections.

Electron Current Flow

The electron flow view of electric current holds that current flows from the source, from its terminal with an abundance or excess of electrons around the circuit and back to the opposite source terminal with a deficiency of electrons. The source terminal with the excess of electrons is called the negative terminal, and the source terminal with the deficiency of electrons is called the positive terminal. So electron flow holds that current flows from negative to positive.

Conventional Current Flow

The conventional current method describes current as the movement of positive charge carriers from a more positive point to a less positive point. Conventional current flow is actually in the opposite direction to that of electron current flow. The basic idea is that the flow of negative charge carriers (electrons) in one direction is exactly equivalent to the flow of positive charge carriers in the opposite direction. Early investigators into the poorly understood phenomenon we call electricity assumed the charge carriers were positive. As it turned out, they were wrong. Most electrical current is the flow of electrons, or the flow of negative charges. This text will use the conventional current flow method when indicating direction of current flow in a circuit. Conventional current is the electrical current standard used in the engineering world. Figure 3.2 shows each view of current flow.

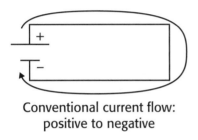
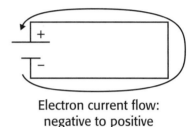

Conventional current flow:
positive to negative

Electron current flow:
negative to positive

FIGURE 3.2 Two views of current flow.

3.2.3 Resistance

Resistance is the opposition that a material offers to DC current in a circuit. It is the bulk property of a material that depends on the material's dimensions, electrical resistivity, and temperature. The resistance of the material determines the amount of current produced by a given voltage.

The standard unit of resistance is the *ohm*, named in honor of a German scientist, Georg Simon Ohm, 1787–1854. The symbol used to indicate resistance is the uppercase Greek letter omega (Ω). In equations, the symbol for resistance is *R*.

The resistance of wire is usually measured in ohms per unit length. The common length used in wire tables is 1,000 feet. Silver offers the lowest resistance of any commonly used metal wires. Copper and aluminum also have very low resistance per unit length. The direct current (DC) resistance of a wire depends on its diameter. Table 3.3 shows the DC resistance of copper wire in different diameters. At least in the United States, wire diameter is indicated by its American Wire Gauge (AWG) rating.

TABLE 3.3 Copper Wire Table. Reprinted with Permission from *The 2001 ARRL Handbook for Radio Amateurs*

Copper Wire Specifications
Bare and Enamel-Coated Wire

Wire Size (AWG)	Diam (Mils)	Area (CM¹)	Enamel Wire Coating Turns / Linear inch² — Single	Heavy	Triple	Feet per Pound Bare	Ohms per 1000 ft 25∞ C	Current Carrying Capacity Continuous Duty³ — at 700 CM per Amp⁴	Open air	Conduit or bundles	Nearest British SWG No.
1	289.3	83694.49				3.948	0.1239	119.564			1
2	257.6	66357.76				4.978	0.1563	94.797			2
3	229.4	52624.36				6.277	0.1971	75.178			4
4	204.3	41738.49				7.918	0.2485	59.626			5
5	181.9	33087.61				9.98	0.3134	47.268			6
6	162.0	26244.00				12.59	0.3952	37.491			7
7	144.3	20822.49				15.87	0.4981	29.746			8
8	128.5	16512.25				20.01	0.6281	23.589			9
9	114.4	13087.36				25.24	0.7925	18.696			11
10	101.9	10383.61				31.82	0.9987	14.834			12
11	90.7	8226.49				40.16	1.2610	11.752			13
12	80.8	6528.64				50.61	1.5880	9.327			13
13	72.0	5184.00				63.73	2.0010	7.406			15
14	64.1	4108.81	15.2	14.8	14.5	80.39	2.5240	5.870	32	17	15
15	57.1	3260.41	17.0	16.6	16.2	101.32	3.1810	4.658			16
16	50.8	2580.64	19.1	18.6	18.1	128	4.0180	3.687	22	13	17
17	45.3	2052.09	21.4	20.7	20.2	161	5.0540	2.932			18
18	40.3	1624.09	23.9	23.2	22.5	203.5	6.3860	2.320	16	10	19
19	35.9	1288.81	26.8	25.9	25.1	256.4	8.0460	1.841			20
20	32.0	1024.00	29.9	28.9	27.9	322.7	10.1280	1.463	11	7.5	21
21	28.5	812.25	33.6	32.4	31.3	406.7	12.7700	1.160			22
22	25.3	640.00	37.6	36.2	34.7	516.3	16.2000	0.914		5	22
23	22.6	510.76	42.0	40.3	38.6	646.8	20.3000	0.730			24
24	20.1	404.01	46.9	45.0	42.9	817.7	25.6700	0.577			24
25	17.9	320.41	52.6	50.3	47.8	1031	32.3700	0.458			26
26	15.9	252.81	58.8	56.2	53.2	1307	41.0200	0.361			27
27	14.2	201.64	65.8	62.5	59.2	1639	51.4400	0.288			28
28	12.6	158.76	73.5	69.4	65.8	2081	65.3100	0.227			29
29	11.3	127.69	82.0	76.9	72.5	2587	81.2100	0.182			31
30	10.0	100.00	91.7	86.2	80.6	3306	103.7100	0.143			33
31	8.9	79.21	103.1	95.2		4170	130.9000	0.113			34
32	8.0	64.00	113.6	105.3		5163	162.0000	0.091			35
33	7.1	50.41	128.2	117.6		6553	205.7000	0.072			36
34	6.3	39.69	142.9	133.3		8326	261.3000	0.057			37
35	5.6	31.36	161.3	149.3		10537	330.7000	0.045			38
36	5.0	25.00	178.6	166.7		13212	414.8000	0.036			39
37	4.5	20.25	200.0	181.8		16319	512.1000	0.029			40
38	4.0	16.00	222.2	204.1		20644	648.2000	0.023			
39	3.5	12.25	256.4	232.6		26969	846.6000	0.018			
40	3.1	9.61	285.7	263.2		34364	1079.2000	0.014			
41	2.8	7.84	322.6	294.1		42123	1323.0000	0.011			
42	2.5	6.25	357.1	333.3		52854	1659.0000	0.009			
43	2.2	4.84	400.0	370.4		68259	2143.0000	0.007			
44	2.0	4.00	454.5	400.0		82645	2593.0000	0.006			
45	1.8	3.10	526.3	465.1		106600	3348.0000	0.004			
46	1.6	2.46	588.2	512.8		134000	4207.0000	0.004			

Teflon Coated, Stranded Wire
(As supplied by Belden Wire and Cable)

Size	Strands⁵	Turns per Linear inch² UL Style No. 1180	1213	1371
16	19×29	11.2		
18	19×30	12.7		
20	7×28	14.7	17.2	
20	19×32	14.7	17.2	
22	19×34	16.7	20.0	23.8
22	7×30	16.7	20.0	23.8
24	19×36	18.5	22.7	27.8
24	7×32		22.7	27.8
26	7×34		25.6	32.3
28	7×36		28.6	37.0
30	7×38		31.3	41.7
32	7×40			47.6

Notes

¹A circular mil (CM) is a unit of area equal to that of a one-mil-diameter circle (π/4 square mils). The CM area of a wire is the square of the mil diameter.

²Figures given are approximate only; insulation thickness varies with manufacturer.

³Maximum wire temperature of 212°F (100°C) with a maximum ambient temperature of 13°F (57°C) as specified by the manufacturer. The *National Electrical Code* or local building codes may differ.

⁴700 CM per ampere is a satisfactory design figure for small transformers, but values from 500 to 1000 CM are commonly used. The *National Electrical Code* or local building codes may differ.

⁵Stranded wire construction is given as "count"×"strand size" (AWG).

Resistance is often placed in a circuit deliberately to limit the current or to provide various voltage levels required by a component. This is done with devices called *resistors*. Resistors will be described in detail in Chapter 6, "Resistors." The resistance inside a component or device is known as *internal resistance*. This term is commonly used to denote the resistance inside a voltage or current source such as cells and batteries.

3.2.4 Power

Power in a DC circuit is the product of voltage and current. It is the rate at which energy is expended or dissipated. *Dissipated* means energy changed from one form to another, such as from electric energy into heat energy. Power is expressed in joules per second, more often called *watts*, in electronics work.

In a DC circuit, power is simply the product of the voltage and current. A source of *E* volts, delivering *I* amperes to a circuit, produces *P* watts. More on this is presented in Section 3.7 on Ohm's law.

Power is also defined as the rate of doing work. Electrical power is used to perform all kinds of useful work for us every day. Operating our PC, watching TV, drying our clothes in an electric dryer, listening to a CD on our home stereo system, all consume, or dissipate power. In general, the more power an electrical appliance consumes, the larger it is physically. A high-powered appliance or system is almost always larger than a low-powered one. Since current is the rate of charge flow, or the amount of charge carriers moving past a given point per unit of time (coulombs per second), then power, which is the product of current and voltage, also involves time.

The electric utility company charges us based on our consumption of electrical power. Customers are charged a price each month based on kilowatt-hours of power use. This is the price for energy, since 1 W = 1 joule/second. In general then, it does not matter if a great deal of power is expended over a very short time, or a very small amount of power is expended over a long time. It is the product of current flow and voltage for which consumers ultimately pay.

3.2.5 Efficiency

Efficiency is a way of relating how well a device or system converts energy into a new, more desirable form. In Chapter 2, "Basic Concepts of Electricity," the concept of conversion efficiency was introduced. Efficiency usually refers to output power divided by input power. The result is then multiplied by 100 to give the answer as a percent. Percent efficiency is a good general method of figuring how well a device or system is performing. Except for resistors and electric heaters (which use resistive wire), no device is 100 percent efficient. Efficiencies are being constantly improved as newer designs, using newer materials, are created. Most solid-state (using transistors and other semiconductor devices) electronic equipment is roughly 50% efficient. This idea will become more important to you as a PC owner/builder/upgrader as you must decide on exactly what wattage capacity power supply you need.

Example 3.10

The rear panel label on a common stereo receiver lists the input power requirements as being 250 watts. The receiver is also rated as capable of producing 60 watts per channel of music power to each of the two stereo channels. What is the efficiency of the receiver? Refer to Equation 3.1.

Solution

$$(P_{out} \div P_{in}) \times 100 = \% \: Efficiency \tag{3.1}$$

$$(120W_{Out}) \div (250W_{In}) \times 100 = (2 \text{ channels} \times 60 \text{ watts each})$$

$$= 48 \text{ percent efficiency}$$

As you can see from the previous example, most common electronic systems are far from 100 percent efficient. Where does the rest of the power go? Much of it is wasted as heat energy, an unwanted byproduct of the conversion process. It is not difficult to prove this concept to yourself. Simply place your hand on top of any electronic system, such as a PC, a stereo receiver, or television set, that has been operating for an hour or so, and you will find that the case of the unit is warm to the touch. This is a direct result of the less than 100 percent efficiency of the system. The losses are manifested as heat. Heat also destroys electronic components. Solid-state devices (diodes, transistors, and microprocessors) are especially sensitive to excess heat. Any large computer has several fans mounted on it. Even small personal computers have fans on their power supplies and microprocessors, and newer, higher-powered PC power supplies include dual fans. Add-on fans are often used in the front and rear of a PC case to increase airflow, and cool the parts inside. Simply placing a fan on the power supply heat sink of most solid-state electronic units will greatly increase the useful life of that unit. A *heat sink* is a piece of metal, called a *thermal mass*, attached to the heat-producing device. The heat sink absorbs excess heat from the device, and then air blown over the fins of the heatsink by a small fan moves the heat from the sink to the air by convection. This action is similar to the way a fan and radiator in your car protects the engine from excess heat. The thermal efficiency of a heat sink, that is how well it moves heat from the device to be protected into the surrounding air, can be greatly increased with the use of a fan, to increase the rate of airflow, and thereby the rate of convective heat transfer. Heat sinks using attached fans are called *active* heat sinks, while those lacking fans are known as *passive* heat sinks. Newer active heat sinks use copper for the sink material, rather than aluminum, because copper is better at transferring heat from the chips to the air. Many PC users who enjoy playing games written for their machines opt to modify or "mod" the cases. One popular case mod is to cut out the sheet metal sides and top and mount larger fans. Often, fans are mounted on the side to direct a cool air stream onto the high-end (and hot) video cards favored by gamers. These video cards have their own fast microprocessors and RAM memory, and often cost as much or more than a complete low-end PC system. Another popular fan position is on the top of the PC case, to help suck out the hot air inside the case. Figure 3.3 shows some heat sinks used to cool microprocessors.

FIGURE 3.3 Microprocessor heat sinks.

Other forms of wasted energy, often called *power losses* or simply losses, will be considered in other chapters of this text.

3.2.6   Ohm's Law

Ohm's law is a mathematical way of stating the relationship between voltage, current flow, and resistance to current flow in an electric circuit. It is an essential concept that should be thoroughly memorized in order to solve common electronic circuit problems. In simplest form, Ohm's law states that for a given voltage applied to an electric circuit, the current flow that results is equal to the voltage applied divided by the circuit resistance. Equation 3.2 shows this:

$$I = E \div R \qquad (3.2)$$

where E is in volts, R is in ohms, and I is in amps.

Two more useful relationships can be derived by simple algebraic manipulation of this basic equation. The first, Equation 3.3, gives the voltage when the values of current and resistance are known.

$$E = IR \qquad (3.3)$$

The second, Equation 3.4, gives resistance when the voltage and current are known.

$$R = E \div I \qquad (3.4)$$

A convenient way of memorizing these relationships is to examine the Ohm's law "thumb wheel." By simply placing your thumb over the desired unknown quantity, the wheel will indicate the appropriate formula used to find the quantity. Figure 3.4 shows the Ohm's law thumb wheel.

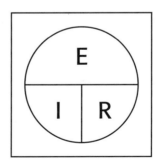

FIGURE 3.4   Ohm's law thumb wheel.

Example 3.11

Find the value of current in the circuit of Figure 3.5.

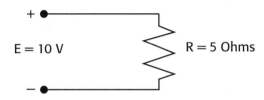

FIGURE 3.5   Schematic circuit diagram.

Always write down the given values in any circuit problem first. The applied voltage, E, is given as 10 V. The resistance value is given as 5 Ω.

Next, write the appropriate Ohm's law formula to solve the unknown value. In this case, use Equation 3.2:

$$I = E \div R$$

Then substitute the given circuit values into the formula.

$$I = 10 \ V \div 5 \ \Omega$$

Finally, solve the equation.

$$I = 2 \ A$$

Another circuit that requires using scientific notation skills follows.

Example 3.12

Solve for the current value in Figure 3.6, using Equation 3.2.

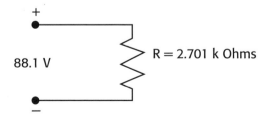

FIGURE 3.6   Schematic diagram.

$$I = E \div R$$
$$I = 88.1 \ V \div 2.701k \ \Omega$$
$$I = 88.1 \ V \div 2.701 \times 10^3 \ \Omega$$
$$I = 3.26175 \times 10^{-2} \ A$$

First, convert this answer to standard form. The number with the least amount of significant digits is 88.1 volts. So the answer must have no more than three significant digits, and use a common metric prefix, as explained earlier in this chapter.

$$I = 32.6 \ mA$$

Example 3.13

Solve for the applied voltage, E, in the circuit of Figure 3.7, using Equation 3.3.

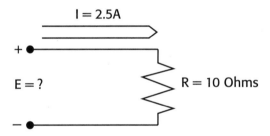

FIGURE 3.7   Schematic diagram.

$$E = IR$$
$$E = (2.5 A)(10 \ \Omega)$$
$$E = IR$$

Now try another problem using scientific notation.

Example 3.14

Solve for the value of applied voltage, E, in Figure 3.8, using Equation 3.3.

$$E = (25 \ mA)(4.7 \ k\Omega)$$
$$E = (35 \times 10^{-3} A)(4.7 \ k\Omega)$$
$$E = 0.16555 \ V$$

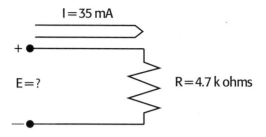

FIGURE 3.8 schematic diagram.

Changing to standard engineering form and rounding off to 3 significant digits gives:

$$E = 166 \ mV$$

Ohm's Law Power Formulas

Another useful Ohm's law formula concerns calculating electric power. For DC circuits, power is the product of current and voltage. Stated as a formula (Equation 3.5),

$$P = IE \qquad (3.5)$$

Two other versions of this basic equation can be obtained (Equations 3.6 and 3.7):

$$P = I^2 R \qquad (3.6)$$

and

$$P = E^2 \div R \qquad (3.7)$$

Example 3.15

Find the power consumed by a circuit that draws 4.5 amps from a voltage source that produces 36 volts. Use Equation 3.5.

$$P = IE$$
$$P = 4.5\ A \times 36\ V$$
$$P = 162\ W$$

Example 3.16

How much power does a 45 Ω resistor that has 35 mA of current flowing through it dissipate? Refer to Equation 3.6.

$$P = I^2 R$$
$$P = (35 \times 10^{-3}\ A)^2 (45\ \Omega)$$
$$P = (1.225 \times 10^{-3}\ A)(45\ \Omega)$$
$$P = 0.055125\ W$$
$$P = 5.51\ W$$

Example 3.17

How much power is dissipated by a device that has 400 volts applied across it, and that has 47 $k\Omega$ resistance? Refer to Equation 3.7.

$$P = E^2 \div R$$
$$P = (400\ V)^2 \div (47\ k\Omega)$$
$$P = (160,000\ V)^2 \div (47 \times 10^3\ \Omega)$$
$$P = 3.4042553\ W$$
$$P = 3.40\ W\ \text{expressed in the text's standard form}$$

3.3   SUMMARY

- *Scientific notation* is a convenient shorthand method for dealing with both very large and very small numbers.
- For electronics and engineering work, a system called *engineering notation* is normally used. It is similar to scientific notation, but uses numbers in the range of 1 to 1,000 times a power of 10 in steps of 3.

■ The decimal part of a scientific notation expression may have any number of significant figures, depending on the required accuracy.

■ When mathematical operations are performed between or among numbers, the resultant cannot have more significant digits than the original numbers themselves.

■ If the original values have different numbers of significant figures, then the resultant must be *rounded off* to have the smallest number of significant figures consistent with the original values.

■ In a decimal number, all nonzero digits are significant.

■ All zeros between significant digits are significant.

■ A zero following a nonzero digit may or may not be significant, depending on whether it is accurate or whether its only function is to help place the decimal point in an approximate number.

■ For electronics work, the metric units of milliamps, millivolts, microamps, microvolts, kilohms, kilovolts, and megohms are commonly used.

■ When rounding off a number to a certain number of significant digits, look at the next least significant digit, that is, the digit to the right of the last significant digit required. If the next least significant digit is 5 or greater, round up the digit to the left. If the next least significant digit is 4 or less, do not round up the next more significant digit.

■ The standard form used throughout this text is engineering notation with standard metric prefixes.

■ *Voltage* is proportional to the strength of the electric field and is the force on an electron tending to make it move in a circuit (electric current).

■ Electric *current* is a flow of charge carriers past a point in an electric circuit. The charge carriers may be any charged particles, but generally are electrons.

■ The conventional current method describes current as the movement of positive charge carriers from a more positive point to a less positive point.

■ This text will use the conventional current flow method when indicating direction of current flow in a circuit.

■ *Resistance* is the opposition that a material presents to direct current in a circuit.

■ *Power* is the product of voltage and current in a DC circuit. It is the rate at which energy is expended or dissipated. *Dissipated* means energy changed from one form to another, such as from electric energy into heat energy. Power is expressed in joules per second, more often called watts, in electronics work.

■ *Efficiency* is a way of relating how well a device or system converts energy into a new, more desirable form.

■ Heat is a destroyer of electronic components. Solid-state devices (diodes, transistors, and microprocessors) are especially sensitive to excess heat. This is the reason for the growing number of fans installed inside a PC's case as new, more powerful (and hotter) processors are developed.

■ *Ohm's law* is a mathematical way of stating the relationship between voltage, current flow, and resistance to current flow in an electric circuit. It is an essential concept that should be thoroughly understood in order to solve common electronic circuit problems.

■ A convenient way of memorizing these relationships is to use the Ohm's law thumb wheel.

■ Power is the product of current and voltage. Stated as a formula, $P + IE$. Two other versions of this basic equation can be obtained:

$$P = I^2R \text{ and } P = E^2 \div R.$$

3.4   KEY TERMS

Scientific Notation	Engineering notation	Efficiency
Voltage	Significant digits	Ohm's law
Current	Rounding off	Power losses
Coulomb	Standard form	Heat sink
Resistance	Metric prefixes	
Internal resistance	Power	

3.5   EXERCISES

1. Convert the following numbers to scientific notation form:
   a. 0.00345
   b. 1,452,599
   c. 23,455
   d. 7,999,333,000

2. Convert the following numbers to engineering notation with standard metric prefixes.
   a. 0.02472
   b. 0.000034567
   c. 39,978,213
   d. 6,731,002

3. Perform the indicated operations and write the answers in engineering notation, using standard metric prefixes.
   a. $(0.00142) \times (123,844)$
   b. $(1,688) \times (.00047)$
   c. $(488.002)/(82.9314)$
   d. $(666,923)/(23.0427)$

4. Compute the percent efficiency of a radio that requires 10 watts of power to operate. The radio produces a maximum of 5.6 watts of audio (electrical energy at sound frequencies) output.

5. Compute the percent efficiency of a microwave oven that uses 290 watts from the house AC power outlet, and produces a maximum of 166 watts of heat in the food.

6. Express 2.356 using engineering notation and a standard metric prefix.

7. Express 145,390 using engineering notation and a standard metric prefix.

8. Express 34,788 in scientific notation form using three significant figures.

9. Express 71,516 in scientific notation form using two significant figures.

   For the rest of the exercises, give answers in engineering notation, using standard metric prefixes.

10. Compute: $34.5\ mA \times 17\ \Omega =$

11. Compute: $1.7\ kV/14\ A =$

12. Compute: $88\ mV/2.7\ k\Omega =$

13. Write the Ohm's law expression used to find voltage, if current and resistance values are given.

14. Write the Ohm's law expression used to find current, if voltage and resistance are given.

15. Write the Ohm's law expression used to find resistance, if voltage and current are given.

16. A circuit has 240 V applied. The resistance is 300 Ω. Find the current in the circuit.

17. A circuit has 9 V applied. The resistance is 3.3 $k\Omega$. Find the current.

18. A circuit draws 2 *A* from a voltage source of 48 *V*. What is the circuit resistance?

19. A circuit draws 23 ΩA from a source providing 5 *V*. Compute the resistance of the circuit.

20. A light bulb draws 250 *mA*, and offers 150 Ω of resistance. What is the value of applied voltage?

21. A CD-ROM drive installed in a PC draws 3.4 *A* and offers 3.63 Ω resistance. What is the voltage supplied by the PC's power supply to the CD-ROM drive?

22. What is the amount of power consumed by the CD-ROM drive in question 21?

23. A 100 *W* light bulb operates at 120 *V*. What is the hot (operating) resistance of the light bulb?

24. A car stereo receiver draws 12 *A* from the battery, which supplies 12.6 *V* of DC. The stereo produces a maximum of 25 *W* per channel for each of the two stereo channels. What is the efficiency of the receiver?

25. An electric dryer is rated to operate on 240 *V*. The dryer draws 15 *A* of current. The dryer produces 3,550 *W* of heat to dry clothes. Compute the efficiency of the dryer.

4

The Electric Circuit and Voltage Generation

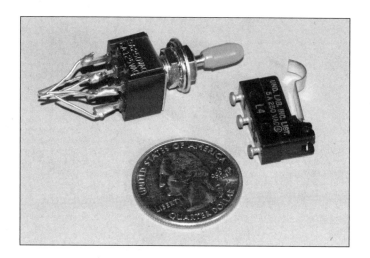

4.1   INTRODUCTION

This chapter presents the concept of a complete pathway for current, which is an essential requirement for electrical charges to flow and allow useful work to be done.

4.2   THE NEED FOR A COMPLETE PATH

Electric current can flow only through a conducting medium that forms a complete electrical pathway around a closed circle, or *circuit*. The path starts at one electrical connection to the source of potential difference, called a *terminal*, runs through all parts of the circuit, and returns back to the opposite polarity terminal of the *voltage source*. Electrons flow from the negative terminal of the voltage source, through the circuit, and return to the positive terminal of the source. As shown in Figure 4.1, for every electron produced by the voltage source, one electron returns back to the source.

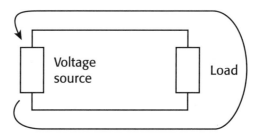

FIGURE 4.1 Electrons flow from a voltage source through the circuit and back to the source.

4.2.1 Apparent Speed

Although it seems as if each electron travels the entire circuit pathway almost instantaneously, the actual speed of any given electron is very much slower. The reason is that each electron travels only a short distance to another neighboring atom, where it may drop back into an empty valence shell (hole). Another electron is pushed a short distance on toward the end of the circuit, before it too falls back into another hole. It is said to recombine with a positively charged atom. This *recombination* releases energy. This idea of moving charges can be visualized by imagining a long row of pool balls on a pool table. Figure 4.2 shows how, if the end of the row of pool balls is struck by the cue ball, only the ball on the opposite end of the long row of pool balls will immediately move. The energy is simply transferred from one ball to the next, in line, until it finally pushes the end ball off in the direction of travel of the cue ball. All this happens extremely quickly, but the cue ball is not the ball that ultimately moved off down the pool table. Figure 4.2 shows this effect.

Electron movement in a conductor acts in a very similar way, although instead of mechanical force at work, it is electrical force that drives the electrons in an overall or net direction. So, despite the apparent rapid speed of the electrons, it takes a considerable amount of time for a single electron to travel the complete distance around an electrical circuit. Nevertheless, when you turn on a switch a nearly instantaneous effect is seen.

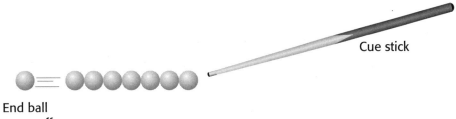

End ball
moves off

FIGURE 4.2 Balls on a pool table.

4.3 BASIC CIRCUIT ELEMENTS

All electric circuits must have at least the basic circuit elements required for current to flow. The next sections look at these basic circuit elements.

4.3.1 The Source

The force that drives the current around the electric circuit was once known as *electromotive force* or *EMF*. That accounts for the designation of *E* for the devices that produce electric potential difference, measured in volts. The device that produces potential difference or provides a voltage to drive a circuit is called the voltage source, or simply the *source*, as shown in Figure 4.3.

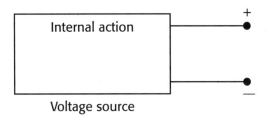

Voltage source

FIGURE 4.3 A voltage source.

4.3.2 The Control Element

Nearly all electric circuits have a means of stopping the flow of current at will. This device is called the *control* element. Most controls are switches that allow for connecting or disconnecting the current pathway around the circuit. A common example of a switch is the one used to turn room lights on and off. The switches mechanically interrupt the circuit in the *off, open,* or *break* position, or complete a connection between conductors—called switch contacts—in the *on, closed,* or *make* position.

Switches

There are many different types of switches, each designed to perform a particular type of control in a circuit. In general, there are mechanical switches and electronic switches. The mechanical types will be presented in this chapter. Mechanical switches are further described as being either linear (those that act in a straight-line mechanical operation) or rotary (those that operated by turning or twisting a handle). Figure 4.4 shows some typical switch types.

FIGURE 4.4   Mechanical switches.

Switch Designations

Toggle switches, such as those used in most homes for control of lights and many appliances, are linear types. There are some simple rules for designating the function of a switch in a circuit. Most linear switches are given a four-letter abbreviated designation according to how they function. The way these types of switches are designated follows two rules:

- Each separate, distinct circuit a switch can control at one time is called a *pole* (abbreviated "P" in circuit diagrams).
- Each switch position that actually causes a connection to be made in a circuit is called a *throw* (abbreviated "T" in circuit diagrams). The "off" position of a switch does not count as a valid position according to this designation scheme.

As an example, the switch that controls a light bulb from a single location in a room can either make (connect) or break (disconnect) a single circuit line to the light bulb. Connect and disconnect are also commonly called the *closed* and *open* switch positions, respectively. Since this type of switch can control only one circuit at a time, it is considered to be a single pole (SP) type switch. Its two physical positions are "on" (connect) and "off" (disconnect). Since only one switch position actually connects a circuit, it is considered a single throw (ST) type. The complete designation for this type of one-circuit, on/off switch is *single pole, single throw* or *SPST*. In the PC world, small gold-plated metal pins are used as switch contacts, and a small plastic covered metal strip makes contact with the two pins. This is called a *jumper*. Jumpers are used in many places in a PC, on the motherboard, drives and so on, to serve as very inexpensive, seldom-changed switches. Figure 4.5 shows the schematic symbols for an SPST switch.

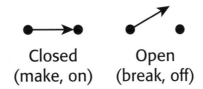

Closed
(make, on)

Open
(break, off)

FIGURE 4.5   Schematic symbol of an SPST switch.

Figure 4.6 shows SPST jumpers on a motherboard.

FIGURE 4.6 Motherboard SPST jumpers

If a switch can connect only one circuit at a time, but has two possible positions that can connect circuits, it is known as a *single pole, double throw (SPDT)* switch. Figure 4.7 shows the schematic symbols for an SPDT switch.

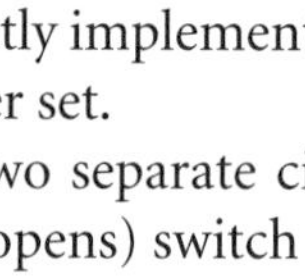

FIGURE 4.7 Schematic for an SPDT switch.

This type of switch is frequently implemented with jumpers on PC equipment. Figure 4.8 shows an SPDT jumper set.

A switch that can connect two separate circuits (but not to each other) and either makes (closes) or breaks (opens) switch contacts at the same time is known as a *double pole, single throw* or *DPST*-type switch. This type of switch is essentially two SPST switches mechanically connected together that operate simultaneously. Figure 4.9 shows the schematic symbols for a DPST switch.

FIGURE 4.8   SPDT jumper set.

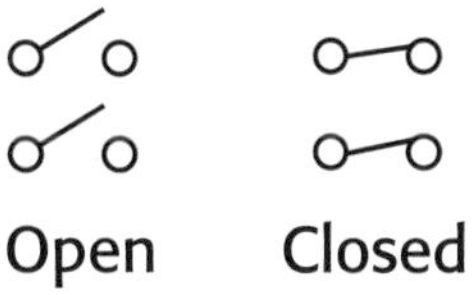

Open   Closed

FIGURE 4.9   Schematic symbol for a DPST switch.

Two SPDT switches connected together that act simultaneously make a *double pole, double throw* or *DPDT* switch. Figure 4.10 shows the schematic symbols for a DPDT switch.

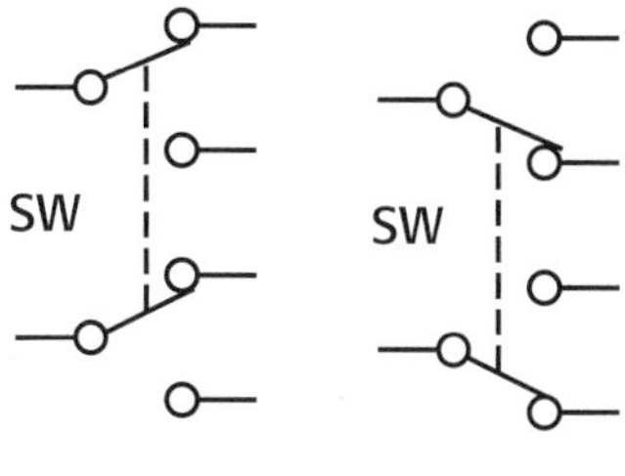

SW   SW

FIGURE 4.10   DPDT switch schematic.

A switch operated by a turning or twisting operation, called a *rotary* switch, is designated a bit differently. An example of a rotary switch is the one used to control a 3-way light bulb. Each of three different switch positions turns on a progressively higher-powered filament in the light bulb. The fourth switch position turns the light off. An example of a two-section rotary wafer switch used to control communication circuits is shown in Figure 4.11.

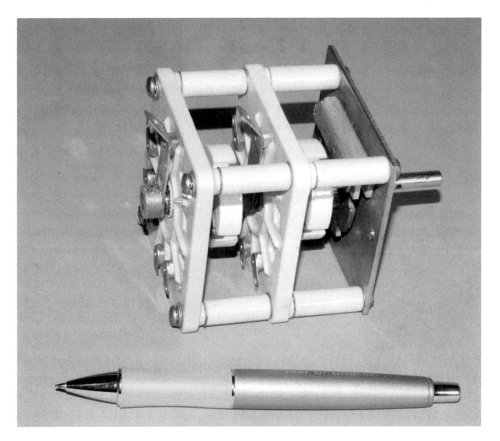

FIGURE 4.11   A two-section rotary wafer switch.

Another example of a rotary switch is the distributor cap in an automobile engine. In the case of a four-cylinder engine, the switch has one center switch contact, called the *rotor*, or more exactly, the *armature*, and four equally spaced contacts around the edge of the inside surface of the insulating material comprising the bulk of the cap. The purpose of the distributor cap is to connect high-voltage electricity, sequentially, to each of four different spark plugs in the engine, one at time. The center conductor (rotor) receives the high voltage energy from the engine's *coil* (actually a high-voltage transformer), and by rotating by each contact for the four spark plug wire connections, distributes the voltage to the respective spark plugs. The rotor's rotation is driven by a shaft connected to the engine, usually the camshaft. Since only one spark plug at time may be connected by the switch, the switch is designated single pole. But there are four possible circuits that connect to the spark plugs. So these are called *positions* (*Pos*) rather than throws. The complete electrical designation for this type of switch is *single pole, four position*, or an *SP4-Pos* rotary switch.

The distributor cap in a six-cylinder car engine would be known as an SP6-Pos. The distributor cap and one spark plug wire for a six cylinder engine is shown in Figure 4.12.

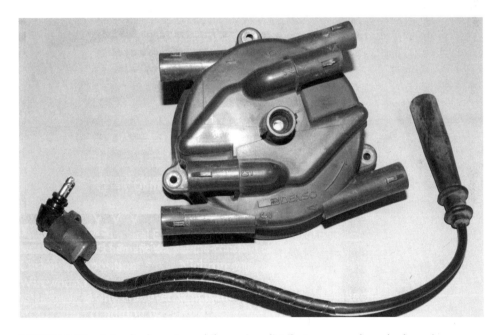

FIGURE 4.12   A 6-cylinder automobile engine distributor cap and spark plug wire.

Switch Ratings

Switches are made to work at specified maximum voltage and current levels. Most switches will have a maximum voltage and maximum current rating clearly marked on the switch body. These ratings should never be exceeded or else damage to the switch may result. In general, the higher the ratings for a given switch, the more the switch will cost. Also, in general, the higher the ratings, the physically larger the switch will be. Figure 4.13 shows switch ratings.

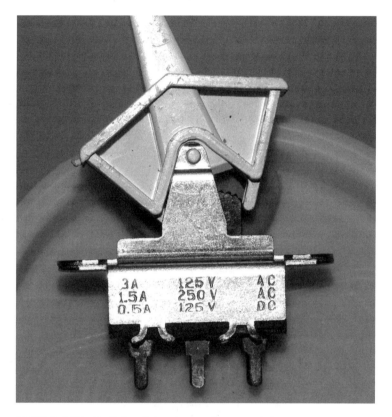

FIGURE 4.13 Switch ratings stamped on a switch.

Switch contacts are further described as to function. All the types presented so far remain in the position where last set until intentionally moved to the opposite position. They never change position on their own. A special-type switch is spring-loaded to return to a certain position after being actuated. This type is called a *momentary-contact* switch. One type of momentary-contact switch connects a circuit or circuits when actuated, or "turned on," and then returns to the off position by itself when released. Other switch types disconnect a circuit when actuated, and then return to the "on" position when released. A switch that is not making a connection until actuated is considered to be a normally open, or *NO* momentary-contact switch. An example of an NO switch is a doorbell pushbutton switch. Figure 4.14 shows examples of NO pushbutton switches used on PCs.

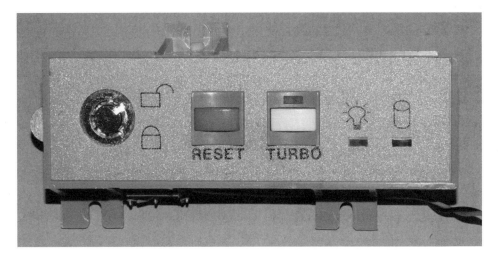

FIGURE 4.14 NO PC pushbutton switches.

A switch whose contacts are connected until the switch is actuated, causing the contacts to disconnect is considered to be a normally closed, or *NC* switch. In an NC switch, the contacts return to the open position once the switch is released. An example of an NC switch is a switch used in a burglar alarm system. The switch has two parts: one stationary part on the door or window side molding, and one that is attached to the door or window edge itself. Door movement causes the two parts of the switch to separate, opening the switch, which triggers the alarm. Although not spring mounted, since the door or window is normally closed, and the switch is closed in its normal position, it can still be considered an NC switch.

Relays

Not all switches directly control a circuit because a person mechanically moves a lever or handle. One exception is the distributor cap presented earlier, which is actuated by a machine (car engine). Sometimes we must control one electric circuit with another circuit, but we do not want any electrical interaction between the two circuits. In this case we use a special type of switch called a *relay*.

During a track and field event, called a relay race, one runner hands off the baton to the next runner, without actually touching the next runner. An electromechanical relay operates in an analogous way. The basic relay consists of a coil of wire wound around a metal form. The metal in the form is specially formulated to concentrate magnetic fields. Recall that whenever current flows in a conductor, a magnetic field is created. One way to increase the magnetic effect in a small area is to wind the wire into a coil. If we place a material inside the coil, which further concentrates the magnetic field lines, called the magnetic *flux lines*, we create a very powerful magnetic field around the coil. This magnetic field will be present whenever the coil is energized by current flow. Whenever the current through the coil stops, the magnetic field will disappear. Close to the coil's end is a moveable piece of metal, called the armature. Recall that the armature in a rotary switch is the moving contact. This armature is also part of a switch. The armature is spring-loaded. When the relay coil is energized, the magnetic field produced by the coil flows through the armature, and it is magnetically attracted to the end of the coil, and "pulls-in." This provides the movement required to operate the switch. This switch can be used to control another circuit. By this method, we can use a small, inexpensive switch to actuate the relay coil, and have the relay control a circuit operating at very high voltage, current, or both (power). Figure 4.15 shows an automobile starting relay.

FIGURE 4.15 Automobile starting relay.

One example of the use of a relay is the starting relay in an automobile. The circuit that actuates the relay coil (the load) consists of the car's battery (the source), ignition switch (the control), and connecting wires (the conductors). Turning the ignition switch to its start position energizes the relay coil and brings into contact two very large, copper switch contacts. These heavy contacts connect the starting motor (*starter*) to the positive battery terminal through the short, heavy-gauge wire. The negative side of the starter circuit is already completed by the metal mounting brackets on the starter. The negative side of the car battery is always connected to the engine and chassis metal. The starting relay allows use of a relatively small and inexpensive switch that operates at approximately two or three amps to control the starter, which operates at perhaps 150–200 amps for a large V-8 engine in cold weather. The ignition switch uses only small-gauge wires to make the relatively long run from the car battery, through the ignition switch, and back to the starting relay. The wire connecting the starter to the battery, however, is many times heavier. By using the relay system, we can remotely place the starting relay in the engine compartment.

Some designs incorporate the starting relay into the starter motor itself. This system is called a *solenoid*. A solenoid is actually a relay that also operates a gear (called a pinion gear) that mechanically links the starter motor to the flywheel (called the ring gear) on the end of the car engine's crankshaft. At the same time the starter motor is turned on, it is coupled by this gear to the flywheel in order to turn over the car engine. When the engine starts, and we release the ignition key from the starting position, it disconnects the solenoid (relay), turning off the starter, while a spring simultaneously returns the starter motor gear to the resting position, uncoupling the starter from the engine. This prevents the starter motor from being driven by the car engine, and being quickly destroyed. Figure 4.16 shows a starting relay and circuit schematic, and Figure 4.17 shows the details of a starter motor.

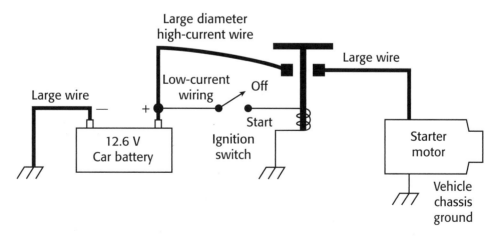

FIGURE 4.16 Automobile starting relay and circuit schematic.

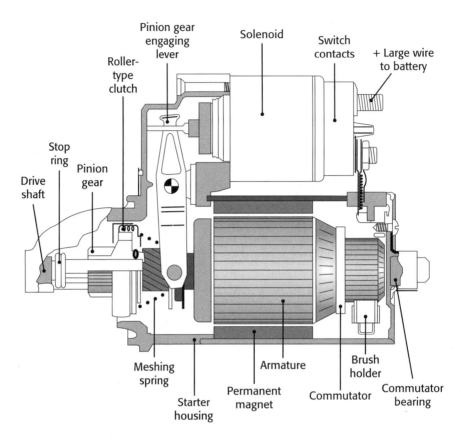

FIGURE 4.17 Automobile starter motor construction.

The remotely mounted relay or solenoid scheme allows us to keep the expensive, heavy current wire to the starter motor short, thereby decreasing cost. Copper used to make wires is relatively expensive. Cars are extremely cost competitive worldwide, so every method of lowering production costs is employed. It is also unsafe to directly switch such high current on the dashboard with one's hands.

Relays are also used to control high, dangerous voltages in radio transmitters, and electric motors that power industrial machinery. They allow us to actuate circuits that would be dangerous for us to directly control by hand. Some relays are very large, in order to control large currents or high voltages, and are called *contactors*. They have heavy metal contacts, which are often silver or gold plated. Figure 4.18 shows a contactor circuit used to control a large electric motor.

So, just as there was no direct physical contact between the runners in a relay race, the only connection between circuits in an electromechanical relay is by magnetism.

FIGURE 4.18   A large contactor.

Relay Ratings

Relays are rated in terms of the coil and the contacts. Relay coils are rated in terms of voltage and current. The ratings should be adhered to for the relay to operate reliably. A relay will usually actuate, or pull-in, at a somewhat lower voltage and current applied to its coil than the rated voltage and current. The extra amount is used to ensure reliable operation. The ratings should never be exceeded, however, or the coils will likely overheat or burn out.

Relays designed to operate at a specific coil voltage are known as *voltage* relays. Such relays are described as being a "6-volt relay" or a "24-volt relay," and so on.

Relays designed to operate at a specific current through them are termed *impedance* relays. The main coil rating for this type of relay is the coil's DC resistance value, or its AC impedance value. Relays such as these would be described as being a "200 ohm" coil resistance relay, for example. In general, the impedance relays are small, draw little coil current, and are used to control small amounts of voltage or current in a circuit. Often they are actuated by a sensor. The relay then controls another circuit, which the sensor is incapable of operating directly. More on impedance will be presented in Chapter 12, "Alternating Current Circuits."

The other main relay rating is for the contacts. Contacts are the small pieces of conducting metal used to make or break during relay actuation. Normally, solid copper is used. The copper is formed into small, flat, round pieces, similar to a coin, although usually much smaller in diameter. One sixteenth to one quarter inch diameter is approximately the range of most medium size relay contacts. These contacts become separated (open) or pressed together (closed) by the armature of the relay. Silver or gold plating is sometimes used to ensure an even lower resistance pathway when the contacts are touching. Silver is an even better conductor than copper, but is subject to tarnish. Gold is a slightly worse conductor than copper, but it does not corrode. Figure 4.19 shows a close-up view of relay contacts.

FIGURE 4.19   A close-up photo of relay contacts.

The contacts are rated for normal amperage value. This value should not be exceeded or the contacts will quickly become pitted and worn. The switching arrangement of the contacts is also important. Remember, a relay is simply a magnetically operated switch, so the switching arrangement is critical. Relays are available with SPST contacts, DPDT, or any other desired arrangement. They are also available with NO, NC, or combinations of these. A special-type relay known as a *stepping* relay was used extensively by the telephone company to select the dialed telephone circuit. One stepping relay for each digit in the dialed number enabled the caller to become connected to only the desired distant phone. Stepping relays were open to the air, and therefore prone to wear and tear, and the contacts became dirty. Most telephone systems have long since been converted to using newer solid-state replacements for the relays.

4.3.3 The Load

The part of the circuit where work, in the desired form, is done is called the *load*. A load can be a light bulb, an electric motor, or a heating element in an appliance that works by producing heat, such as a toaster, heater, or clothes dryer. Other loads are radios, TVs, and computers. Anything that is connected to a source of electrical potential difference—a voltage source—and that has a complete conducting pathway to and from the source, serves as a load in a circuit.

All practical loads have some amount of resistance. As you learned in Section 3.2, when current flows through a resistance, power is dissipated. All real loads dissipate some amount of power in a circuit. As was presented in Section 3.2.5, some types of loads are more efficient at converting energy to its new form than others.

Load Ratings

Most loads are rated in terms of voltage, current, and power. These ratings are important to know, as a load operated beyond its rated specifications (specs) will have a much increased likelihood of premature failure. If a load does not fail outright, it will surely have a decreased life span due to operation at overrated voltage and/or current (power). As shown in Figure 4.20, a simple *light-emitting diode* (*LED*), commonly used for panel indicators in all sorts of electronic equipment, normally operates at around 20–30 milliamps at about 2 volts. A resistor is normally used in such a circuit to limit the current to the LED. Under these conditions, it will last virtually a lifetime. If, however, the current is allowed to increase to around 75 mA, the diode will burn very brightly, but it will last only a few minutes to a few hours at most.

FIGURE 4.20 A light emitting diode (LED).

4.4   VOLTAGE RISES AND VOLTAGE DROPS

A circuit has voltage rises and voltage drops. The next sections present each of these concepts.

4.4.1   Voltage Rises

A voltage, which comes about because of energy transformation from one form to another, is known as a *voltage rise.* Voltage rises are designated by the letter E in a circuit. Voltage rises are measured in volts. There are presently six common methods of producing a voltage rise. These methods are *electromagnetic, electrochemical, photoelectric, thermoelectric, piezoelectric,* and *triboelectric.*

Electromagnetic Voltage Generation

By far, the largest single method of producing electricity on a large scale is *electromagnetism.* This involves the relationship between an electric current and a magnetic field. There are two parts to this relationship. The first part is this: whenever current flows in a conductor, a magnetic field exists around the conductor. The strength of the magnetic field is directly proportional to the strength of the current flowing in the conductor. This means the greater the current, the greater the strength of the field. The polarity of the field is dependent upon the direction of current flow. There is a simple way to remember this relationship. It is called the *right-hand rule for conductors.* It states that if you grasp an insulated conductor (never place any part of your body on a bare wire carrying current) with your thumb pointing in the direction of conventional current flow (positive to negative) then your fingers will indicate the direction of the magnetic field around the conductor. It is assumed that the magnetic field travels from the north magnetic pole to the south magnetic pole. If the conductor is formed into a coil, there is another right-hand rule to use to remember the relationship of current flow and magnetic polarity. If the fingers of the right hand wrap around the coil in the direction of conventional current flow, (from positive to negative), the thumb of the hand will point in the direction of the North magnetic pole. The right-hand rules for coils and conductors are shown in Figure 4.21.

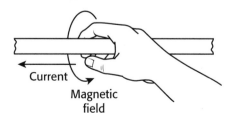

FIGURE 4.21   Right-hand rules for conductors and coils.

The second part of the electromagnetic relationship states that whenever there is relative motion between a conductor and a magnetic field, there will be a voltage induced (created) in the conductor. The value of this *induced voltage* depends on four things:

The strength of the magnetic field acting on the conductor: A stronger field produces a stronger induced voltage.

The length of the conductor subjected to the magnetic field: A longer conductor in the field will have more induced voltage than a shorter one. The most common way of doing this is forming the wire into a coil.

The angle between the magnetic field and the conductor: The greatest amount of induced voltage will occur when the magnetic field is at right angles (90 degrees) to the conductor.

The speed of interaction between the field and the conductor: The greater the speed, the greater the induced voltage.

Electromagnetic generation of power is the method by which large electric generators operate. The commercial power companies use the interaction of powerful magnetic fields and large coils of wire to produce electricity. There are two types of electric generators: one with a fixed field coil and a rotating armature, and one with a fixed armature and a rotating magnetic field. The rotary motion is supplied in mechanical form from coal- or nuclear-fired boilers, moving water, or even wind power.

The largest amount of commercial power generation uses the system of burning coal to heat water, turning it to steam to drive a turbine. The rotating turbine blades drive a shaft that is connected to a large generator. Large power stations may have a dozen of these large boiler-turbine-generator units. Refer to Figure 4.22 for a picture of large commercial power turbines.

In a hydroelectric power station, water is built up behind large dams that harness the water's work potential. That energy is given up, or changed to mechanical energy (work is done), as the water drops to a lower level. The falling water is forced through tubes and made to turn the turbine blades that convert the water's energy into mechanical rotation of a shaft. The shaft turns a generator. The generator spins coils of wire inside a strong magnetic field (or vice-versa). The result is generation of a voltage rise due to the voltage induced in the coils cutting through the magnetic field. It's the water's potential energy (high up, with the capability to fall and release energy) harnessed to produce electric energy.

Nuclear power stations use the heat produced by the radioactive decay of certain elements to produce steam. The steam is then used to turn a turbine. The turbine turns the generator, as in the water-powered and coal-fired systems.

FIGURE 4.22   A large commercial power generator. Photo courtesy of
Pacific Gas and Electric Company.

Geothermal power stations use pipes sunk deep into the ground over active
heat sources in the Earth's crust. Water circulates through the pipes, and is turned
into steam by the heat deep in the earth. The steam is then used to turn turbines,
and so on. Figure 4.23 shows a geothermal electric power genrating site with multiple gererators (each escaping cloud of steam represents a single generator site).

FIGURE 4.23   Geothermal electric power generating sites. Photo
courtesy of Pacific Gas and Electric Company.

Wind generators are driven by the huge blades that operate the generator. This system is extremely clean and quiet, generating no harmful byproducts. Current wind generators are being used to supplement coal-fired power generators in many places in the world. Electricity produced in this way still costs more than electricity produced by the burning of fuel. As world prices for fossil fuels increase, as the supply surely dwindles, wind-driven electric power generators will become evermore competitive. Figure 4.24 shows some wind-powered electric generators.

FIGURE 4.24 Wind-powered electric generators. Photo courtesy of Pacific Gas and Electric Company.

So, even though many people consider these systems "alternative energy sources," they all work because of the relationship of mechanical movement between conductors and a magnetic field. A car engine uses mechanical rotation of the crankshaft to turn a fan belt, which drives an alternator (AC generator), to power the electric equipment and recharge the battery. It is another form of electromechanical generator.

Electric motors operate in the reverse of electric generators. They use the interaction of magnetic fields and conductors to produce torque, or rotating force. An electric motor consists of a *field coil* that is energized by the applied voltage to produce a magnetic field. The rotating armature is also wound with coils of wire having many layers. The *armature* is also energized by the applied voltage. The motor is manufactured to cause a difference in the polarity of the field produced by the field coils and the armature. It is this difference in magnetic fields that drives the motor. Figure 4.25 shows a simplified motor/generator classroom model. More detail on motors and generators will be presented in Chapter 10, "Magnetism and Electromagnetism."

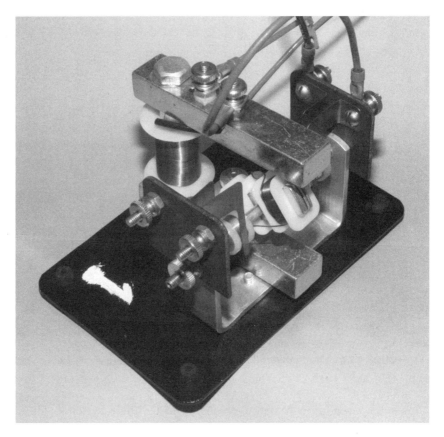

FIGURE 4.25 Simplified motor/generator model.

Although not as easily understood, the interaction of moving charges and conductors is the principle by which radio, radar, and television operate. A transmitter causes a very high-frequency (rapidly changing in polarity between positive and negative) current to flow in an antenna, and the electromagnetic field produced by the antenna current is transmitted a long distance away. This principle is known as *radio wave propagation*. When the propagated electromagnetic waves strike a distant antenna, a minute voltage is induced in the receiving antenna. This minute voltage is amplified by the receiver, and the information contained within the transmitted signal is reproduced in the form of radio sound, radar images, or television sound and pictures. A simple radio link is shown in Figure 4.26. See Chapter 16, "Communications," for more on electromagnetic energy propagation.

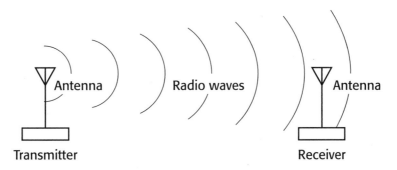

FIGURE 4.26   A simple radio link.

Electrochemical Voltage Generation

Generation of a voltage rise by electrochemical means is currently the second largest method by which electricity is generated. *Electrochemical* voltage generation is done by *cells*. A cell is a single unit that consists of two different *electrodes* and a substance that chemically acts on the electrodes, called an *electrolyte*. A simple electric cell is shown in Figure 4.27.

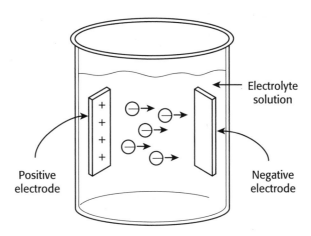

FIGURE 4.27   A simple electric cell.

There are two general categories of electrochemical cells: primary cells, which cannot be recharged, and secondary cells, which can be recharged. A connection of two or more cells to act together in a circuit is called a *battery*. Sometimes, several cells are connected together inside a common case, and the entire unit is known as a battery. Most common units used in standard flashlights, such as the "AA," "C," or "D" size, are actually cells. Many people mistakenly call them batteries, however. The 9-volt rectangular battery used in many toys, smoke alarms, many digital multimeters (DMMs), and other forms of small hand-held test equipment is, in fact, a connection of six cells in series inside a metal case, and actually is a battery. Also, the unit used in automobiles is made of six separate cells connected together in series-aiding form (the individual cell voltages add up to the total voltage) inside a common case, and is correctly called a battery. A simple series-connected battery is shown in Figure 4.28.

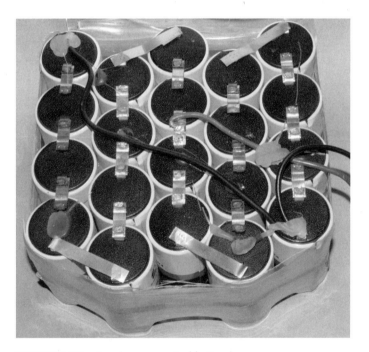

FIGURE 4.28   A series-connected battery.

Primary Cells

Of the primary type cell, there are many different types in use today. One that was formerly the most popular for general service applications is the carbon-zinc "dry cell." A dry cell is not really dry. The electrolyte solution is in the form of a thick, gooey paste. This is the stuff that leaks out of the cells if they are left in equipment too long. It usually results in damage to the cell connectors and may ruin the equipment. In fact, if the electrolyte solution does dry out, the cell stops working. A "D"-sized 1.5-volt carbon-zinc dry cell and a 6-volt lantern battery are shown in Figure 4.29.

FIGURE 4.29   A carbon-zinc dry cell lantern battery and a D cell.

The voltage developed at a cell's terminals is known as its *terminal voltage*. The terminal voltage is different for different types of cells. It is determined by the materials used to construct the cell; both the electrodes and the electrolyte. Terminal voltage also depends on the state of cell discharge. The amount of discharge depends on whether the cell is connected to a load that draws current from the cell, and the amount of that current draw. Terminal voltage also depends on temperature. Some types of cells lose their capability to produce voltage at very high or low temperatures. Table 4.1 gives a good comparison of some of the more popular types of cells.

So-called alkaline cells are very popular for applications that require the load device to sit unused for long periods. Alkaline manganese-dioxide cells use amalgamated zinc as the negative plate, manganese dioxide for the positive plate, and potassium hydroxide for the electrolyte. Terminal voltage is about 1.55 volts. They

TABLE 4.1 A Comparison Chart of Different Types of Electric Cells

Cell Type	Materials	Relative Age	Advantages	Disadvantages	Rechargeable?
Carbon-Zinc	Carbon rod positive electrode, zinc can negative electrode, sulfuric acid electrolyte paste.	Old	Cheap, relatively long shelf life.	Poor discharge curve. Electrolyte corrodes through zinc can and can cause burns and damages equipment. Causes land-fill pollution.	No
Alkaline	Manganese-dioxide positive electrode, zinc can negative electrode. Alkaline zinc-chloride electrolyte.	Newer	Last about 5 times longer than Carbon-zinc cells, have a flatter discharge curve.	Rather poor discharge curve.	Generally no, but some newer rechargeable types have been developed, but can only be recharged about 25 times.
Nickel-Cadmium (NiCd)	Nickel positive electrode, cadmium negative electrode, zinc-chloride electrolyte.	Newer	The first widely available recharge-able small cell type. Relatively flat discharge curve. Cheaper than other types of rechargeables.	Contains highly-toxic cadmium. Suffers from "memory effect" if not fully recharged. Will go bad if left discharged for long periods.	Yes, up to about 1,000 times if properly cared for.
Nickel	Metal-Hydride NMh	New			
Lithium-Ion	Carbon positive electrode, cobalt oxide negative electrode, lithium solution in ether electrolyte	New	Very light weight, very high energy density. Don't suffer form memory effect.	Lithium is explosive in contact with water.	Yes
Lithium Polymer (PLB)	Aluminum positive electrode, copper negative electrode, polymer electrolyte.	Very new	Extremely thin packaging forms. They don't like heat.	Expensive, lower energy density than lithium-ion type.	Yes
Sealed Lead-Acid (SLA)	Two types of lead electrodes. Sulfuric acid-water electrolyte.	Old	Does not leak electrolyte, offers good energy density, can be recharged many times, "mature" technology.	Heavy, prone to failure from over-charging, or left for long times without charging.	Yes, up to perhaps 250 times.

can operate continuously at high discharge rates over a wide temperature range. They provide low cost per hour of use and have a shelf life of over three years. Several alkaline cells and an alkaline 9-volt battery are shown in Figure 4.30.

FIGURE 4.30   Alkaline cells and 9-volt battery.

Secondary Cells

Secondary cells also produce a chemical reaction between the electrodes and the electrolyte during discharge. However, this reaction may be reversed when the cell is recharged. Many types of secondary cells produce a gas as byproduct of the charging phase. In the case of the wet-cell such as a motor-vehicle battery, which uses an acid-water solution electrolyte and lead plates, the gas produced is hydrogen. Hydrogen is highly explosive, and great care must be used when recharging this type of cell to avoid flames or sparks near the battery. Lead-acid car batteries operate by the dissolution of lead plates into dilute sulfuric acid electrolyte. The positive plate is made from lead peroxide while the negative plate is made from pure lead. When the battery is discharging, the lead plates dissolve into lead sulfate and water. Figure 4.31 shows the internal construction of a lead-acid battery.

When a certain state of discharge is reached, the battery should be recharged. As it recharges, the dissolved lead sulfate and water in the acid solution turn back into lead and lead peroxide, which are redeposited onto the plates. For a given size battery, there is a rated specific rate of recharging to ensure long cell life. A too-rapid charge results in the lead clumping on the plates rather than being evenly deposited in a smooth even layer. The lead clumps can be easily dislodged by

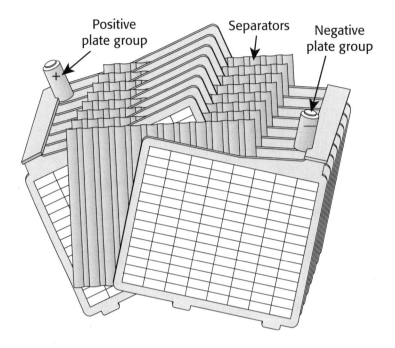

FIGURE 4.31 Lead-acid battery internal construction.

vibration. When this sloughed off lead material builds up on the bottom of the battery; it can eventually short out one or more cells, making the battery useless. This type of battery should be kept clean and not be overfilled, or the electrolyte solution can become overly diluted or contaminated. It is essential to keep the terminals clean and the terminal lead clamps tight to ensure a low-resistance connection. At a normal starting current of 100 amps in automotive use, it doesn't take much terminal resistance to result in enough voltage drop at the terminal connections to leave insufficient starter voltage for the engine to start.

A lead-acid battery is rated in terms of capacity, specified by an ampere-hour (AH) rating. A 200 amp-hour battery is capable of delivering 200 amps of current for 1 hour, 100 amps for 2 hours, 50 amps for 4 hours, and so on. Lead-acid batteries are affected by temperature and the condition of the electrolyte solution. Electrolyte is checked for proper level, and proper specific gravity, using a hydrometer. A *hydrometer* measures the amount of water in solution. As the battery becomes discharged, the amount of water in the electrolyte increases. For this reason, most battery manufacturers recommend adding only clean, distilled water to a cell with low electrolyte when the cell is in a fully charged condition. Figure 4.32 shows a hydrometer used to check wet cells.

FIGURE 4.32   A hydrometer used for checking the specific gravity of wet cells.

A new type of lead-acid cell uses an electrolyte in a gelled state. The cells can be operated in any position. They are often used as backup power supplies for burglar alarms and computers. This type is called a *gel-cell*, and is available in many popular ampere-hour ratings and sizes. They generally cost more than a traditional wet-cell battery of equivalent ratings. Figure 4.33 shows several gelled electrolyte, sealed lead-acid batteries.

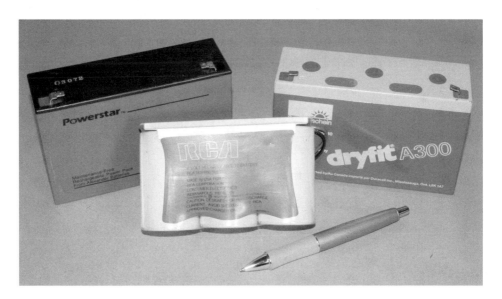

FIGURE 4.33   Examples of sealed lead-acid (SLA) batteries.

A common form of secondary cell used in cameras, calculators, toys, and other electronic equipment is the *NiCd* (pronounced Ni-cad) cell. The electrodes are forms of nickel and cadmium, and the electrolyte is potassium hydroxide. The positive electrode is actually made of nickel hydroxide, and the negative electrode is made of cadmium hydroxide. This type of cell lasts a long time unused. This quality of maintaining a capacity to produce current over a long time is called *shelf life*. NiCd cells may be recharged hundreds to thousands of times, if properly done. Each specific size of cell has a recommended charging current and time that should be closely followed. If overcharged, or allowed to discharge too much, the life of the cell will be greatly shortened. Refer to Figure 4.34 for a picture of NiCd cells.

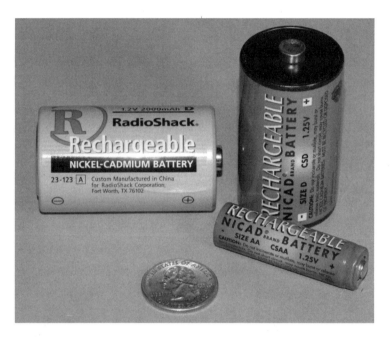

FIGURE 4.34   NiCd cells.

Early forms of nickel-cadmium cells were expensive, used a wet electrolyte, and could be used only in upright positions. Modern NiCd cells use a paste form of electrolyte, and can be used in any position. NiCds should normally be allowed to discharge only to a specified voltage, and then recharged. If allowed to discharge below the recommended voltage, they sometimes exhibit a "memory." That is, they will not recharge fully after that, resulting in short running time between recharges. Other causes of memory include using a cell before it has been charged to its full capacity, and recharging before it has been discharged down to its recommended voltage. For instance, it is recommended to wait until a cordless phone stops working due to a low battery before placing it back on its cradle. If cells exhibit memory behavior, connecting the cells to a load and allowing them to fully discharge can sometimes remedy this condition. The cells are then recharged and tested. It sometimes takes several discharge/charge cycles to repair the cell. If the proper use conditions are met, this of type cell offers one of the best cost-to-performance ratios of any type cell.

Other Forms of Rechargeable Cells

Thomas Edison invented a type of secondary cell that uses nickel and iron for the electrode plates and potassium hydroxide for the electrolyte solution. It has a terminal voltage of 1.4 volts. They were originally used to power telephone systems, and there is still a limited use of Edison cells today.

Figure 4.35 shows several lithium cells (which are *not* Edison cells) used on PC motherboards to backup the CMOS configuration data.

FIGURE 4.35 Lithium cells used to maintain CMOS configuration data.

A newer type of rechargeable cell is the Nickel-Metal-Hydride, or NiMH cell. It overcomes most of the problems with NiCds, such as the so-called memory effect. Figure 4.36 shows several AA size NiMH cells.

Photoelectric Voltage Generation

Photoelectric cells, commonly called *photocells*, are used to produce electric current directly from light. The term *photocell* is used actually to refer to a number of different devices that convert light into electricity. There are two kinds of photocells: those that generate a current by themselves when subjected to light, and those that

FIGURE 4.36   NiMH cells.

change their resistance to current due to a change in light striking their surface. We will consider the type that generates a voltage when it is struck by light. This type is often called a *photovoltaic cell* or *solar cell*.

Solar cells are made of semiconducting material such as silicon or gallium arsenide, and metal conducting surfaces. Low-current solar cells are used as sensors or to power small loads such as pocket calculators. This type of cell is usually packaged in a metal case with glass windows or in transparent plastic packages. Figure 4.37 shows a two-panel bank of photovoltaic cells.

Cells used for large-scale solar energy converters require large surface areas to provide maximum current capacity. Solar cells are also available in flat strip form for efficient coverage of available surface areas.

There have even been recent attempts to build solar-powered electric cars. Races have recently been held in Australia, where entrants' cars have raced across the outback. Some of the more efficient cars averaged 40 miles per hour, through rain, shine, and clouds. The limitations of the electric cars are mainly the weight and cost of the batteries for use during darkness or cloudy weather, and the limited number of miles per battery charge. Efficiencies in solar-cell voltage and current output have steadily risen, however.

FIGURE 4.37 A two-panel bank of photovoltaic cells that produces 19.5 VDC at 1 A in full sun.

Thermoelectric Voltage Generation

A device that develops a difference of potential when heated is called a *thermocouple.* It consists of two different types of metal wires, twisted together at the end. When the junction of the two dissimilar wires is heated, the opposite ends of the wires show a voltage across them. This is called the *Seebeck effect.* The value of voltage produced by a thermocouple depends on the kinds of metal used to make the wires and the amount of heat applied to it. Some common examples of thermocouples include copper-constantan, iron-constantan, Chromel®-Alumel®, and Chromel-constantan. These cover the approximate temperature range of 371°C (700°F) to 1,200°C(2,300°F). Chromel is an alloy of 90 percent nickel and 10 percent chromium; Alumel is an alloy consisting of 95 percent nickel with traces of other metals, and Constantan is 55 percent copper and 45 percent nickel. Other thermocouples are chosen for use at higher or lower temperature ranges.

Thermocouples are very useful for measuring temperatures in areas too remote or too hazardous for humans to directly measure with a thermometer. One such application is the temperature of the inside of a jet aircraft engine. It would be impossible to directly measure the temperature, but if thermocouples are placed inside the engine compartment, and several are wired together, the output voltage from all the sensors can be averaged to get the engine temperature. A *thermopile* is a combination of several thermocouples connected in series. The output of the thermopile is equal to the output of each thermocouple multiplied by the total number of thermocouples in the entire assembly. Another application for a thermocouple is to monitor the temperature of the flue gasses of a wood stove. A thermocouple could also be used to monitor the temperature of a microprocessor chip. And don't forget the food-temperature-sensing thermocouple used in a home oven. The typical output voltage of a thermocouple is less than several tenths of a volt. A Chromel-constantan junction, for example, will produce 70 millivolts at 1,000°C. Figure 4.38 shows a thermocouple used as an oven temperature sensor.

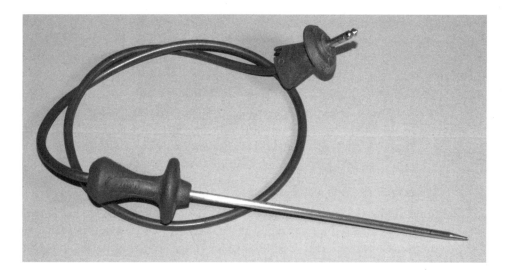

FIGURE 4.38 Thermocouple used to sense oven temperature.

Piezoelectric Voltage Generation

There is class of materials that, when mechanical stress is applied, produce a voltage. The inverse is also true; when a current is allowed to pass through these materials they change their physical dimensions. If the voltage applied is alternating, that is, changes from positive to negative and repeats, the materials will vibrate with the alternating current. This is known as the *piezoelectric* effect, pronounced pee-ay-zoh.

Certain crystalline or ceramic substances can act as transducers at audio and radio frequencies. There are three groups of substances that exhibit the piezoelectric effect: quartz crystals, Rochelle salts, and tourmaline. Some materials offer a high-voltage output (2 volts or so), but are subject to cracking, while others are very strong, but offer low-output voltages. Quartz is usually chosen as the best compromise between high-output voltage and durability.

Piezoelectric devices are employed at audio frequencies as pickups for phonograph records, crystal microphones, earphones, and buzzers. Examples of these include ceramic phonograph cartridges and piezo-buzzers. Many newer loudspeaker systems employ a small, high-frequency sound transducer, which uses a crystal to reproduce the high musical notes. These are called piezo tweeters. Figure 4.39 shows a piezoelectric PC clock crystal.

FIGURE 4.39 Piezo PC clock crystals.

At radio frequencies, the piezoelectric effect makes it possible to use crystals and ceramics as oscillators and tuned circuits. Refer to the chapters on communications and computers (Chapters 16 and 18) for more information. Piezoelectric sensors are also used as mechanical strain gauges, producing a voltage proportional to the stress. Figure 4.40 shows a piezo speaker, the round object in the lower right corner of the modem card.

Triboelectric Voltage Generation

When certain materials are rubbed together, an electrostatic charge develops on each object. One object becomes positive because electrons are transferred to the other object, which becomes negative. This is called the *triboelectric* effect.

FIGURE 4.40   A piezo loudspeaker on a PC modem card.

The most common example of the triboelectric effect is the transfer of charge from a carpeted floor to the shoes of a person walking across the floor. This action charges the person. If that person then touches a metal doorknob or other conducting object, the person will be discharged. This often causes a painful electrostatic discharge in the form of an electric arc or spark.

The Triboelectric Series

Materials can be classified as far as their tendency to give up electrons and therefore become electrically positive, or accept extra electrons and become electrically negative. A list of common materials classifying their tribolectric tendencies is known as a *tribolelectric series.* A technician should be aware of this list, as materials widely separated on the list pose very real ESD problems for computers, boards, and chips. Refer to Table 4.2.

Pay particular attention to human skin and silicon. Most ICs are based on silicon, and when touched by human hands, there is a danger presented to the chips by discharging electricity from the two materials. Another common danger is wearing clothing made from synthetic materials such as plastic around ICs, boards, and PCs. Notice that the plastics are near the negative end of the triboelectric series and aluminum (used in most ICs to make internal connections) is toward the opposite

TABLE 4.2 The Tribolelectric Series

More Positive (+)	Aluminum	Platinum, gold
Dry air	Paper	Sulfur
Dry human hands	Cotton (neutral)	Acetate, rayon
Leather	Steel (neutral)	Polyester
Asbestos	Wood	Celluloid
Rabbit fur	Lucite	Polyurethane
Glass	Sealing wax	Polyethelyne
Human hair	Amber	Polyproylene
Mica	Polystyrene	Vinyl (PVC)
Nylon	Polyethelyne	Silicon
Wool	Rubber balloon	Teflon
Lead	Hard rubber	Saran™ Wrap
Cat fur	Nickel, copper	More Negative (−)
Silk	Silver, brass	

end of the series. Even walking around a few feet away from an exposed mother-board while wearing clothes made from synthetic materials can induce sufficient voltage in the board's chips to damage them!

A Van de Graaff generator uses the triboelectric principle to generate millions of volts of electricity for high-voltage laboratory research. The electric charge is stored on a hollow metal sphere supported by an insulating column. A moving endless belt of silk or rubber is also housed inside the column, as shown in Figure 4.41. The belt, passing over an idler pulley at the top, is moved up the column by a motor-driven sheave. An electrode in the shape of a comb with pointed teeth is positioned near the bottom of the moving belt. It is connected by wire to a source of high voltage with respect to ground, such as a high-voltage transformer. The high-intensity electric field on the points produces inductive charge separation in the belt, removing electrons to the points by corona discharge and leaving the belt positively charged.

A second comb electrode, attached to the inside of the hollow sphere with a conductive rod, is positioned close to the top of the belt to collect the charge. Inductive charge separation in the metal sphere creates a very strong electric field strength near the tips of the comb, and electrons travel to the belt by corona discharge action. This discharges the belt and leaves the outside of the sphere positively charged because positive charges always reside on the exterior of a conductor.

A Charge Accumulator

The charge energy eventually built up on the sphere comes from the work done by the motor in driving the charged belt upwards against the repulsive electric forces exerted by the similarly charged sphere. If the charge on the sphere reaches a saturation level with respect to the size of the sphere, it will repel additional charges, and the outside charge will leak off into the air. Figure 4.41 shows a Van de Graaff generator.

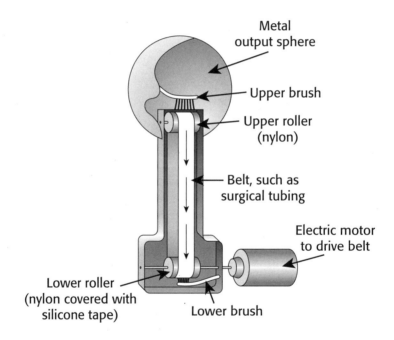

FIGURE 4.41 A Van de Graaff generator.

A large research laboratory-sized Van De Graaff generator can have a sphere with a five-foot diameter mounted on a 50 foot insulating column. A machine of this size is likely to be equipped with a DC voltage generator capable of generating 10,000 volts (10 kV) output. These large machines are used to simulate lightning and to operate high-voltage x-ray machines and high-energy particle accelerators in nuclear fission experiments. Smaller benchtop units several feet tall are used in physics classroom demonstrations of electrostatics. The Boston Museum of Science houses one of the nation's largest Van de Graaff generators, and regularly presents public demonstrations.

Measuring Voltage Rises

Any of the six methods of producing a voltage rise presented in this chapter can be measured with a device called *voltmeter*. Voltmeters are always connected *across* (in parallel with a device that generates the voltage), never in series with the device. *In parallel* means positive-terminal to positive-terminal, and negative-terminal to negative-terminal, such as when jump-starting a car. *In series* means that the positive terminal of one device is connected to the negative terminal of the next device, such as cells in a flashlight. There are two general types of voltmeters: analog and digital. All analog meters have a moving pointer or needle, which can move only in one direction when the voltage is measured. They are polarized instruments, and must be correctly connected across the voltage rise in order to prevent the pointer from trying to move backwards, damaging the meter. Most types of digital voltmeters are not polarized; the display simply indicates if the polarity of the conection is in reverse. Voltmeters will be presented in more detail in Chapter 5, "Using a Digital Multimeter."

4.4.2 Voltage Drops

A *voltage drop* is a voltage that is developed as current flows through opposition. In a direct current circuit, this opposition is called resistance. Resistance is present in all types of devices that are connected in an electrical circuit. We call these devices loads, as presented in Section 4.2. Typical circuit loads are light bulbs, relays, buzzers, resistors, and all sorts of specialized circuits used in radios, TVs, communications equipment, computers, and so on. In fact, we can speak of these specialized circuits as being a load in themselves, even though they may contain hundreds of separate components. As far as the voltage source is concerned, the entire circuit connected to the source becomes the load. Figure 4.42 shows loads connected to a voltage source.

Measuring Voltage Drops

Voltage drops are measured in the same manner as are voltage rises. The difference is that a voltage drop will disappear as soon as the current in the circuit stops. A voltage rise is not dependent on a complete circuit in order to be present, or measured. Voltage drops are measured with analog or digital voltmeters.

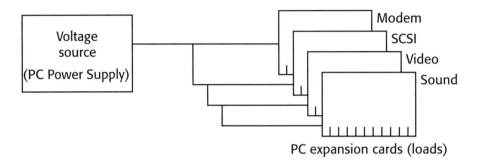

FIGURE 4.42 Loads connected to a voltage source.

Making voltage drop measurements is a good troubleshooting method for determining if a load is acting normally. If a load is burned out, it will usually have an increased voltage drop across it. If a load is shorted, it will have no voltage drop across it. See Section 4.6 for more information on shorts and opens. Figure 4.43 shows voltage drop measurement connections.

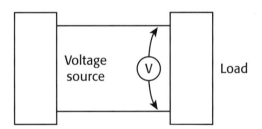

FIGURE 4.43 Measuring a voltage drop.

4.4.3 Summary of Voltage Rises and Drops

A voltage rise is a potential difference developed by one of the six methods of producing voltage. A voltage rise is not dependent on the source being connected to a circuit, causing a resultant current flow. A voltage rise is measured with a device called a voltmeter. Voltmeters are connected across a voltage rise. To avoid meter damage, analog voltmeters should be connected across the voltage rise with respect to its polarity.

A voltage drop is a potential difference resulting from electric current flowing through a load. A voltage drop is not present when there is no current through the load. Voltage drops are measured with voltmeters, connected across the load.

4.5   THE CONCEPT OF GROUND

One of the most misunderstood words in electronics work is *ground*. A ground can mean many things, each slightly different, depending on the particular application. The idea central to all uses of the word ground in electronics is that of a common connection between two or more circuit components.

4.5.1   Common Ground

A *common ground* is a connection in an electrical system or unit between several discrete components. A common ground usually means the leads of all components connected to it are at the same electrical potential, or voltage. Refer to Figure 4.44 for picture of a common ground.

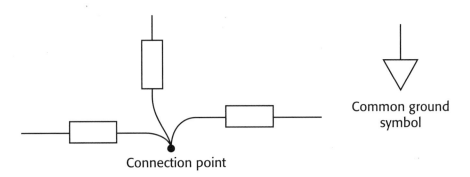

Connection point

FIGURE 4.44   A common ground.

4.5.2   Chassis Ground

Early in electronics, it was the practice to secure all the separate electrical components to a rigid surface. Legend has it that one enterprising electrical experimenter first used his mother's wooden board for cutting bread to solidly mount these components. That is the origin of the word *breadboard*, still in use today to describe assembling a circuit for the first time to test its functionality. Figure 4.45 shows a breadboard circuit.

With the advent of metal boxes used to mount the components, it was discovered that the metal chassis itself could very conveniently be used as a common connection between components, thereby saving extra wires. *Chassis ground* refers to the use of the metal chassis to provide this common electrical connection. A chassis may or may not be connected to another form of ground, such as to the earth (earth ground). Figure 4.46 shows a metal chassis.

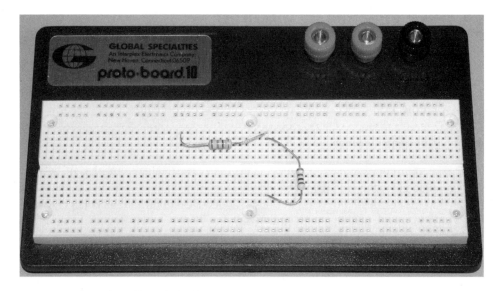

FIGURE 4.45   A breadboard circuit.

FIGURE 4.46   A metal chassis.

4.5.3   Earth Ground

An *earth ground* is a connection to the earth, usually via the shortest direct route possible. The earth ground conductor should be a large gauge (large diameter) cable securely fastened to both the appliance and circuit to be grounded, and to a conducting rod driven into the earth. Normally, copper rods or copper-plated steel rods are used, eight or more feet in length, driven into the ground for most of their length. The connection to the ground wire must be made with the proper type of clamp, usually cast brass, and be of the required size for the size ground rod used. All this is covered by local and national electrical codes.

The commercial power companies require earth grounds installed at the service entrance points to all residential and commercial buildings (see Chapter 17, "Building Power Wiring"). All large electrically operated machines have an earth ground connected to the metal chassis. A large number of electric power tools and other electrical and electronic devices are also earth grounded by using the third wire on the power cord. This green wire connects directly back to the earth ground rod at the service entrance of the building in which the tool is used. Earth grounding helps prevent electrical shock hazards, and may save a person from death by electrocution should the wiring inside the appliance or tool fail and contact the metal housing. As mentioned in the safety section of Chapter 2, "Basic Concepts and Electricity", the human body is very sensitive to electric current passing through it. Death may result from as little as 6 milliamps, if the current pathway passes through the heart. Figure 4.47 shows the shock hazard of poorly grounded equipment.

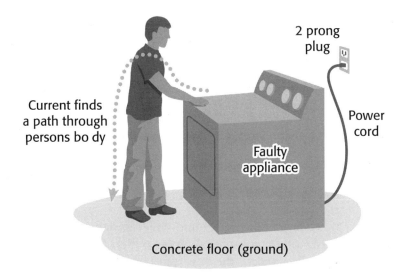

FIGURE 4.47   Poorly grounded equipment presents a real shock hazard.

An earth ground is designed to offer the current an easier path to ground than through someone's body. Never defeat the earth ground by removing the ground lug or prong from a 3-wire power cord. Always make certain that all electrical outlets are the three-wire earth grounded type. Often, people will use an adapter to allow the 3-wire cord of the power tool to be plugged into an old-style 2-wire outlet. The problem is the third wire of the adapter is a short green wire with a connector that is supposed to be hooked under the small screw that holds the outlet cover on. This screw is supposed to complete the ground connection to the metal enclosure (called a box) housing the 2-conductor outlet. Often, there is no actual connection of the box to earth ground. So the use of such an adapter only defeats the purpose of the third wire, thereby subjecting the appliance or tool user to potential shock or death. Some outlets are miswired, having the earth ground and the "hot" wires, or the hot and neutral wires reversed. A simple plug-in tester, shown in Figure 4.48, can be used to test for proper outlet wiring. Three small indicator lights will show the status of the wiring to a particular outlet when the tester is plugged into the outlet. The lights indicate if the outlet wires are reversed or missing. These testers are available for only a few dollars at hardware and electronics supply stores, and are a worthwhile addition to anyone's toolbox.

FIGURE 4.48   An AC outlet tester
used to check for proper outlet wiring.

A new safety device called a "ground-fault circuit interrupter" or *GFCI*, may be installed in place of a conventional power outlet. Internal circuitry senses, and nearly instantly disconnects, the power to the outlet when a short to ground—called a ground fault—is detected. These GFCIs can be purchased for as little as ten dollars at many electrical supply stores and should be used in all bathrooms, laundry rooms, and wherever electrical devices are are operated in wet conditions, such as on the lawn. A person in a bathroom may be using an electrical appliance such as a hair dryer or electric shaver. They may also have water running. If there was a problem with the electric appliance causing electrical contact with the person's skin through the metal housing on the appliance, and the person comes into contact with the water, or the water pipes, the person could be severely shocked or electrocuted. Water pipes are buried in the ground outside the building, and are usually connected to the power company's ground system. A GFCI would sense the current flow, and almost instantaneously interrupt the power to the appliance, preventing disaster. They are cheap insurance, and are quite valuable in preventing electrocution. Many local electrical building codes require the use of these devices in new construction. Figure 4.49 shows a GFCI.

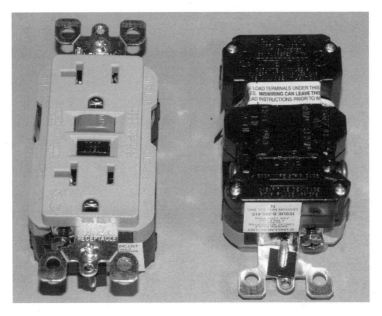

FIGURE 4.49   A ground-fault circuit interrupter (GFCI).

4.5.4   Ground Bus

A *ground bus* is a large diameter metal conductor, such as a strap, braided wire, or section of tubing, to which all of the common connections in an electrical system are made. In commercial power systems and home wiring systems the ground bus is routed to earth by the most direct path possible. The size of the conductor used for a ground bus depends on the number of pieces of equipment in the system, and on the total current the system carries. Typical ground-bus conductors are made from copper tubing having a diameter of ¼ to ½ inch. For radio frequency applications, the larger the conductor size, the better. The distance to the earth ground connection should be as short as possible. Figure 4.50 shows a ground bus made from a piece of copper tubing to safety ground all the equipment on an amateur (ham) radio station table. The main purpose of this type of ground is to help prevent electric shock and RF burns. However, it is also necessary for proper operation of the antenna

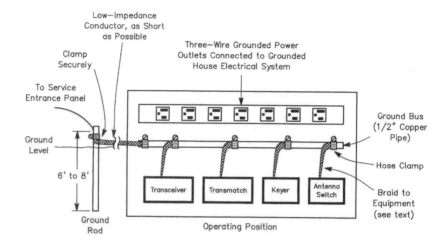

FIGURE 4.50 A ham radio table ground bus. Reprinted with permission from The 2001 ARRL Handbook for Radio Amateurs.

A ground bus used inside a metal chassis is often simply a length of solid tinned-copper wire, usually bent into a "U" shape, which follows around the inside of the chassis, and may be connected to the chassis. Another example of a ground bus is the slightly larger-width copper trace on a printed circuit board. Again, the ground trace is usually on the perimeter of the board. Many times the ground bus trace is connected to the chassis ground, which, in turn, is connected to earth ground through the power cord's third (green) wire.

4.5.5 Floating Ground

A *floating ground* refers to a common ground connection that is at a different voltage potential than some other ground used in a large system. For instance, a floating ground can signify a connection of several components, but that connection is not tied to the equipment chassis or earth.

4.5.6 Ground Shields

A *ground shield* is a conductor, usually in the form of a metal plate, enclosure, braid, or foil, which electromagnetically isolates one section of a circuit from another.

Ground shields, in the form of little metal walls, are used between sections in communications equipment, even TVs, to keep the electromagnetic energy from one circuit from unintentionally coupling to another circuit. Michael Faraday, an English scientist interested in electric research, found that placing an enclosure around a static electric generator would cause the electric field to be developed on the surface of the shield, rather than passing beyond it, as the field normally would. For that reason, a ground shield in the form of a complete conducting metal enclosure is today called a *Faraday shield*.

A ground shield is used in coaxial cable to prevent the radiation of electromagnetic energy from escaping from the center conductor. It also prevents the unwanted pickup of spurious (undesired) signals by the cable, such as in the cable TV industry, ham radio, CB, and military communications systems (see Chapter 16). Figure 4.51 shows examples of shielded and unshielded EIDE cables.

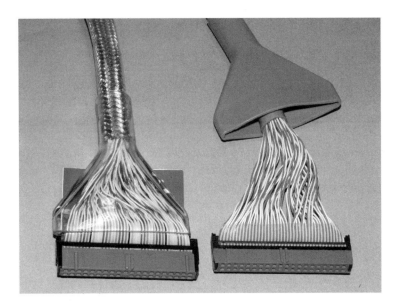

FIGURE 4.51   Shielded and unshielded EIDE cables.

4.5.7   Ground Summary

To summarize, when someone tells you a circuit is "grounded," unless it is obvious what they mean by that, you should ask the very intelligent question, "What type of ground do you mean?" There are many different forms of grounds, each used for a slightly different purpose, but all grounds connect several components together.

4.6 CIRCUIT PROBLEMS

All electrical or electronic circuits are prone to failures at one time or another. These failures can be categorized into one of three possible types: shorts, opens, or intermittents. It is the job of the electronics technician to diagnose and repair each of these circuit problems, using a technique called *troubleshooting*.

4.6.1 Shorts

A *short* is an undesired connection in a circuit. A short usually causes destruction or damage by increasing the total current in a circuit, thereby overheating components. Recall the simple circuit presented in Section 4.1. If a conductor completes the path across the source, it will offer a very low opposition to current. Current will always be greatest in the path that offers the least resistance. The majority of current will flow through the undesired connection, taking a "shortcut" and bypassing the intended circuit path. Since the short offers less resistance than the original circuit was designed to accommodate, the total current supplied by the source is increased. The original wires connecting the source to the point where the short happened were, however, not designed to handle the increased current value, and will likely overheat and burn out. If the wires themselves don't burn out from the excess current, the source itself may be destroyed due to its internal resistance, which dissipates heat. This is why fuses and circuit breakers are used. Fuses and circuit breakers open the circuit when a critical amount of current flows through them, preventing damage. Figure 4.52 shows two types of shorts. It is also possible for a circuit component to short internally, and therefore not be visually detected until the damage to the circuit occurs.

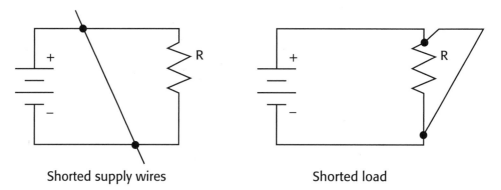

Shorted supply wires   Shorted load

FIGURE 4.52 Examples of two types of shorts. In the first case, the load has developed an internal short; in the second case, the wires leading to the load are shorted.

Before the use of current-sensitive protection devices such as fuses and circuit breakers, many homes and factories were lost to fires resulting from people exceeding the current capability of electrical outlets. You should never replace a fuse or breaker with a unit rated higher than the one that failed. The gauge (diameter) of wires used in a particular circuit is chosen based on the load current required. The fuse or breaker must burn out or trip at a lower value of current than the value that causes excessive heating in the wires. Figure 4.53 shows fuses, and Figure 4.54 shows circuit breakers.

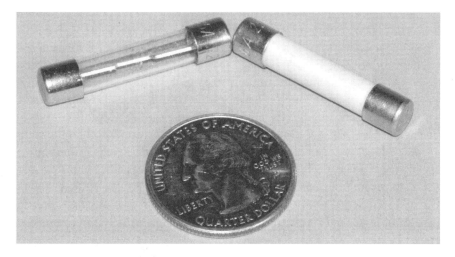

FIGURE 4.53 Cartridge fuses.

FIGURE 4.54 Circuit breakers.

4.6.2   Opens

An *open* refers to a current pathway that has been interrupted. When this happens, there is no current flow. Voltage rises will usually increase due to no current demand by the circuit, and voltage drops will disappear (go to 0 volts). Opens can be found by making voltmeter measurements. A load that measures zero volts across it either has no current flowing through it or it is completely shorted (zero ohms resistance).

Opens can be caused by a circuit that first had a short, thereby burning out a part or device in the circuit due to excessive current. Opens can also be caused by poor electrical connections such as dirty contacts, broken wires, or mechanical damage. One of the first checks a troubleshooter must do is to visually inspect a circuit for obvious signs of damage such as loose wires. If nothing is found, voltmeter tests will usually help to pinpoint the exact location of the open. Any component with no current through it will measure 0 volts drop across it, unless it is shorted, or it is the only load in the circuit. If it is the only load in the circuit, then the voltage will be that of the source. If a component is shorted, then the source will measure less than normal voltage. So, one of the first things to measure is the source voltage. If it is OK, then measure the drops across all components. If they are all zero except one, and it measures the same as the source voltage, you must unhook and de-energize the circuit, and test individual components for resistance. The component that is open will measure infinite Ω with a DMM resistance test.

Another easy test for an open is to use a simple continuity tester, as shown in Figure 4.55. If the component or circuit under test is open, the lamp will not light. If the circuit or component has a continuous current path, the lamp will light. It is necessary to ensure that the voltage of the tester does not exceed the voltage that will cause damage to the device or circuit being tested. Low-voltage cells or batteries are usually used in continuity testers.

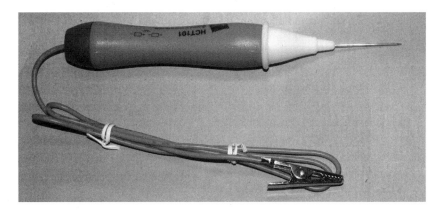

FIGURE 4.55   A continuity tester.

4.6.3 Intermittents

An *intermittent* is a problem that happens sporadically. It could be the result of a short or an open, or a combination of the two. Many techniques are used to locate the cause of an intermittent. Mechanical movement will frequently reveal the source of an intermittent. Caution should be used in order to prevent contact with a live circuit. Wooden handles or small sticks can be used to gently move a component or circuit board while looking for the problem. Paint stir sticks and even chop sticks work well for this purpose. Figure 4.56 shows examples of such sticks. Be careful not to use a wooden pencil to perform this job as the graphite center is a semiconductor and can result in a nasty shock if touched.

FIGURE 4.56 Wooden sticks used to mechanically probe a circuit.

Often, a temperature change causes the value of a component to change. Blowing on a circuit component with compressed air or the use of a pressurized can of coolant will usually cause the component to change value. A soldering iron can be brought near (never touching) a component to heat it up. Portable hair dryers are sometimes effectively used to heat components in order to force them to become intermittent when heated.

Changing applied voltage values will cause certain faulty components to display their problems. By the use of a variable power transformer, called an autotransformer, different values of voltage can be applied to a circuit. Voltage-sensitive components can be made to give up their secrets. An increase or decrease of perhaps 10 percent applied voltage can often trigger the problem in an intermittent component. Normally, an adjustable DC power supply or an autotransformer is used to change the applied voltage (see Chapter 14, "Power Supplies").

4.7 INTRODUCTION TO TROUBLESHOOTING

Troubleshooting, as explained in Chapter 1, "PC Overview, Safety, and Tools," is the art and science of diagnosing problems, pinpointing the exact component involved, and repairing the problem. The repairs can be in the form of removing and replacing individual components, replacing an entire circuit board, or performing repairs without replacement. Often, simply resoldering a suspect soldered connection will cure the problem. In any troubleshooting situation, there is a general approach that will increase chances of success.

4.7.1 The Logical Approach

Clear thinking and attention to detail are the marks of a successful troubleshooter. If a problem shows up immediately after changing something in a circuit, it is almost certainly the replaced part that is the culprit. Often components of the improper value are placed in a circuit, especially during school lab exercises, when the circuit builder may feel rushed, and is yet unfamiliar with electronic component color codes.

4.7.2 Note Taking

Develop the habit of always recording the exact problem encountered with a circuit, and all steps taken to repair it. This prevents "spinning your wheels," simply trying various cures without thoroughly thinking out what the trouble is. For instance, simply replacing each component in turn until the circuit performs is time-consuming and not very practical with a large circuit. Some less-skilled troubleshooters employ the "shotgun approach." They simply replace everything in the circuit, and charge the customer for all the parts. Unfortunately, this happens all too often in TV, automotive, and unhappily, even PC repair shops.

A few minutes checking the normal expected values from a data book, lab manual, or motherboard document, and then thinking out what could cause the measured values to be different, is much more efficient.

Most lab exercises encountered in classroom work will provide a space to record all measurements taken. You should also record any unexpected problems on the lab report, and the steps that are taken to correct the problem. No one has a perfect memory, and it may be necessary to refer to the lab notes later. Notes describing problems and fixes are invaluable during later reading.

4.7.3 Documentation

The best source of reference for any circuit problem is the manufacturer's data book or manual that comes with a piece of equipment. The system manual that comes with a new PC or the motherboard booklet that comes with a separate motherboard contains a wealth of useful information. Often such references contain essential schematic diagrams. Often included with the schematics are typical voltage values measured at various points in the circuit. The legend area of the schematic will have a notation of the specific type of meter used for the measurements, as well as the position any switches are in during measurements. This is necessary in order to know whether any difference in circuit values measured are due to problems with the circuit, or differences in switch setting or measuring instruments used.

Classroom lab exercises are most beneficially done after first thoroughly reading the lab manual and text used for the class. It is an excellent idea to completely read the lab exercise before coming to class. This will help familiarize you with what the circuit does, the inputs and outputs, and normal circuit values. Some lab manuals will also include a discussion section, which will give an overall plan of attack for the lab work. Be sure to examine all charts or graphs that must be filled out or drawn, to become familiar with what is required. Fervently involved in trying to get a lab exercise to function, you may become frustrated trying to become familiar with all the details of a lab circuit. Figure 4.57 shows examples of motherboard manuals.

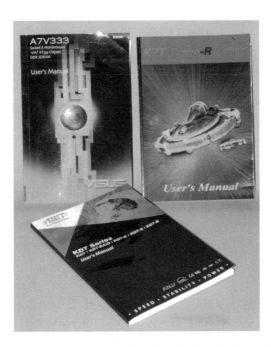

FIGURE 4.57 Examples of typical motherboard manuals.

The former Heath Company of Benton Harbor, Michigan, once the largest producer of build-at-home electronics products available in kit form, reported that over 90 percent of the problems with returned kits involved poor soldering techniques or wrong parts used during construction. The key idea is to be prepared by reading reference material before starting circuit work.

4.7.4 Inputs and Outputs

An *input* is a signal applied to a circuit, or the point in a circuit where a signal, usually in the form of a voltage or current, is applied. A circuit may have more than one input. For example, a circuit may require two different signals in order to produce a desired output. In the case of a very simple circuit, one with no signals applied to it, the input can be thought of as simply the connection to the battery or power supply. Typical inputs to a PC include the keyboard, mouse, telephone line, and Ethernet cable.

An *output* is a signal produced by a circuit, or the point in a circuit where the output is provided. A circuit may have more than one output. For example, a stereo system has audio outputs for each of two speaker units. Typical outputs found on a PC include those for the speakers, video monitor, printer, telephone line, and Ethernet cable.

Since both the telephone line and Ethernet cables include the capability to both send and receive data, called full-duplex mode, they count for both inputs and outputs.

When troubleshooting, the first thing to be checked should be the output. If the output from the circuit is normal, the circuit is functioning normally. The absence of an output, or an output that is different from normal, will give an indication on the problem. The next thing to be checked, if the output is not normal, is all the inputs. If all inputs to the circuit are normal, the problem must be in the circuit under test. Don't make the common mistake of replacing parts in a circuit having an abnormal output without first checking all inputs. As an example, don't replace the light bulb in a defective flashlight without first checking all the cells for proper voltage.

Figure 4.58 shows examples of typical flashlights used for PC troubleshooting.

4.7.5 The Signal Path

The *signal path* in a circuit is the path taken, starting at the input to the circuit, and passing through each part of the circuit to the output. In a simple circuit such as a flashlight, the input is the voltage produced by the cells. The next component through which the current passes is the mechanical on/off switch. The output is the light bulb. In order to recognize where the problem is, it is necessary to think of the signal path.

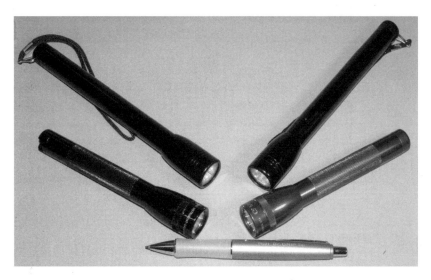

FIGURE 4.58   Typical flashlights used for PC troubleshooting.

In complex electronic systems, the signal path can become quite complex. Some schematic diagrams cover several pages, each diagramming a separate subsystem of the whole system. One useful technique is to use a colored pen or pencil to lightly mark the signal path on the schematic. Systems as complex as a television or communications receiver usually include a short description of overall signal paths of the various signals present. For instance, a television receiver has several signals present: picture, sound, horizontal and vertical synchronization, color, intensity, and a reference signal. If it is a stereo TV broadcast, there are even two different channels of sound signals present. PCs also have signals paths, and knowing them is essential to troubleshooting. Figure 4.59 shows the different signal pathways in a computer.

4.7.6   Recognizing Faults

The first place to start, in order to know if a circuit is not performing correctly, is to become very familiar with what the circuit should do when functioning correctly. This may sound like an obvious and simple thing to do, but as the circuit complexity increases, so does the difficulty in knowing correct circuit operation. You must know the correct position of all switches and controls, the normal operating parameters (measurement values) of the piece of equipment, and have the ability to detect any substantial differences. It should be noted, however, that most electronics equipment has tolerances in most specifications. That is why you must refer to the manufacturer's specifications in order to recognize abnormal performance. Most

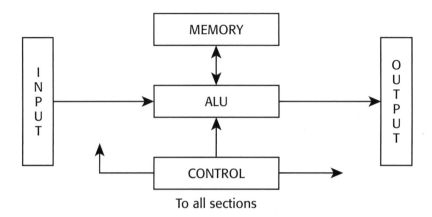

FIGURE 4.59   Signal pathways in a computer.

inexpensive equipment has tolerances of plus or minus 20 percent (+/20%) of all components. It is cheaper to produce the equipment if these wide tolerance components are used rather than the higher-priced precision components. The difference between a 20 percent resistor, for example, and a 5 percent one, is usually at least twice the price of the 20 percent resistor. In a typical piece of equipment, there may be hundreds of resistors. There is, of course, much more to making a high-quality circuit than simply using close-tolerance components. Good engineering design is essential to making a good quality, long working product.

4.7.7   Parts Replacement

After the troubleshooter has pinpointed the component responsible for the malfunction of the circuit, it must be repaired or replaced. Most components in modern electronic circuits are not repairable. The replacement part must meet several criteria: it have the same voltage, current, power, and resistance ratings as the original part, it must physically fit into the same location as the original, and it must be affordable.

Most components are soldered into place on a circuit board. The old part must first be unsoldered, without damage to the tiny traces of copper on the board. Some components use sockets, and are easily removed and replaced. Much of the digital integrated circuits (ICs) use sockets. A new form of IC uses very short connecting leads, in order to reduce performance degradation due to stray electronic effects. First developed for military applications, this new class of ICs is called surface mount technology, or SMT. Precision soldering skills are required in order to replace these devices.

Many of the ICs in modern electronic gear such as PCs are extremely static sensitive. The general device category is called Metal Oxide Semiconductors, or MOS. They will be destroyed by casual touching by one's fingers, or even from using ungrounded soldering irons. They must be shipped in special static-proof cases, and are sometimes wrapped with a small wire that connects all their pins together. The wire is removed after mounting the IC. The wire grounds all pins to the same potential. Chips, including microprocessors, are sometimes shipped in conductive black foam for the same reason. Special safety precautions must be used when working with these devices. Soldering irons must be earth grounded, the workbench surface must be grounded, and even the repair worker's wrist is connected to ground through a high-resistance path (usually a 1 M resistor), in order to drain off any body static charge. Figure 4.60 shows an ESD wrist strap and conductive mat.

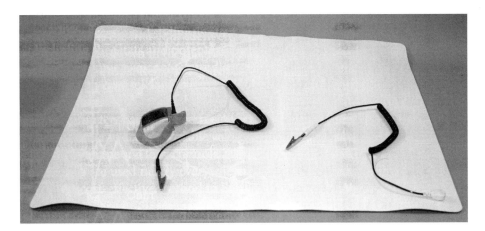

FIGURE 4.60 A typical ESD wrist strap and conductive mat.

4.8 SUMMARY

- An electric circuit requires a complete path to function.
- The basic circuit elements are the source, conductors, control, and load. Frequently, safety components, such as fuses and circuit breakers, are used as well.
- For most circuits encountered by PC and network techs, the voltage source is the commercial wall outlet, a power supply, or a cell or battery.
- Conductors can be wires, cables, traces on a circuit board, or the connecting wires inside ICs.

■ The control elements most used in circuits are switches. Jumpers, a type of switch, are commonly used in PCs since they are very small and extremely inexpensive.

■ Safety items include over-voltage or over-current protection. Fuses and circuit breakers are the most common over-current devices used.

■ Switches are rated according to their switch function, rated voltage and current, and type of mechanical action.

■ Relays are used for one circuit to control another without the direct interconnection of the two. Examples are automobile starting relays and the small relays used to switch the telephone line on a PC modem card.

■ A circuit load is where the intended work is done. Examples are the circuits inside a PC or network component. Integrated circuits, drives, fans, and LEDs are all examples of circuit loads.

■ A voltage rise results from the conversion of one form of energy to another, such as in a generator, cell, battery, or photocell. With the exception of a short circuit or very excessive current draw, a voltage rise always produces a voltage, as long as it is in good condition.

■ A voltage drop results from current flowing through resistance or impedance. No current through a component results in no voltage drop across that component.

■ There are many different types of grounds, but the common theme is a common electrical connection. Often voltages are given with respect to an understood ground reference point.

■ A short can be either desired or undesired. An intentional connection of test probes on a test meter in order to check its reading is an example of a non destructive short. An undesired electrical connection that causes excess current to flow, and almost always causes problems, is an example of an unintentional, destructive short circuit.

■ An open circuit results in no current and no voltage drops. It is an interruption of the complete circuit. It can be either intentional, such as turning off a control switch, or unintentional, such as when someone trips over a power cord, pulling it from the wall outlet.

■ Troubleshooting involves recognizing wrong circuit behavior, using good documentation, using test instruments, note taking, knowing the signal path, and taking a logical, common-sense approach to things.

■ The signal path must be understood in order to perform troubleshooting.

■ Quality tools and test equipment are vital to performing troubleshooting.

■ One of the best troubleshooting skills is simple common sense.

4.9   KEY TERMS

Circuit	Load	Photocell
Terminal	Voltage rise	Thermocouple
Voltage source	Electromagnetic	Voltmeter
Recombination	Electrochemical	Across
EMF	Photoelectric	Voltage drop
Control	Thermoelectric	Ground
Poles	Piezoelectric	Breadboard
Throw	Triboelectric	Chassis
Position	Right-hand rule	Short
NO	Induced voltage	Open
NC	Field coil	Intermittent
Relay	Cell	Input
Flux lines	Electrode	Output
Solenoid	Electrolyte	Signal path
Contactor	Battery	ESD

4.10   EXERCISES

1. Indicate which components of a flashlight are considered the source, the control, and the load.

2. A stereo sound system consists of a CD player, an amplifier, and a loudspeaker system. Which component of this system would be considered the input? Which component of the system would be considered the output section?

3. Draw a simple block diagram of the stereo system described in Question 2. Use arrows to indicate the signal flow path.

4. Describe what precautions must be used when replacing a device that is soldered to a printed circuit board.

5. If the device to be replaced in Question 4 is a MOS-type component, what additional precautions must be taken when replacing it?

6. What is the first thing a troubleshooter should check when testing a circuit?

7. What usually happens as a result of a short circuit?

8. What devices are used to protect against damage caused by shorts?

9. Describe how to determine if a component that is soldered into a circuit, and cannot be removed easily, can be tested to determine if the component is open.

10. Describe how the third wire of a 3-wire safety ground system works to protect people from electrocution.

11. What device should be installed in power outlet locations where power tools are to be used in wet conditions, or where appliances may be used within reach of water?

12. A circuit sometimes overheats when energized. Sometimes it operates normally. What name is given to this type of circuit problem?

13. When attempting to repair an unfamiliar piece of equipment that was commercially manufactured, what should be your first course of action?

14. When troubleshooting a circuit, what should you do while performing tests to determine the cause of the problem?

15. Describe the major contents of a standard 1.5 volt dry-cell.

16. Can a primary cell be recharged?

17. What type of cell is used on a PC motherboard?

18. Should water ever be added to a lead-acid cell when the cell is in a discharged condition? Why or why not?

19. A certain circuit experiences problems only when it is cold. You are working in a repair facility that is always kept at about 70°F. What could you do to help diagnose the problem?

20. What name is given to a small buzzer that uses a piece of quartz to produce a sound?

21. Describe the basic component parts of a commercial geothermal electric power station.

22. Draw the schematic symbol for an SPST switch.

23. Draw the schematic symbol for a DPDT switch.

24. If you needed to build a circuit requiring a switch to turn on and off a single load, and you had only a DPDT switch, how many different combinations of switch terminals could you use to control the circuit?

25. Draw a schematic diagram showing a relay with a 6-volt coil that is controlled by a 6-volt battery and a switch. The relay contacts are SPST, NO. The relay controls a circuit consisting of a 120-volt bell, connected through the relay switch contacts to a 120-V supply.

26. Jumpers on a hard drive can be placed to set the drive for master, slave, or cable select (CS). What type switch function does a single jumper serve? Refer to Figure 4.61.

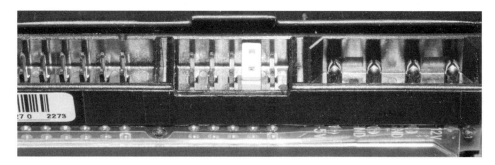

FIGURE 4.61 Hard drive M/S jumpers.

5 | Using A Digital Multimeter

5.1   INTRODUCTION

A diagnostic and testing tool common to the electronics industry is the digital multimeter or *DMM*. Any decent PC tech should carry one in his toolbox. Prices for new DMMs suitable for general PC troubleshooting can be as low as $15.

5.2   A BRIEF HISTORY OF METERS

In the early days of electronic devices and circuits, electrical meters were very expensive and were manufactured to measure a specific electrical parameter, such as voltage, current, or resistance. In addition, there was a single maximum value on the meter scale that could not be easily altered. So it was necessary to purchase a large variety of each type of meter in order to cover the needed ranges of voltage, current, and resistance. In addition, all the meters used a moving pointer, usually called a needle, to indicate the measured value. Various schemes for suspending the armature and needle assembly were used in competing mechanical designs. Some used a taut-band of metal to serve as the suspension; others used a spring mechanism similar to those used in mechanical pocket watches. The design most used in analog electrical meters has been the moving coil, field magnet type credited to a Frenchman named D'Arsonval. It uses a magnet in the shape of a horseshoe and a rotating coil wound on an armature suspended in the center of the horseshoe magnet's two magnetic poles. The fixed magnet provides a magnetic field that passes through the armature. When the meter is connected to a circuit and electric current flows through the armature coil, a second magnetic field is set up around the armature that interacts with the field magnet's fixed magnetic field. The result is torque, or rotational movement of the armature. Attached to the armature is the needle or indicator, which rotates with the armature. The position on the indicator on the paper scale behind the needle shows the value being measured. These meter assemblies are known as *basic meter movements*, to reflect the fact that there is a moving part, the needle, used as the indicator. Figure 5.1 shows some dedicated analog meters.

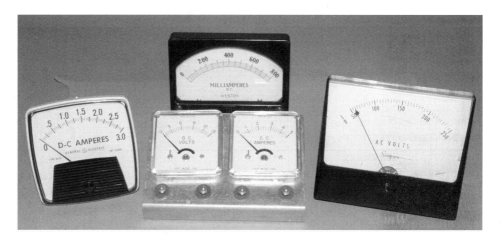

FIGURE 5.1   Dedicated analog meter movements.

5.2.1   Multiple Ranges

Scientists next figured out how to use a single-range meter to measure different maximum amounts of an electrical parameter, specifically voltage and current. In the case of voltmeters, the use of series resistors called *multiplier resistors* allows a single range meter to be used for several different maximum, called *full-scale*, values.

To measure multiple ranges of current using a single ammeter, resistors placed in parallel with the meter, called *shunt resistors*, will bypass the excess current that the basic meter movement can not accommodate without damage and allow measurement of current values in excess of what the meter movement by itself can handle (called the full-scale current value).

5.2.2   Multiple Functions

Next in the development of meters, a single meter movement was used with switches that allowed for connecting either multiplier resistors or shunt resistors of different values, so as to allow the measurement of either voltage or current at various maximum values. Then the meter circuits were further modified to allow deriving resistance values from knowing the voltage across and the current through the meter (Remember Ohm's law?). This allowed the same meter movement to also indicate resistance. This is the first occurrence of what has come to be called a *multimeter*, since a single meter movement can be used to measure multiple electrical parameters. The problems remaining were sensitivity and ruggedness.

5.2.3   Increasing the Sensitivity

By the early 1900s, the invention of the vacuum tube allowed the first electronic *amplifier* to be produced. In simple terms, an electronic amplifier allows a small electrical signal to appear on the output of the amplifier with a much greater magnitude. Think of common stereo amplifiers used today. They allow a small electrical signal, such as that produced from a radio receiver or CD player to be built up to a point that it can drive large loudspeakers. Actually it is not the small signal itself that is made larger, it is the amplifying process that produces a faithful reproduction (that's the real meaning of *high fidelity*) of the original signal. So by using a vacuum tube amplifier, early multimeters were made much more sensitive. They are called vacuum tube volt meters or VTVMs. They can still be purchased new today. The major drawbacks to a VTVM are the lack of the capability to measure current and the need to be plugged into a power source of 120 volts AC in order to power the vacuum tube, which requires relatively high voltage to operate.

5.2.4   Transistors to the Rescue

By the late 1960s, transistors, able to run on relatively low voltage and current, were being cheaply produced. They were incorporated into the new meters, and a special type of transistor, called a *field effect transistor*, or *FET,* was selected to replace the less rugged and much more demanding vacuum tube. These transistors enabled the production of *field effect volt ohm multimeters* or *FETVOMs*. They were very popular and replaced many of the older VTVMs for all but the most demanding testing.

5.2.5   Digital Meters

Digital meters first appeared in the 1940s in any great number. They had no moving parts, but used vacuum tubes for the amplifiers as well as a special type of indicating tubes called *nixie* tubes. They were large, bulky, and were current-hungry beasts, but were preferred by the military and large commercial users for increased accuracy. Typically 19-inch rack-mounted affairs, they generated so much heat that they required cooling fans. Figure 5.2 shows this type of voltmeter, a Hewlett-Packard® Model 405 CR "Automatic Digital Voltmeter" that weighs roughly twenty pounds! It was considered in its day to be the "Cadillac" of laboratory-quality voltmeters.

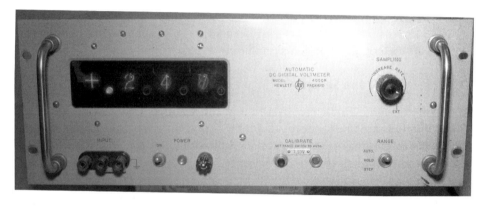

FIGURE 5.2   A rack-mounted voltmeter with nixie display tubes.

By the 1970s, the first small, handheld digital meters appeared. They were very expensive and rather finicky. Today they are smaller, very rugged, and very reliable. The integrated circuits that make up the bulk of the meter circuit are so inexpensive that a digital multimeter, or DMM, can be purchased for less than $20. Every PC tech should own one and carry it in his toolbox. Figure 5.3 shows an example of an inexpensive DMM.

5.2.6   Increased Versatility

Most DMMs today can measure AC and DC voltage, AC and DC current, and resistance. Most also include a dedicated diode test setting, and all can test a circuit for

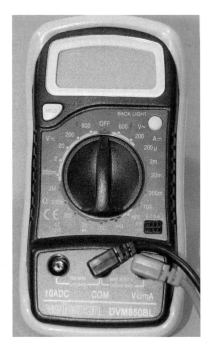

FIGURE 5.3   An inexpensive DMM
suitable for PC troubleshooting use.

continuity. In addition, some can also test transistors, measure temperature using a
special probe, and some can even measure frequency.

5.3   MEASURING VOLTAGE

Voltage measurements are the most commonly done tests using DMMs. Voltage
measurements are done across, meaning in parallel with, the component under
test. Typical voltage tests are performed on PC power supply output leads while the
supply is connected to at least a minimum number of components, in order to
draw at least the minimum current from the supply required to ensure the proper
regulated voltage. More information on power supplies is presented in Chapter 14,
"Power Supplies." For now, it is enough for you to know that PC power supplies are
of a special type known as *switching supplies*. Switching power supplies, or *switchers,*
get their name because they don't always conduct current through the control device, as do the older *linear* supplies. In a switcher, a control or current-pass device,
usually a heavy-duty transistor, is turned on and off at some rate. The rate is varied

according to what the load current demand is. The switching rate is constantly changed and updated by monitoring the load voltage to produce a close-tolerance output voltage from the supply. In the modern ATX +12 PC power supplies, there are several voltages produced; these voltages are +12 VDC, + 5 VDC, + 3.3 VDC, –12 VDC, and –5 VDC. In general, the +12 VDC runs the main microprocessor, drive motors, and fans, and the +5 VDC runs the other ICs in the PC. The negative voltages are remnants of the first original PC and are no longer used. The older ATX power supply specification called for the main processor to run off the + 3.3 VDC supply line.

A Minimum Load is Required

There are a couple peculiarities of switching power supplies. They need to be connected to at least a minimum load in order to draw sufficient output current, and excess current will cause them to "fall out of regulation," meaning they can no longer produce the required voltage levels. Different switching supply manufacturers use slightly different specs for their supplies' output voltage percentages, but in general, a switcher should provide the specified voltage regulation, usually +/–5%, when the output current demand placed on it is between 30% and 70% of the total rated current. Most PC supplies are actually rated in terms of total power output rather than current, but you should know the current demand is what is really crucial. Connecting the motherboard with a normal amount of RAM and the processor installed, and at least one hard drive to the power supply can establish the minimum power supply demand. If this minimum load is not connected, most low-quality power supplies will fail outright as soon as they are turned on. Others will produce no output whatsoever, and simply shut down. Some of the better-quality supplies include internal load resistors to ensure that the minimum output power is drawn from the supply at all times. To be safe though, always assume you are dealing with a PC supply that will be damaged if you forget to connect at least the minimum load to it.

Don't Overload the Supply

Switchers will also tend to come out of regulation when more than approximately 70% of their total rated power is drawn from them. This underrating is necessary since power supplies cannot supply as much power when they get hot. So to ensure they can still produce the minimum power to run the PC when warm, the 70% of total rated power figure is used. Different manufacturers tend to "fudge the numbers" more or less, so the old adage, "you get what you pay for" applies here. A typical example is a PC that locks up or randomly reboots after several hours of use, or under heavy load, or when the room temperature is abnormally high. Another reason for underrating of the output power is because drive motors draw more starting current than running current. The supply must allow for this extra initial motor power demand when the PC is first turned on. SCSI hard drives can be easily set for a delayed turn-on for this reason; it helps to spread the initial power demand on start up. Most EIDE drives do not offer delayed startup options, but some modern BIOSs allow for doing so.

With a populated motherboard, meaning all normally installed devices are connected and working and at least one hard drive is connected, the easiest way to measure the power supply output voltages is to use the so-called "back probing" method. This entails placing the meter probes into the back plastic connector shell openings that house the discrete wires in the power supply connector attached to the motherboard. This is known as the "P1" connector on modern ATX or ATX +12V power supplies, and the "P8" and "P9" connectors on the older AT-style power supplies. Most DMMs come with a set of probes that end in long, narrow metal pins. These are placed into the proper wire holes in the back of the connector shell in order to measure the voltage. Refer to Figure 5.4 for a picture of the back-probing measurement technique.

FIGURE 5.4 Back-probing the power supply connector in order to measure power supply voltages.

The measured voltages should be checked against both the connector pinout chart and the power supply manufacturer's published voltage tolerance specifications. In other words, refer to the power supply connector diagram or chart in order to anticipate the proper voltage value that should be present at certain connector pins, and refer to the supply's tolerance ratings in order to tell if the measurements fall within the manufacturer's specified ranges. A diagram of the signal or voltage that is present on each pin or wire is known as a *pinout*. Table 5.1 shows an illustration of a PC power supply connector pinout. Table 5.2 shows the P1 power connector pinout.

TABLE 5.1   AT and ATX power supply connector pinouts

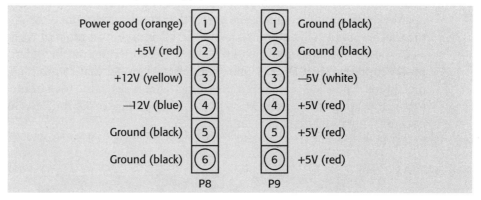

TABLE 5.2   The P1 power connector pinout used by ATX and NLX PCs

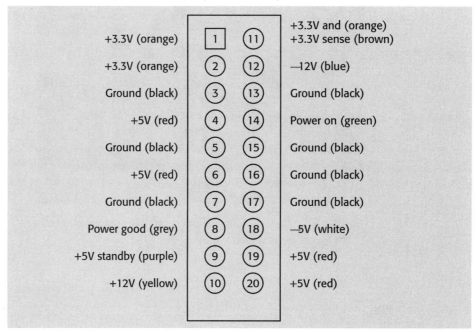

Voltage measurements can also be made at drive power connectors. Just because the voltages check out at the P1 connector does not ensure that the proper voltages will also be present at other places in the PC. Other places that voltage measurements can be made are the three-pin fan connectors on the motherboard. Most modern high-quality motherboards come with three or more 3-pin cooling fan power connectors. One is dedicated to the main processor fan, and two are for optional chassis fans. One chassis fan draws cool air in from the case front, and the other pushes hot air out the rear of the case. Some chipsets even come with a dedicated fan for the main chipset component, the "north bridge" in Intel's chipset language. Some motherboards come with a special monitor chip that reports temperature of the processor, inside case temperature, and the speed of several fans, usually the processor fan and at least one other fan. It also monitors the power supply output voltages. In motherboards without such a monitor chip, making voltage measurements at each of the fan connectors is a useful troubleshooting technique. This is especially true if system overheating is a problem. Low fan voltage equates to low fan RPM and consequently, insufficient airflow. Figure 5.5 shows an example of a motherboard hardware monitoring chip.

FIGURE 5.5   Motherboard hardware monitor chip.

5.3.1   Using the Proper Meter Settings

Always set a meter's switches and connect the meter ends of probes to the proper meter jacks before connecting the meter to a circuit. To adjust a DMM to measure voltages in a PC, use the 20 VDC scale. Be sure to set the meter to DC if it uses a separate function control, rather than a single combined function/range switch. Be sure to switch the DMM on if it uses a separate power switch. All these settings should be made before the meter is connected to the circuit.

5.3.2 Using the Proper Meter Jack Connections

Most DMMs are set up to use the Common connector jack for the negative, black lead and a jack labeled Volts/Ohms for the positive, red colored lead. The Volts/Ohms may be labeled by the symbols "V/Ω" instead of the words. These connections are absolutely critical to both get the proper reading and to avoid meter damage. Incorrectly connected leads can blow out fuses, replacements for which can be difficult to find, or permanent damage can occur to a meter at worst. Refer to Figure 5.6 for a picture of correct DMM voltage measurement lead connections.

Key Ideas

Whenever possible, power down the circuit before connecting the meter. Always set the proper meter function range and lead connections before connecting the meter to the circuit. Then re-energize the circuit and take the meter reading.

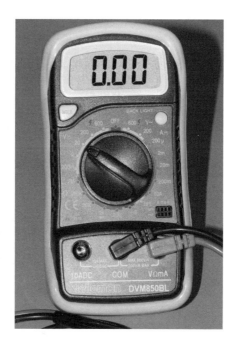

FIGURE 5.6 DMM voltage measurement lead connections.

5.4 MEASURING CURRENT

Current is measured through a component, which means in series with it. It can be mechanically difficult to make such a series connection. This is why current measurements are not as common as voltage measurements. In order to place the meter in series with a component, one end of the component must be disconnected from the normal circuit and the test meter inserted in series with it. For components soldered in place to a board, this is a real problem, since desoldering and re-soldering a component from a board requires special skills. In addition, the process of removing and reinstalling a component will almost certainly leave telltale marks on a circuit board. There still are certain times, however, where a current measurement is advantageous. There are two rules that usually apply to making current measurements with a DMM. First, select the correct type of current, either DC or AC, by setting the function control, and second, one lead of the meter must be moved into a new jack position. The function control may serve a dual purpose as a range selector as well. Be sure to set the control to DC A, DC mA, AC A, or AC mA as required. Some DMMs also require the operator to select the proper range. The common, black colored lead is still used to connect to the common, negative meter jack, and the red lead is moved to the meter's appropriate current jack. The meter leads are then connected as follows: red lead to the positive power connection and the black meter lead to the device under test's normal positive connection. Refer to Figure 5.7 for a close-up picture of a DMM set to measure up to 10 A of DC current.

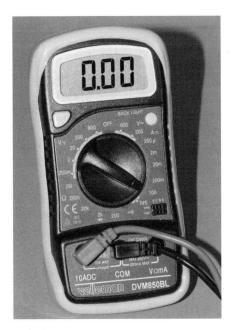

FIGURE 5.7   DMM showing leads and range positions to measure up to 10 A DC current.

Current is measured through a component; you must connect the ammeter in series with the component through which current is to be measured. The positive amme-ter lead should be connected to the more positive side of the circuit where the break is made; the negative ammeter lead should be connected to the more negative side of the break in the circuit.

5.5   MEASURING RESISTANCE

Resistance measurements are made with no power applied to the circuit or device to be tested. This is because doing so will almost certainly damage the meter. Meters capable of making resistance measurements include their own internal power supply. Hand-held DMMs include a small battery to allow for resistance measurements. In the case of a component mounted to a circuit board, or wired into existing equip-ment, at least one end of the component to be measured must be disconnected from the circuit in order to isolate it. The device under test is connected to both DMM leads after first setting the DMM for the proper resistance range, or in the case of an "autoranging" DMM, at least setting the function control to resistance. Select the range that provides the greatest accuracy by ensuring that the most digits are used. This means you must first have at least a general idea of the expected value in order to properly set the range. Figure 5.8 shows a DMM set to measure resistance.

FIGURE 5.8   A DMM set to measure resistance.

Key Ideas

At least one end of a component must be disconnected from the circuit when making resistance measurements. Never attempt to measure resistance in a powered circuit.

5.5.1 Diode Testing

Diodes are tested with a DMM by supplying them with current and measuring their resistance. Diodes are "one-way devices" for current, meaning they normally allow an easy current pathway in the forward biased direction and a very high resistance in the *reverse biased* direction. *Forward biased* means connecting the positive voltage source to the anode (positive lead) and the negative of the voltage source to the cathode (negative lead). This test is essentially the same as a resistance test. Most DMMs include a special switch position for diode testing, usually marked with the diode symbol. Refer to Figure 5.9 for a close-up picture of the diode test switch position on a DMM. Refer to figure 5.10 for a picture of a diode; note the white band indicating the diode's negative (cathode) end.

FIGURE 5.9 DMM measuring a forward-biased diode. The meter reading indicates the diode's forward bias voltage.

FIGURE 5.10 A close-up photo of a diode.

5.6 METER LOADING

The effect the meter itself has on the circuit under test is known as *meter loading*. The following sections detail how this affects the readings of different DMM measurements.

5.6.1 Voltmeter Loading

Ideally, the measuring device should not change the circuit or device being tested. In reality, measuring something does change it to some degree. In the case of voltage measurements, the voltmeter is connected in parallel with the device being measured. This creates a circuit with the device under test and the voltmeter connected in parallel to it. As you will learn in more detail in Chapter 8, "Parallel Circuits," when a device is placed in parallel with another device, the total circuit resistance decreases, since there is now another current path. So in order to not change the voltage across a component under test, a voltmeter should not appear to be present at all. What this means electrically, is a perfect voltmeter should have infinite resistance, as with an open circuit. Of course, if the circuit were truly open there would be zero current. Real voltmeters are not an open circuit; some amount of current must flow through the voltmeter to cause a reading. Just how little a voltmeter influences the device under test is a function of its degree of circuit loading. This depends on the relative resistance range of the item being tested and the internal impedance of the meter. More on impedance will be presented in Chapter 12, "Alternating Circuit Currents." For now, think of impedance as similar to resistance. Loading a circuit means connecting more devices that increase the total circuit current. Refer to Figure 5.11 for a picture of a simple parallel circuit.

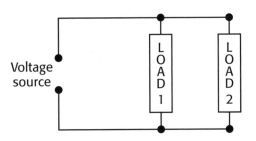

FIGURE 5.11   A simple parallel circuit.

Various methods have been used in the evolution of voltmeters in order to minimize the current drain from the circuit. We discussed the design of analog meters in Section 5.2, and more on this subject will be presented in Chapter 10, "Magnetism and Electromagnetism." One technique used to make a more sensitive voltmeter was using a finer gauge and more turns of the wire used to wind the armature. This resulted in a greater magnetic field set up by the small current flowing through the armature coil. Better methods of suspending the armature helped as well, which reduced meter bearing friction.

5.6.2   Amplified Meters

The invention of the vacuum tube meant an even smaller amount of current was required to make useful voltage measurements, as the tube could amplify the current drawn from the test device. Semiconductor replacements for the vacuum tube, such as field effect transistors (FETs), are used for the same reason. The first circuit inside a DMM (called the *front end*), to which the device under test is connected, presents very high impedance to current, thereby helping to isolate the meter from the device under test. Refer to Figure 5.12 for a picture of a DMM making a voltage measurement.

5.6.3   Meter Loading Examples

A few examples will help illustrate the idea of meter loading.

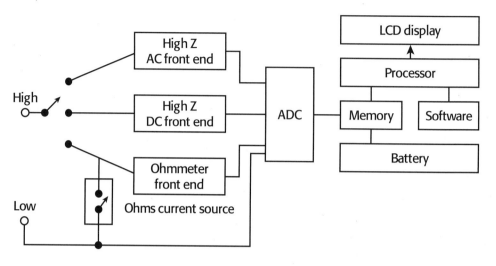

FIGURE 5.12 DMM block diagram for voltage measurement showing high-impedance front end circuit.

Example 1

In the first case, the voltage across one of two series-connected 10 kΩ resistors is being measured. The total voltage applied is 10 volts. What should the meter indicate? Using the voltage-divider formula from Chapter 7, "Series Circuits," the value of voltage that should exist across this resistor is one half of the applied voltage, or 5 volts. Assuming a meter that presents an internal resistance of 10 kΩ is used, how much voltage will be indicated by the meter?

First, find the equivalent resistance of the resistor under test and the voltmeter:

$$R_{eq} = (R_2 R_{Meter}) \div (R_2 + R_{Meter}) \qquad (5.1)$$
$$= [(10k\Omega)(10k\Omega)] \div (10k\Omega + 10k\Omega)$$
$$= 5k\Omega$$

Next, use the voltage-divider formula to find the voltage across this equivalent resistance.

$$V_{Meter} = (R_{Eq} \div R_{Total}) \times E \qquad (5.2)$$

Then substitute the values for applied voltage and equivalent resistance.

$$V_{Meter} = (5k\Omega \div 15k\Omega) \times 10V$$
$$= (1 \div 3) \times 10V$$
$$= 3.33V$$

5.6.4 Ammeter Loading

When measuring current, the meter must be placed in series with the circuit. Any extra resistance offered by the ammeter will affect the total circuit current value. So the ideal ammeter would have 0 ohms resistance; it should look like a perfectly conducting short. Real ammeters do have a small amount of resistance however. In general, an ammeter should have a relatively small amount of resistance compared to the circuit under test. One tenth or less is a good starting point for the amount of ammeter resistance compared to the resistance of the circuit under test. If the circuit under test has a very low value of resistance, then the insertion of the ammeter will significantly change the total circuit current and seriously degrade the ammeter reading accuracy.

5.6.5 Ohmmeter Loading

When using an ohmmeter, the problem encountered is the requirement to first isolate the component being tested from the circuit into which it is connected. Otherwise, the ohmmeter will read the total resistance of the device under test and the rest of the circuit in parallel with it. So resistance readings of components installed in a circuit are usually not done, as it would involve desoldering and later resoldering components, and that is just too costly in time and worker expense.

Know the specifications and limitations of the meter you are using. Know the consequences of meter loading.

Key Ideas

5.7 METER PRECISION

Meter precision is an often-debated topic among technicians, and is one that should be considered before buying or using a digital meter.

5.7.1 Sensitivity

Sensitivity refers to the smallest sample of something that can be measured; how accurate the reading is another thing, however.

5.7.2 Accuracy and Resolution

Meter accuracy means how close to the actual value present in the circuit being measured the meter is indicating. Many analog meters are rated accurate to a certain percentage of the full-scale value. One problem with an analog meter is that this percentage of full-scale is applied to any reading on the scale. For example, if an analog voltmeter is rated at +/– 10% accuracy, and the scale being used at the moment is 100 volts, then the meter can be off from the indicated value by plus or minus 10 percent of 100 volts, or 10 volts. So if you are measuring exactly 100 volts, the true voltage value could be 100 volts minus 10 volts, or 90 volts on the low side, and 100 volts plus 10 volts, or 110 volts on the high side. This is better than expected for an inexpensive analog meter. But what if this same meter is used to measure a voltage that is far less than the meter's full-scale value? Assuming a value of 10 volts is applied to the meter, and still using the 100 volts full-scale setting, the +/– 10 % of full-scale value is still the rule for this meter, so it is possible the meter would actually display somewhere between 0–20 volts. This is not very accurate at all. So the general rule when using an analog meter is to select a meter scale that gives as close to a full-scale reading as possible. The number of different analog meter scales has a real effect on analog meter accuracy for this reason. It is desirable to use a meter scale where the actual reading will be as close to the full-scale value as possible *without* exceeding it. Of course, more expensive analog meters are available. It is possible to buy 5% analog meters, and even 1% meters, for a lot more money, of course. Another problem with analog meters is they use mechanical bearings of one sort or another, and the bearings have a certain amount of friction. It was commonplace to see engineers and techs gently tapping on an analog meter's glass cover to help "settle" the movement before taking the reading. Fortunately, digital meters don't suffer from this problem, as there are no moving parts involved in the display.

5.7.3 Digital Meter Accuracy

Digital meters instead are rated for a certain percentage basic accuracy of the reading and +/– the least significant digit error. Most DMMs are rated for +/– 0.5% accuracy or better, and +/– the LSD. In addition, the accuracy is usually given separately for DC, AC, and individually for current versus voltage as well. Refer to Table 5.3 for an example of a typical DMM specifications chart.

5.7.4 Resolution

DMMs are rated for a certain number of digits resolution, such as 3 or 4. Like accuracy, resolution can be expressed as a percentage. A 2-digit meter has 1% resolution (1 count out of 100) while a 3-digit meter has a 0.1 % resolution (1 count

TABLE 5.3   A DMM specification chart.

Digital Multimeter Specifications, Velleman Model DVM850BL

Features:	AC voltage: 200/600 V ranges
Auto-polarity and 3 Ω digit LCD panel	**Maximum input:** 600 V
DC current up to 10 A max	
AC and DC voltage up to 600 V max	**DC Current:** 200Ω/2m/20m/200m/10A
Resistance up to 2 MΩ	**Overload protection:** 10A fused
Diode, transistor and continuity tests with buzzer	**AC Current:**
Data-hold function and backlight	**Overload protection:** 10A fused
Protective plastic holster	**Resistance:** 200/2k/20k/200k/2M_
Dimensions: 2.7 x 5.4 x 1.2 inches	**Over range indicator:** "1"
Power: one 9 V battery	**Maximum display:** 1999
Specifications:	
DC Voltage: 200m/2/20/200/1000V ranges	
Maximum input: 600 V	

out of 1,000). Resolution is not the same as accuracy, however. A 3-digit meter has 0.1% resolution but may have 0.5% full-scale accuracy. One should read the meter manufacturer's meter specifications carefully. Often it is the meter's resolution that is better than the meter's accuracy. In other words, you can not simply believe a meter reading without knowing how to interpret it.

Understand the limitations of your meter. Don't rely on it being absolutely accurate.

Key Ideas

5.8   HOW TO SELECT THE CORRECT METER

There are many types of meters from which to choose when deciding which to use for a PC toolkit. In general, a DMM offering DC and AC voltage and current scales, resistance, and diode/continuity testing is about the minimum one should consider. The DC voltage scales should include a 20-volt as well as a 5- or 10-volt scale, in order to measure the common PC power supply lead values. In order to measure the line voltage in the United States, a 200-volt AC scale should be included. Since many other countries use an AC line voltage in excess of 200 volts AC, a higher voltage AC scale would be required as well. The capability to measure a transistor's gain is a nice added feature, even if seldom used in normal PC troubleshooting electrical work. Some DMMs also measure frequency, but their range is rather limited.

5.8.1 PC Interface

Some of the most sophisticated digital multimeters offer even more features. With a built-in computer interface, some can be directly connected in a PC via the serial port. With a data logging feature, it's possible to monitor any system with a variable sampling rate for extended periods of time without supervision. The data is instantaneously displayed on the computer with its sampling rate and measurement in chart form. The continuous samples are automatically charted for graphical analysis. DMMs that offer this feature come with monitoring software included, as well as the PC interface cable. A good PC test using this capability would be a long-term voltage measurement of the power supply outputs, covering times of varying PC power load, in order to see how well it maintains regulation. Figure 5.13 shows a DMM with a PC interface cable and software.

FIGURE 5.13 A DMM with its PC interface cable and companion software.

5.8.2 A Sturdy Case

A sturdy case and an added shroud or meter holder are good ideas, as unprotected DMMs can be easily damaged if thrown into a toolbox with sharp tools inside. The author likes to carry his DMMs in small padded bags such as those used to carry a small notebook computer. These can often be had very inexpensively at electronics flea markets.

5.8.3 Affordability

A DMM should not "break the bank," and many good-quality DMMs can be had for $20 or so. More expensive models usually offer more features, and the better known brand names command higher prices. Some of the features found in higher-priced DMMs include a bar graph to act as a faster-response replacement for an analog meter, the capability to measure temperature using a special probe, and others. Potential purchasers must analyze their real measurement needs before buying a DMM.

5.9 COMMON PC TROUBLESHOOTING USING THE DMM

DMMs can perform several important electrical tests on a PC. The following sections illustrate the most common examples of these.

5.9.1 Voltage Tests

The AC line voltage can be checked with either an AC line tester or a DMM. A DMM can be used to check all of the PC power supply output voltages as well as the voltages at each drive or fan connector. Even voltages to power LED indicators for on/off, hard drive activity, and so on can be checked with a DMM. Using a low AC voltage scale, the audio output to a PC's speaker can even be checked.

5.9.2 Continuity Tests

A continuity check can be made on any wire or cable if both ends of the wire or cable are within reach of the DMM's test leads. Although time-consuming, a continuity check of a PC's drive cables can locate an open wire, although this isn't the most efficient way to do that. For instance, checking each and every wire in an EIDE 80 wire cable would be quite time consuming and tedious. Swapping the suspect cable with a known good one would be much faster and easier. Virtually any DMM or ohmmeter can be used to perform a continuity check. Simply test the unpowered circuit or wire for a low resistance (close to zero ohms) reading.

5.9.3 Current Tests

It is possible to perform DC current measurements on PC subsystems such as drives, but doing so would require a special cable fixture to allow inserting the DMM set up as an ammeter in series with the device. Such an adapter can be custom made by any tech with soldering skills. AC current measurements can be performed using a clamp-on type ammeter, but the power cable must be split so that only one conductor passes though the pickup coil, which functions as a one-turn primary transformer winding. Refer to Chapter 13, "Transformers," for more on this subject. Normally, no one would tear an AC line power cable apart in order to do a current test using a clamp-on ammeter.

5.10 SUMMARY

- The first meters were analog, with moving pointers, and operated on the electromagnetic motor principle.
- The first analog meters were dedicated to a single function and range; later improvements allowed for multiple functions and ranges.
- The invention of the amplifier vacuum tube allowed for much better meter sensitivity and less meter loading, resulting in more accuracy. The need for high voltage and current kept these vacuum tube voltmeters (VTVMs) to be tied to the tech bench.
- Semiconductor replacements for vacuum tubes allowed smaller, battery-operated versions of the VTVM, which allowed for high-accuracy measurements to be made in the field as well as on the bench.
- The invention of rack-mounted digital meters made for easier reading.
- Portable DMMs offer the best of both worlds: digital readouts and portability.
- Modern DMMs are also affordable, rugged, and reliable.
- Many basic troubleshooting tests that can be made with a DMM include voltage, current, resistance and continuity, among others.
- The best value for most PC techs is a DMM with a rugged case and an affordable price.
- Understand a DMM's accuracy and resolution specifications.
- A low-priced, good-quality DMM should be part of every PC tech's toolkit.
- When using a meter, always first determine the expected range of what you are to measure, and then set the meter's function and range knobs before connecting the meter to the circuit.
- Whenever possible, set up the meter and connect it to an unpowered circuit before reenergizing the circuit and making the meter reading.
- Read the owner/operator's manual that is provided with your DMM. Understand the measurement limitations of the meter.

5.11 KEY TERMS

Voltmeter

Multiplier resistor

Resolution

Accuracy

Meter loading

Clamp-on ammeter

Ammeter

Shunt resistor

Ohmmeter

Continuity

5.12 EXERCISES

1. The proper order of operations when using a DMM to measure a voltage is . . .
   a. connect the meter to the device, turn on the circuit power, turn on the meter, set the meter function and range, and then take the reading
   b. turn on circuit power, turn on meter, set the meter function and range, connect the meter to the device, and then take the reading
   c. turn on the meter, connect the meter to the device, turn on the circuit power, adjust the meter function and range, and then take the reading
   d. determine the expected current type (DC or AC) and range of voltage, and set the meter function and range controls accordingly. Then connect the meter to the circuit and turn on the meter, turn on the circuit power, and then take the reading
   e. take the reading, connect the meter to the circuit, turn on circuit power, and then turn on the meter

2. When used as a voltmeter, a DMM should be connected how?
   a. In series with the device under test
   b. In parallel with the device under test
   c. In series-parallel with the device under test
   d. Voltmeters are never used in live circuits
   e. None of the above

3. A DMM used as a DC ammeter should be connected how?
   a. In series with the device under test
   b. In parallel with the device under test
   c. In series-parallel with the device under test
   d. Ammeters should never be used in live circuits
   e. None of the above

4. A DMM used as an ohmmeter should be connected how?
   a. In series with the device under test
   b. In parallel with the device under test, with at least one of the device's leads isolated from the circuit
   c. In series-parallel with the device under test
   d. Ohmmeters should never be used in live circuits
   e. Both a and d

5. Most analog meters are rated for accuracy how?
   a. As a plus or minus percentage of the reading
   b. As a plus or minus percentage of the full-scale meter value
   c. Analog meters are all 100% accurate
   d. Analog meters are accurate only on the lowest ⅓ of the scale
   e. None of the above

6. A voltage rise is . . .
   a. developed by current flowing through a device
   b. developed by current flowing around a device
   c. developed by a change from another form of energy, such as in a cell or battery
   d. what happens in a magnet
   e. none of the above

7. A voltage drop is . . .
   a. produced when current flows through a resistive device
   b. produced when current flows around a resistive device
   c. produced at the terminals of a power source, such as a generator
   d. produced by a perfect conductor
   e. both b and c

8. A method used to increase the range of an analog voltmeter is to connect a . . .
   a. shunt resistor in series with the voltmeter
   b. shunt resistor in parallel with the meter
   c. multiplier resistor in series with the meter
   d. multiplier resistor in parallel with the mete
   e. voltage rise resistor in parallel with the meter

9. DMMs are specified as having what typical value of input impedance?
   a. 10 kΩ
   b. 100 kΩ
   c. 1000 kΩ
   d. 20 kΩ/V
   e. 10 MΩ

10. Which of the following are desirable DMM features a PC tech would use?
    a. Relatively inexpensive price
    b. Rugged case and/or shroud-type protective holder
    c. The capability to read voltage, current, and resistance
    d. The capability to read both DC and AC values
    e. All of the above

11. A 4-digit DMM is used to measure the output of a PC power supply. The indicated reading is 12.31 volts DC. The basic meter accuracy for DC voltage is specified as 0.05%, and +/– the LSD. What range of actual voltage could be present in the circuit being tested?
    a. 10.0–13.5 volts
    b. 12.38–12.24 volts
    c. 11.69–12.92 volts
    d. 11.68–12.94 volts
    e. 11.0–13.0 volts

12. DMMs have no effect on the circuit under test.
    a. True
    b. False

13. DMMs include their own power source, in the form of a battery, in order to measure what parameter?
    a. Voltage
    b. Current
    c. Resistance
    d. Temperature
    e. RPM

14. It is common to have to relocate a DMM's red lead plug from one meter jack to another when changing from making voltage tests to current tests.
    a. True
    b. False

15. Most DMMs contain one or more fuses to help guard against over-current damage.
    a. True
    b. False

16. How does a DMM display an over-range condition?
    a. All 1s displayed
    b. All 0s displayed
    c. A blank display
    d. A single 1 displayed as the leftmost digit
    e. None of the above

17. DMMs are absolutely accurate; you can always totally trust the indicated value in all instances.
    a. True
    b. False

18. Clamp-on ammeters are commonly used to test a PC's power cable.
    a. True
    b. False

19. A reasonable-quality DMM can be purchased new for as little as how much?
    a. $100
    b. $75
    c. $50
    d. $20
    e. $5

20. Hand-held DMMs are powered by . . .
    a. solar power
    b. commercial power
    c. internal batteries
    d. external batteries
    e. only the circuit or device under test

6 ▪ Resistors

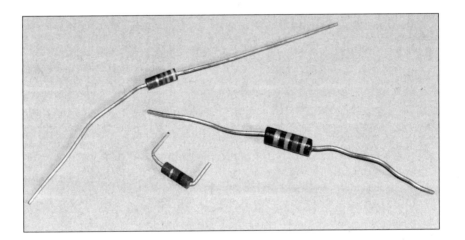

6.1   INTRODUCTION

One of the basic electric circuit elements is the resistor. This relatively simple device is used in many circuits to control the flow of current. In this chapter you will learn many of the different styles and uses of resistors and resistive devices. In general, resistors can be described as components that tend to oppose the flow of electric current. They differ from insulators in that the specific amount of current opposition (resistance) can be designed into the device during manufacture. Most insulators, however, are intended to have the maximum resistance possible. That amount of resistance depends on the material from which the insulator is made, and the temperature and the amount of voltage they are subjected to.

Several styles of resistors have evolved over the years due to the need for increased reliability and wider ranges of values. Some resistors are designed to always present a constant and (hopefully) unchangeable resistance value. Others are designed to be adjusted. A special-type resistor, called a *thermistor*, actually changes resistance value with changes in temperature. Resistive devices are classified as either *fixed* or *variable*.

6.2   FIXED RESISTORS

Fixed resistors are made to have a specific value of resistance, which is not supposed to change, and they can not be manually adjusted to some new value of resistance.

6.2.1   Carbon Composition Resistors

Beginning in the early days of electronics, and up until the mid 1970s, the most widely used type of resistor was the *carbon-composition* type. Relatively inexpensive and available in a wide number of different standard resistive values, the carbon composition type has a rod-shaped body, composed of a mixture of finely-ground carbon granules and a ceramic powder called the *binder*, a sort of glue. The binder is also a good insulator. The ratio of carbon granules to binder determines the exact resistance of the mixture. The more carbon used, the less the resistance. The carbon/binder mixture is pressed and formed into the desired shape, and then coated with a varnishlike coating, to help prevent moisture absorption.

The advantages of the carbon-composition resistor were its low price, the availability of many popular resistance values, and relatively constant resistance value over a wide frequency range. The disadvantages were the tendency for the resistor's value to vary over time, and with changes in temperature and humidity. They are not usually manufactured with power dissipation capabilities beyond 5 watts. Relatively fragile components, they are easily broken if strained or overheated. Resistors of this type even momentarily overheated would often change resistance value, with the final resistance value different than the original design value. Figure 6.1 shows the construction of a carbon-composition resistor. Color codes, in the form of colored bands painted around the outside of the resistor, signified the resistance value, tolerance in percent of indicated value, and sometimes additional data such as expected failure rate.

6.2.2   Package Styles

The typical package style for carbon-composition resistors is the common rod shape with one electrode (lead) exiting each end of the device. This is called an *axial-lead* arrangement. Other styles of carbon-composition resistors include metal end caps attached to each end of the cylinder shape for installation into spring-contact mounts, and small leadless resistors known as *chip resistors*, for attachment

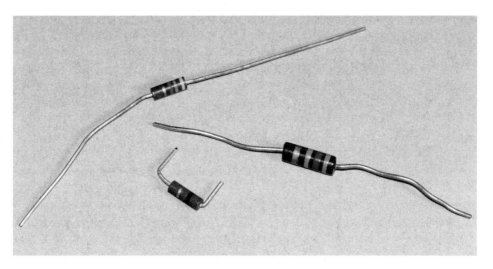

FIGURE 6.1   Carbon composition resistors.

directly onto the traces of a printed circuit board. Another package arrangement includes many individual resistors, usually of all the same value, in a single package that is designed to mount in a vertical fashion on a printed circuit board. It is called a *single inline package*, or *SIP*.

6.2.3 Wattage Sizes

The most popular wattage (power) rating sizes for carbon resistors are ¼ W, ½ W, 1 W, and 2 W. Occasionally, larger wattage sizes are produced for special applications, but the majority of carbon composition resistors are 2 watts or smaller.

6.2.4 The Resistor Color Code

There are two main methods of indicating the ohmic value of carbon composition resistors: printing the value directly on the body of the resistor, and use of three or more color bands around the body. As shown in Table 6.1, the first color band is the first significant digit, the second is the second significant digit, and the third band indicates the decimal multiplier. Some people like to think of the third band as the number of zeroes to add to the first two digits. The fourth band indicates the tolerance of the nominal ohmic value, expressed as a percentage. It is extremely important for anyone working in electronics to memorize this color code scheme, as the values of many different types of components, including most small resistors, are designated by this method. Table 6.1 lists the standard values for resistors and capacitors.

TABLE 6.1 The resistor color code chart. Reproduced with permission from The 2001 Handbook for Radio Amateurs

Resistor-Capacitor Color Codes

Color	Significant Figure	Decimal Multiplier	Tolerance (%)	Voltage Rating*
Black	0	1	-	-
Brown	1	10	1*	100
Red	2	100	2*	200
Orange	3	1,000	3*	300
Yellow	4	10,000	4*	400
Green	5	100,000	5*	500
Blue	6	1,000,000	6*	600
Violet	7	10,000,000	7*	700
Gray	8	100,000,000	8*	800
White	9	1,000,000,000	9*	900
Gold	-	0.1	5	1000
Silver	-	0.01	10	2000
No color	-	-	20	500

*Applies to capacitors only

6.2.5 Wirewound Resistors

Wirewound resistors overcome the major disadvantages of carbon-composition resistors. They do have other disadvantages that carbon-composition resistors don't, namely, they appear as inductors, which can be a very bad thing when dealing with high frequencies. Refer to Chapter 11, "Inductors and Capacitors," for more on inductors. This type of resistor is made on a hollow ceramic tube, which is capable of withstanding high heat without damage. Wire, usually Nichrome®, a special alloy of nickel and chromium, is wound on the ceramic form to serve as the resistive element. Wirewound resistors are manufactured in power sizes ranging from several watts to several hundred watts. As you will learn in Chapter 14, "Power Supplies," many early PC power supplies included an internal power resistor to serve as a minimal load, something absolutely required by the type of supply used in PCs.

For a given diameter, it takes less length of Nichrome than for a good conductor such as copper to attain the same value of resistance. Because of this, a resistor made of Nichrome can be made smaller than a resistor made with copper. Nichrome is also commonly used to form heating elements in such appliances as toasters and portable room heaters.

TABLE 6.2   Standard Values for Resistors and Capacitors.   Reproduced with permission from
The 2001 ARRL Handbook for Radio Amateurs

+5%	+10%	+20%	+5%	+10%	+20%
1.0	1.0	1.0	3.6		
1.1			3.9	3.9	
1.2	1.2		4.3		
1.3			4.7	4.7	4.7
1.5	1.5	1.5	5.1		
1.6			5.6	5.6	
1.8	1.8		6.2		
2.0			6.8	6.8	6.8
2.2	2.2	2.2	7.5		
2.4			8.2	8.2	
2.7	2.7		9.1		
3.0			10.0	10.0	10.0
3.3	3.3	3.3			

*Applies to capacitors only

The body of the wirewound resistor is wound with Nichrome or other special alloy metal wire. The correct length and turns spacing is required to provide the desired resistance value. The current capability depends on both the gauge (diameter) of wire used and the size of the ceramic tube used to construct the resistor. Larger sized ceramic tubes can dissipate more heat. After winding, the resistor is dipped into a ceramic slurry, and then oven dried. Three popular electrode connection styles are used: axial-lead arrangement, usually employed for units of less than 5 watts dissipation rating; metal end caps to allow insertion into special holders, and metal solder lugs, to allow attachment of wires to the ends, usually employed for high-power units.

Even if the wirewound resistor is overheated for short periods, as long as the wire does not burn in two, the resistance value will not usually change over time. It is never a good engineering practice to exceed the rating of any electronic device. Even so, the wirewound is much more durable than an equivalent size and value of carbon-composition resistor. Because it costs more to manufacture, wirewound units cost more than carbon-composition resistors. This is why their use is usually limited to demanding applications such as machine tool controllers, and space and avionics (aircraft electronics), where high precision as well as high heat capability is required. Figure 6.2 shows various wirewound resistors.

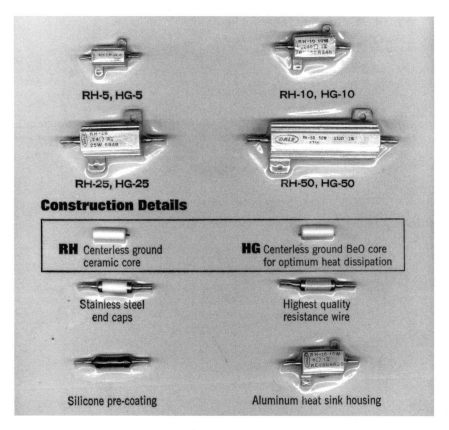

FIGURE 6.2 Precision wirewound resistors.

6.2.6 Film-type Resistors

Carbon-composition and wirewound resistors are the two extremes of device cost compared to reliability and accuracy. Film resistors were developed as a compromise between the two resistor types. Film-type resistors offer advantages of each type, with few of the disadvantages of either. They are used now more than carbon-composition resistors.

Similar to wirewound resistors, a ceramic tube is coated with a thin film of resistive material. Both carbon mixtures and metal films are used. Part of the film is then machined away, in a spiral fashion, which leaves the desired amount of a resistive spiral between the two opposite metal end caps. The more material removed, the greater the resistance.

The film removal is a very precise operation, and the resistive qualities of the film are accurately known. Film resistors can be made with high precision. The resistor package is dipped into a ceramic coating mix, and oven cured. Then it is marked, indicating its resistance value and tolerance. Either the resistor is color-coded in a manner similar to the carbon-composition resistors, or it is printed with ink, indicating the exact value and tolerance in word form, such as 10 $k\Omega$, 1%.

Film-type resistors can be easily identified because of the special construction method used. A bulge from the metal end caps on the film-type resistor is obvious. The film-type resistor has a distinctive "dog-bone" shape. Carbon composition resistors have a straight-sided profile. The resistors also differ in body color. Carbon composition resistors' body color is usually a dark brown, whereas film resistors usually have a light-colored body color, such as pastel tan, green, or blue. Figure 6.3 shows carbon film-type resistors. Figure 6.4 shows metal film resistors.

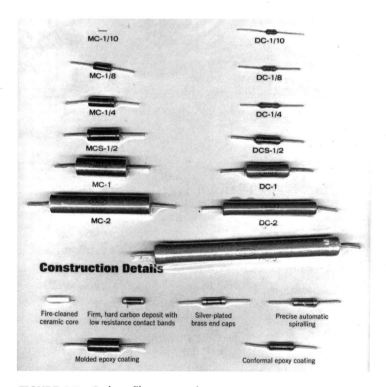

FIGURE 6.3 Carbon-film type resistors.

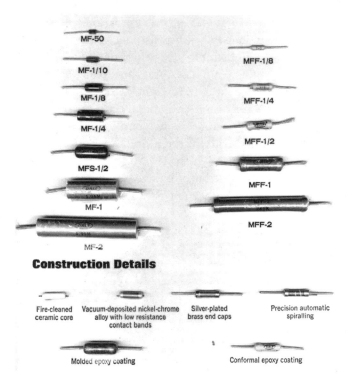

FIGURE 6.4 Metal film resistors.

6.2.7 Comparison of Fixed Resistor Types

In the past, the carbon-composition resistor was nearly always used for low-wattage fixed-resistor applications. Wirewound resistors were used only where their extra cost could be justified by the increased accuracy and reliability they offered. Film resistors are now the low-cost alternative to the carbon-composition resistor, and are the preferred style for all but large (greater than 5 watts) power applications.

Both carbon-composition and film resistors are color-coded using the standard method. Some film resistors, usually high-precision types of 1% tolerance or less and 1 watt or greater, have their values printed on their sides. Most wirewound resistors, because their larger size allows for it, use printed markings.

Wirewound resistors generate less electrical noise (undesired random signals produced by the device), than do the carbon-composition or film types, and are therefore preferred for sensitive measuring equipment such as deep-space satellite receivers, and biomedical devices that monitor such things as heart-rate and brain electrical activity.

Film-type resistors offer good value, availability in all the popular resistance values, and better resistance value stability over time and with changes in humidity and temperature. Prices for film resistors have fallen consistently. It is now possible to purchase film resistors for approximately the same price as the less-desirable carbon-composition type.

The most common wattage sizes for film resistors are ¼ W, ½ W, 1 W and 2 W. Film resistors in power sizes greater than 2 W are rare.

Both carbon-composition and film-type resistors are available in similar wattage sizes, from approximately ¼ W through 2 W in many popular sizes. Wirewound resistors are available in wattage sizes ranging from about 2 W thorough several hundred watt sizes. Wirewound resistors are still the most expensive of the three resistor types used. Carbon composition and film resistors usually have axial-lead style wires, but most wirewound resistors use metal end-caps or solder lugs.

6.3 VARIABLE RESISTORS

For applications requiring the capability to vary the resistance value in a circuit, several styles of *variable* resistors are manufactured. The simplest style is known as a linear variable resistor. *Linear* means in a straight line. The movement of an adjustable contact on this type resistor is in a back and forth, or straight line fashion. The sliding contact allows for choosing different resistance values.

A different form of variable resistor is called *rotary* variable resistor. These are made in two forms, *rheostats* and *potentiometers*. Both types are used to allow the operator of electronic equipment some degree of control over the circuit operation.

6.3.1 Seldom-adjusted Resistors

The simplest type of linear variable resistor looks like a modified version of a wirewound resistor. Most have a metal clamp that wraps around the resistor body, with a small protruding contact point that touches the resistive wire element. When adjusted to the desired point along the length of the resistor, the moveable metal band is tightened with a set screw. The screw may be later loosened to allow moving the sliding metal contact to a new location along the length of the resistor wire. The screw is then retightened, making a positive electrical contact that will not change. Since it is relatively difficult to change positions of the contact, this type resistor is normally used in *prototype* equipment. A prototype is the first unit of a new design produced. Often the values of certain components must be changed to optimize the performance of the circuit. Once the best design is found, the linear variable resistors are usually replaced with cheaper fixed-value resistors. Linear variable resistors are also sometimes used in consumer products, but are designed to be adjusted by trained service workers doing repairs.

There is an old saying in electronics that "If you don't want somebody to adjust it, don't put a knob on it!" The difficult-to-adjust resistor with the moveable contact clamp arrangement fits this situation, as most people unfamiliar with electronics would never attempt to adjust it. Figure 6.5 shows a linear adjustable wirewound resistor.

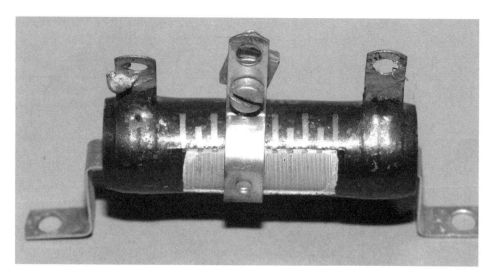

FIGURE 6.5 An adjustable wirewound resistor.

6.3.2 Linear Potentiometers

A special type of linear variable resistor is called a potentiometer. It does have a knob on it, designed for ease of adjustment. These units are often found on such consumer appliances as "graphic-equalizers" used in home and auto stereo music systems. The term "graphic" comes from the fact that a series of these resistive units mounted behind a panel gives the appearance of a line graph or chart. Such linear potentiometers are also used on professional recording mixing panels, TV production-room mixing consoles, and other precision control applications. Figure 6.6 shows a linear potentiometer.

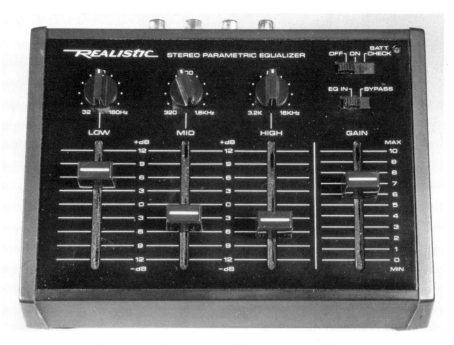

FIGURE 6.6 Linear potentiometers used in a stereo equalizer.

6.3.3 Rotary-variable Resistors

Another style of variable resistor is the rotary variable. They are manufactured in two main styles, the rheostat and the potentiometer. Both styles allow relatively easy adjustment of the resistance value offered by the device between close to zero ohms and a specific maximum amount of resistance.

Rheostats

A *rheostat* is a two-terminal resistor, made in the form of either a carbon-composition element or a wirewound element, shaped in the form of a circle. One end of the circular resistive element has an electrode for connection to the circuit. A second electrode connects to a rotary wiper, or armature, that slides along the resistive element, allowing for about 300 degrees of rotation. The effect of moving the rotary contact is to vary the resistance, from close to zero ohms at one extreme, to the maximum resistance of the resistive element at the other extreme of movement. Originally developed to enable control of circuit current, rheostats have generally been replaced by more modern solid-state devices for current control. Today, they are considered a large, expensive, and an inefficient means to affect "brute-force" load current control. Figure 6.7 shows a rheostat used to control the brightness of a lamp. Adjusting the rheostat's variable contact position will increase or decrease the amount of resistance in series with the lamp, and consequently, determine the lamp's brightness. One original use of rheostats was the control of theater lights. The rheostats were often large, high-power units, with greased-bushings around the moveable contact shaft. Occasionally, overheated rheostats would set fire to the grease or other nearby objects, resulting in more than one theatre fire.

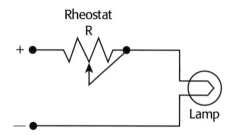

FIGURE 6.7   A schematic diagram of rheostat used to control lamp brightness.

Rheostats are nearly always used in series with a circuit load in order to control the amount of load current flowing. Rheostats are generally made with 5 watts or greater power dissipation ratings, and often use Nichrome resistance wire.

Potentiometers

Potentiometers are three-terminal variable resistors. In general, they are used not as brute-force current-control devices as are rheostats, but are used instead as variable voltage dividers. Refer to Chapter 7, "Series Circuits," for more on voltage dividers. Potentiometers are used in parallel with a load, to allow voltage control across the load device or circuit. A voltage divider is a circuit composed of resistances that provide two or more different values of voltage to a circuit.

Potentiometers are manufactured in one of two styles, composition and wirewound. The composition style has a thin film of carbon-composition material deposited on a circular base material, such as plastic, Bakelite®, or ceramic. The wirewound style uses wire wound around a semicircular form, similar to a wirewound fixed-value resistor.

The composition resistor types are used for low-power applications, and the ceramic wirewound types are used for higher wattage applications.

By far, most potentiometers, or *pots* as they are known, are used for applications that require the pot to dissipate less than two watts of power. Above the two-watt levels, wirewound pots are more common. Some high-quality pots are wirewound types designed for applications requiring less than two watts. Many wirewound pots are still available from military-surplus stores, and are usually preferred to the cheaper composition types for critical applications. The wirewound construction style is less prone to developing high resistance spots in the resistance element from dirt, moisture, and arcing (sparking). In the past, a common use of a carbon-deposited type of pot was the gasoline level sensor in the gas tanks of cars and trucks. The changing resistance as the level of gasoline changed sent a voltage proportional to the level to the gas gauge, an electrical voltmeter. When the carbon resist material on the tank sensor eventually wore off on spots from the constantly sliding contact, the gas gauge would "jump" at certain levels of fuel, and become operable again as the level went down. The only fix was a costly replacement of the sensor in the tank, usually part of the electric fuel pump assembly.

Figure 6.8 shows two pots and a potentiometer wired as a rheostat, used to control a microprocessor's active heat sink fan speed (voltage).

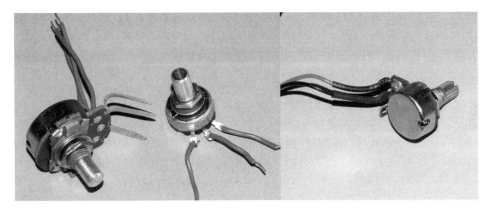

FIGURE 6.8   Two pots and a pot wired as a rheostat, used to control a microprocessor's active heat sink fan speed (voltage).

Other typical potentiometer applications are "volume" and "tone" controls on a radio receiver or home stereo system. Other uses are speed controls for electric appliances, all sorts of controls on electronics test equipment, and even the heart of the joysticks used with every video game or home computer.

Potentiometers are used whenever it is necessary to sample a portion of a voltage, which usually represents a signal. Amplifiers are used to increase the magnitude of a weak electrical signal. The signal usually represents a form of intelligence, such as speech, music, or digital data. Amplifiers are designed to always amplify, or increase the signal applied to them by a certain ratio. For instance, an amplifier that is designed to increase a signal's amplitude by ten times is said to have a *gain* of 10. Amplifiers are classed according to what particular aspect, called a *parameter*, is to be increased. For instance, we hear of *voltage amplifiers, current amplifiers, and power amplifiers.* Approximately 90 percent of all amplifiers are designed to be voltage amplifiers. A good example of potentiometer use is in the amplifier in your home stereo system. The stereo amplifier has one or more signal inputs. These signals are produced by sources such as a radio tuner, a CD player, phonograph, tape deck, microphone, and so on. The signal produced by any of these input sources is very small. Loudspeakers require a relatively large amount of power to produce a useful amount of sound. It is the amplifier's job to increase the magnitude of the input signal produced by the sources until it is sufficient to drive the speakers. The problem arises when we consider how amplifiers are made. In order to produce the best increase in signal strength with the minimum amount of change in what the signal looks like, (an undesired change in signal is called *distortion*), amplifiers are designed to amplify a certain number of times. As just explained, this is called *gain*. The gain is constant. So if a signal is applied to an amplifier, the total amount of gain would always be applied to the signal. In the case of a home stereo system, that would mean the music volume would be maximum all the time. This is obviously not the right way to operate the stereo, as a constant loud volume level would surely create problems with the neighbors, to say the least. What is needed is an easy method of controlling how much of the input signal is applied to the amplifier. By using a potentiometer on the input to the amplifier, it is possible to control how much signal is applied. Figure 6.9 shows a potentiometer connected to the input of an audio amplifier to serve as a volume control. When used in this manner, the potentiometer functions as a variable voltage divider in parallel with the load, in this case, the amplifier.

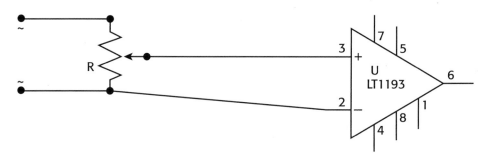

FIGURE 6.9   A potentiometer used as a volume control.

6.4 SAFETY CONSIDERATIONS

The most important safety consideration when using a variable resistor is the power rating of the device. This power rating should never be exceeded or damage to the device and possibly to other circuitry may result. It is good engineering practice to allow for 50% to 100% overrating of a resistive device. This assures safe, cool operation over long periods of time. For instance, a rheostat or potentiometer that is required to dissipate four watts of power should be rated at least for five watts, and if space and cost allow, then a 10-watt unit (the next larger wattage size unit) would be an even better choice. The idea is to stay well below the maximum power rating of an electrical device. Heat is the enemy of electronic devices, especially solid-state semiconductors. Typically these devices are diodes (one-way current valves) and transistors (semiconductor devices capable of switching and amplifying devices). The majority of components inside a PC are solid-state. Underrating resistor wattage or placing components that get hot too close to other devices is a sure recipe for disaster. As PCs run ever faster they generate ever more heat and the cooling requirements get ever more critical. That's why you see more and larger fans used in all "gamer" PCs. You see fans placed in the side and top panels of these high-end PCs to keep the insides cool enough to continue to operate even when the processor is set to run faster than its designed speed, a concept known as *overclocking*. Overclocking is extremely demanding of the electronic components used in a PC. The minimal increase in actual processing power achieved in this way does not really justify the extra problems created by the increased heat and decreased reliability of the PC.

Resistive devices should always be mounted in a manner that allows a good deal of air to circulate around them. They should not be exposed to dirt or moisture, as this will cause premature failure, possibly causing catastrophic short circuits.

When high voltage is applied to a panel-mounted rotary resistor, the control should be mounted on an insulating material, and should always use an insulated shaft coupling between the resistive control and the shaft with the knob on it. This method will ensure that the operator can not be inadvertently shocked by the high voltage potential on the metal control shaft. Panels of plastic, glass, or Teflon® are often used to insulate rotary controls for high-voltage applications.

6.5 THERMISTORS

Thermistors are resistive devices specially manufactured to have a large negative temperature coefficient. One style thermistor has a high resistance when cool, and the resistance decreases with an increase in temperature. This type is commonly used to control the initial current surge when a circuit using vacuum tubes is first turned on. This surge will shorten the life of the circuit components if allowed to happen. A thermistor is chosen that will offer as large a resistance when it is cool as when the circuit is turned off. When the current is applied, it flows through the thermistor, slowly raising its temperature. As the temperature increases, the thermistor's resistance value goes down, allowing more current to the circuit. In this way, the initial current surge is reduced to a safe level. Thermistors are used to control current surges in this way in TVs, stereos, and many electronic appliances.

Thermistors can also be used to sense temperature. The thermistor is placed in an environment where temperature is to be monitored. The thermistor is connected to a sensor circuit that converts the change in resistance of the thermistor into a temperature readout. It is a safe method to electronically sense temperatures that would be hazardous to people, such as temperatures inside blast furnaces or jet engines.

6.6 RESISTORS IN THE PC

Resistors are used to limit current and drop voltage in many places in a typical PC. An example is the resistors used to drop 5 VDC to the approximately 1.2 VDC required by the front panel LEDs used to indicate power on and hard disk drive activity. They are used in many power supplies as a safety load, to provide a minimum current draw to prevent a PC's power supply from self-destructing in case it is operated without its normal load connected. A normal load for a PC power supply is usually considered to be the motherboard with the processor and RAM installed, and at least one hard drive connected. Resistors are to be found on the motherboard and virtually all expansion cards. They are typically used with integrated circuits and can be seen next to them on a circuit board. Figure 6.10 shows a section of a motherboard featuring small, surface-mount technology (SMT) resistor networks, labeled "RN."

FIGURE 6.10 Motherboard resistor networks—marked RN.

6.7   SUMMARY

- Resistors offer opposition to electric current.
- When current flows through a resistance, it produces a voltage drop.
- Resistors are made to have either fixed- or variable-resistance values.
- Three manufacturing types have been used: carbon composition, now considered obsolete, carbon and metal film types, and wirewounds.
- Wirewounds are the most expensive but offer the greatest reliability, especially when used in extremes of temperature or humidity.
- Resistors are marked with either a standard color code or printed values.
- Resistor lead styles include: axial leads, meaning one wire exits the device on each end on the same center line axis, metal end caps, or solder lugs.
- When selecting a resistor, pay attention not to underrate its power capability. Allow for some extra safety margin to allow it to run cool and not overheat.
- Variable resistors include the seldom-adjusted wirewounds, which are adjusted using a set screw and sliding contact, and the made-to-be-easily-adjusted types, including rheostats and potentiometers.
- Rheostats have two contacts and are most frequently used in series with a load in order to control load current. They are typically made in 5 *W* or greater power sizes.
- Potentiometers have three connections. They are typically used in parallel with a load in order to act as a variable voltage divider. A typical application is as an amplifier's volume and tone controls. Most pots are made to handle two watts or less.
- Resistors are used nearly everywhere on a PC's circuit boards and power supply as a cost-effective method to limit current and drop voltage.

6.8   KEY TERMS

Carbon composition	Nichrome	Rheostat
Wirewound	Temperature coefficient	Potentiometer
Film resistor	Linear	Thermistor
Color code	Prototype	

6.9   EXERCISES

1. The most popular style of fixed resistor up until the 1970s was the:

2. A type of fixed resistor that is made by application of a thin film of resistive material to a ceramic tube is known as a:

3. A variable resistor that uses a metal band adjusted by a set screw is known as a:

4. A variable resistor that has two terminals, and is used to control current to a load is called a:

5. A variable resistor that has three terminals and is used to adjust a voltage to a circuit is called a:

6. A special resistor used to control the initial rush of current into a circuit is called a:

7. The method of using colored bands around the body of a resistor is known as the:

8. A resistor that can safely dissipate more than five watts of power would most likely be which type resistor?

9. A disadvantage of a carbon-composition resistor is:

10. An axial-lead arrangement on a resistor means what?

11. List at least two uses of resistors in a PC.

7 | Series Circuits

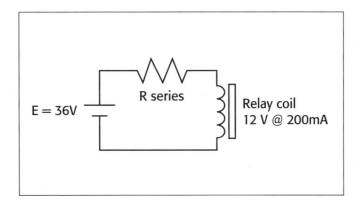

7.1 INTRODUCTION

One of the simplest electric circuits is the *series* circuit. In this chapter, you will learn how to solve series circuit problems to find current, voltage, and power values. You will also learn to quickly interpret a schematic diagram to determine which type of series circuit you are dealing with. During the lab activities you will learn to make measurements in each type of series circuit in order to properly troubleshoot them.

7.2 THE TWO-RESISTOR CIRCUIT

A two-resistor circuit consists of at least a voltage source, connecting wires, and two resistors. The following sections show how to deal with this type of circuit.

7.2.1   Finding Current Values

Figure 7.1 shows a two-resistor circuit consisting of a voltage source in series with two resistors. There is only one complete path for current in this circuit. All the current must flow from the source, through the first resistor, through the second resistor, and back to the source again. The same value of current must flow in each part of the circuit.

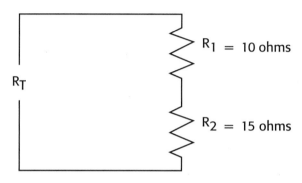

FIGURE 7.1   A two-resistor series circuit schematic diagram.

The value of current is the same in all parts of a series circuit.

Key Ideas

Example 7.1

The Ohm's law equation $I = E \div R$ can be used to find the current in this circuit. In order to use this formula, however, first find the value of *total circuit resistance,* R_{Total}, or simply R_T. Then just add up the values of both resistors to find R_T, as shown in Equation 7.1.

$$R_T = (R_1 + R_2) \qquad\qquad (7.1)$$
$$R_T = (10\ \Omega + 15\ \Omega)$$
$$R_T = 25\ \Omega$$

Then substitute the values for R_T found above and the voltage, E, given in the problem, into the Ohm's law formula. This gives Equation 7.2:

$$I = E \div R_T \qquad (7.2)$$
$$I = 25 \ V \div 25 \ \Omega$$
$$I = 1 \ A$$

This is the value of current that flows through each resistor, since in a series circuit the current value is the same for all parts of the circuit. The 1 amp of current leaves the source, flows first through R_1, then through R_2, and then back to the source. This method of current flow is the same for any series circuit.

The general formula used for finding the total resistance, R_T, for any series resistor circuit is found using Equation 7.3.

$$R_T = R_1 + R_2 + \ldots R_N \qquad (7.3)$$

R_N signifies that you simply add all the values of resistors you have in a series circuit including the highest number resistor you have, R_N. For instance, in a series circuit having five resistors, R_N would be the fifth resistor, so the formula for that circuit would be:

$$R_T = R_1 + R_2 + R_3 + R_4 + R_5$$

Simply use the ohmic values for each of the appropriate resistors in the formula.

7.2.2 Finding Voltage Drop Values

How much voltage is dropped across each of the resistors in Figure 7.1? Since the values of total resistance and current have already been found, use Ohm's law to find the voltage drops. To find the value of voltage dropped across R_1, use the Ohm's law formula for voltage in terms of resistance and current, $E = IR$. A convenient way to designate the voltage drop across resistor R_1 is to use the designation V_{R_1}. See Equation 7.4

$$V_{R_1} = IR_1 \qquad (7.4)$$
$$= 1 \ A \times 10 \ \Omega)$$
$$= 10 \ V$$

Then use the same method to find the value of voltage dropped across R_2.

$$V_{R_2} = IR_2$$
$$= 1 \ A \times 15 \ \Omega)$$
$$= 15 \ V$$

Note that the sum of the voltage drops around a closed circuit must always equal the applied voltage, *E*. This is formally known as *Kirchoff's Voltage Law for series circuits*, or *KVL*. Gustav Robert Kirchoff was a German physicist born on March 12, 1824, in Konigsber, Prussia. His first research topic was the conduction of electricity. As a result, Kirchoff wrote the Laws of Closed Electric Circuits in 1845. These laws were eventually named after their author, which are now known as Kirchoff's Current and Voltage Laws. Since Kirchoff's Voltage Law (KVL) and Kirchoff's Current Law (KCL) apply to all electric circuits, a person studying electronic circuits needs a firm understanding of these fundamental laws.

So, in the circuit just presented, $V_{R_1} + V_{R_2}$ should equal *E*. Check for this:

$$V_{R_1} = 10 \ V$$
$$V_{R_2} = 15 \ V$$
$$E = 25 \ V$$
$$10 \ V + 15 \ V = 25$$

Key Ideas

According to Kirchoff's Voltage Law for series circuits (KVL), the sum of the voltage drops in this circuit equals the applied voltage.

KVL is an easy way to check your answers in a quiz or test on series electric circuits. Always check to make sure the sum of the voltage drops in a series circuit exactly equals the applied voltage. If not, you have made a mathematical error, and should recheck your calculations until you get the applied voltage value and the sum of the voltage drops to equal. Don't overlook excessive rounding-off errors here either. It's best not to round off answers until the final calculation.

Example 7.2

Solve another series circuit for voltage drops. Use Figure 7.2 and Equation 7.5.

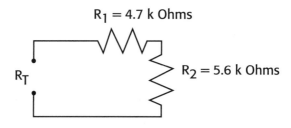

FIGURE 7.2   Circuit schematic diagram for Example 7.2.

Begin by finding total resistance, R_T.

$$R_T = R_1 + R_2 \qquad (7.5)$$
$$= 4.7\ k\Omega + 5.6\ k\Omega$$
$$= 10.3\ k\Omega$$

Using Ohm's law, find the current in this circuit:

$$I = 80V \div 10.3\ k\Omega$$
$$= 7.77\ mA$$

Using this value of current, find V_{R_1}.

$$V_{R_1} = IR_1$$
$$= 7.77\ mA \times 4.7\ k\Omega$$
$$= 36.5\ V$$

Find V_{R_2} in the same manner:

$$V_R = IR_2$$
$$= 7.77\ mA \times 5.6\ k\Omega$$
$$= 43.5\ V$$

It is a good idea to check to see if the sum of the calculated voltage drops for the circuit equals the applied voltage, E.

$$V_{R_1} + V_{R_2} = ?$$
$$= 43.5\ V + 43.5\ V = 80\ V$$

The voltage drop values calculated must be correct because they add up to the applied voltage.

7.2.3   Finding Power Values

To find the amount of power dissipated (converted from electric energy to heat energy) in a circuit, we use the Ohm's law power formulas learned earlier. Return to the circuit studied in Figure 7.1. To find the total power dissipated by the circuit, use Equation 7.6.

$$P = I \times E \qquad (7.6)$$
$$P = 1\ A \times 25\ V$$
$$= 25\ W$$

To find the power dissipated by each resistor,

$$P_{R_1} = IV_{R_1}$$
$$= 1\ A \times 10\ V$$
$$= 10\ W$$

$$P_{R_2} = IV_{R_2}$$
$$= 1\ A \times 15\ V$$
$$= 15\ W$$

This shows another interesting fact about series circuits.

Key Ideas

The sum of the individual power dissipations in a series circuit must exactly equal the total power dissipated. This is consistent with the law of conservation of energy.

To emphasize this point, check the circuit just studied, Figure 7.2, using Equation 7.7.

Example 7.3

$$P_T = 25\ W$$
$$P_{R_1} + P_{R_2} = ?$$
$$10\ W + 15\ W = 25\ W$$
$$P_T = P_{R_1} + P_{R_2} \qquad (7.7)$$

Now find the power values for the circuit of Figure 7.2, using Equation 7.8.

Example 7.4

$$P_T = IE \qquad (7.8)$$
$$= 7.77 \ mA \times 80 \ V$$
$$= 622 \ mW$$

The individual power dissipations are:

$$P_{R_1} = IV_{R_1}$$
$$= 7.77 \ mA \times 36.5 \ V$$
$$= 284 \ mW$$

$$P_{R_2} = IV_{R_2}$$
$$= 7.77 \ mA \times 43.5 \ V$$
$$= 338 \ mW$$

To test for calculation mistakes, see if the sum of the individual powers equals the total circuit power:

$$P_T = P_{R_1} + P_{R_2} \ ?$$
$$622 \ mW = 284 \ mW + 338 \ mW$$
$$622 \ mW = 622 \ mW$$

This check is valuable for checking calculations, and confirms that the sum of the individual power values equals the total circuit power.

7.3   MULTIPLE RESISTOR CIRCUITS

Figure 7.3 shows a three-resistor series circuit with 12 V applied to the series string of resistors. Solve for total resistance, current, voltage drops, and power values.

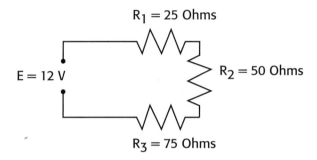

FIGURE 7.3   A multiple resistor circuit schematic.

Example 7.5

Solve to find R_T, I, V_{R_1}, V_{R_2}, V_{R_3}, P_T, P_{R_1}, P_{R_1}, and P_{R_3}, using Equation 7.9. First, find the total resistance, R_T.

$$R_T = R_1 + R_2 + R_3 \qquad (7.9)$$
$$= 25\ \Omega + 50\ \Omega + 75\ \Omega$$
$$= 150\ \Omega$$

Next, solve for circuit current. See Equation 7.10.

$$I = E \div R_T \qquad (7.10)$$
$$= 12\ V \div 150\ \Omega$$
$$= 80\ mA$$

Then find the voltage drops:

$$V_{R_1} = IR_1$$
$$= 80 \ mA \times 25 \ \Omega$$
$$= 2 \ V$$

$$V_{R_2} = IR_2$$
$$= 80 \ mA \times 50 \ \Omega$$
$$= 4 \ V$$

$$V_{R_3} = IR_3$$
$$= 80 \ mA \times 75 \ \Omega$$
$$= 6 \ V$$

Do a quick check on the voltage drops.

$$V_{R_1} + V_{R_2} + V_{R_3} = E?$$
$$= 2 \ V + 4 \ V + 6 \ V = 12 \ V$$

Finally, find the power values.

$$P_T = IE$$
$$= 80 \ mA \times 12 \ V$$
$$= 960 \ mW$$

$$P_{R_1} = IV_{R_1}$$
$$= 80 \ mW \times 2 \ V$$
$$= 160 \ mW$$

$$P_{R_1} = IV_{R_2}$$
$$= 80 \ mW \times 4 \ V$$
$$= 320 \ mW$$

$$P_{R3} = IV_{R3}$$
$$= 80 \; mA \times 6 \; V$$
$$= 480 \; mW$$

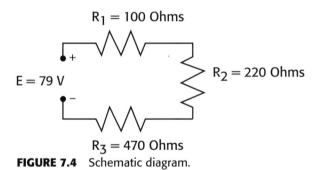

Now check the power values.

$$P_T = P_{R2} + P_{R2} + P_{R3} \; ?$$
$$= 160 \; mW + 320 \; mW + 480 \; mW$$
$$960 \; mW = 960 \; mW$$

You can use the same method to solve for values in series-resistor circuits using any number of resistors.

Example 7.6

Refer to Figure 7.4 for Questions 1–9.

R$_1$ = 100 Ohms

E = 79 V

R$_2$ = 220 Ohms

R$_3$ = 470 Ohms

FIGURE 7.4 Schematic diagram.

1. What is the value of R_T?
2. What is the value of I?
3. What is the value of V_{R_1}?
4. What is the value of V_{R_2}?
5. What is the value of V_{R_3}?

6. What is the value of P_T?
7. What is the value of P_{R_1}?
8. What is the value of P_{R_2}?
9. What is the value of P_{R_3}?

Solutions

Refer to Figure 7.4 for Questions 1–9.

1. What is the value of R_T?
   790 ohms
2. What is the value of I?
   1 *mA*
3. What is the value of V_{R_1}?
   10 V
4. What is the value of V_{R_2}
   22 V
5. What is the value of V_{R3}
   47 V
6. What is the value of P_T
   7.9 W
7. What is the value of P_{R_1}
   1 W
8. What is the value of P_{R_2}
   2.2 W
9. What is the value of P_{R3}
   4.7 W

7.4 VOLTAGE DIVIDERS

In all the previous series circuits, finding the voltage drops across individual resistors required several calculations. First, the total resistance had to be calculated. Next, the current value was calculated, and finally, the voltage drops were calculated. There is an easier method to use when you know the applied voltage, E, and the values of the resistors. This convenient formula is known as the *voltage divider formula*. It requires performing just a single calculation to find the voltage drop across a resistor. The voltage divider formula shows that the drop across a resistor in a series circuit is proportional to the resistor we are interested in finding the drop across, divided by the total circuit resistance, with that ratio multiplied by the applied voltage. Equation 7.11 shows this:

$$V_{R_1} = (R_1 \div R_T) \times E \qquad (7.11)$$

Recall the circuit in Figure 7.1. The voltage drop across R_1 can be found by using the voltage divider formula:

$$V_{R_1} = (10\ \Omega) \div (10\ \Omega + 15\ \Omega) \times 25\ V$$
$$= 0.4 \times 25\ V$$
$$= 10\ V$$

Now solve for the drop across R_2, using a modified Equation 7.11.

$$V_{R_2} = (R_2 \div R_T) \times E$$
$$= (15\ \Omega) \div (10\ \Omega + 15\ \Omega) \times 12\ V$$
$$= 0.6 \times 25\ V$$
$$= 15\ V$$

A quick check will show that these are the same values found before, using three separate calculations. The voltage divider formula saves calculation steps and time as long as the values of all resistors and the applied voltage in a circuit are already known.

Solve for voltage drops in the circuit depicted in Figure 7.4 using the voltage-divider formula:

$$V_{R_1} = (R_1 \div R_T) \times E$$
$$= (25\ \Omega \div 150\ \Omega) \times E$$
$$= 0.166 \times 12\ V$$
$$= 2\ V$$

$$V_{R_2} = (R_2 \div R_T) \times E$$
$$= (50\ \Omega \div 150\ \Omega) \times 12\ V$$
$$= 0.333 \times 12\ V$$
$$= 4\ V$$

$$V_{R_3} = (R_3 \div R_T) \times E$$
$$= (75\ \Omega \div 150\ \Omega) \times 12\ V$$
$$= 0.5 \times 12\ V$$
$$= 6\ V$$

These are the same values found before, but this method required only a single voltage calculation instead of the three calculations required by the Ohm's law method.

7.5   VOLTAGE DROPPING APPLICATIONS

A practical use of voltage-dropping resistors is the use of a load component that requires less operating voltage than what is available from the source. In this case, a resistor is used to drop the extra amount of voltage.

Example 7.6

It is desired to operate a light-emitting diode (LED) that requires 2 volts and that draws 30 *mA* of current. This value of current was selected to give the desired brilliance from the LED as well as a reasonably long life. The LED is to be powered from an available 6-volt source. Refer to Chart 7.1 for the forward-biased *IV* (current, *I* vs. voltage, *V*) curve of the LED. Find the required voltage-dropping resistor value. The required dropping resistor will be designated as V_{Drop}. Equation 7.12 is used to compute a required voltage-dropping resistor value.

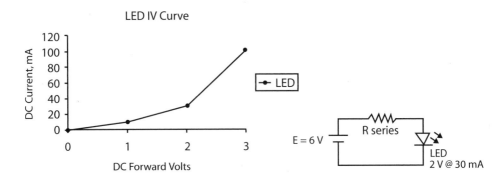

CHART 7.1   LED *IV* curve and schematic.

$$R_{Drop} = (V_{Source} - V_{Load}) \div I_{Load} \tag{7.12}$$
$$= (6\ V - 2\ V) \div 30\ mA$$
$$= 133\ \Omega$$

The amount of power dissipated by the voltage-dropping resistor must also be calculated. Since this is a series circuit, which means that the same current flows through the dropping resistor as flows through the load, then:

$$P_{Drop} = I_{Load} \times V_{Drop}$$
$$= 30 \ mA \times 4 \ V$$
$$= 120 \ mW$$

Since 133 ohms is not a standard value resistor, a 150 Ω unit, being the next larger standard value, should be selected. Although a ⅛ W (125 mW) resistor could be used, the calculated power dissipation is too close to the rating of a 125 mW unit. Good engineering practice and concern for safety dictate that one should always use the next higher standard wattage size resistor, so use a 150 Ω, ¼ W (250 mW) resistor. This will give just slightly less light output but with a greater safety factor.

Example 7.7

It is desired to operate a 12 V relay that draws 200 mA from a 36 V source. Find the resistor required to safely operate the relay in this circuit. Refer to Figure 7.5.

$$R_{Drop} = (36 \ V - 12 \ V) \div 200 \ mA$$
$$= 120 \ \Omega$$

$$P_{Drop} = 200 \ mA \times 24 \ V$$
$$= 4.8 \ W$$

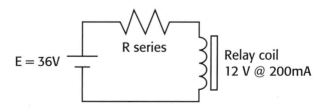

FIGURE 7.5 12-volt relay voltage-dropping schematic.

In this case, a 120-ohm, 5-watt or greater resistor should be used to operate the relay from this voltage source.

7.6 MOVING THE REFERENCE POINT

Up to now, voltage has been expressed as being measured across a resistor, rather than expressing a voltage *at a point* in a circuit. There are two methods of specifying a voltage: between two given points, such as V_{AB}, or voltage at a point, such as V_A. The following circuit will help illustrate the concept. Refer to Figure 7.6.

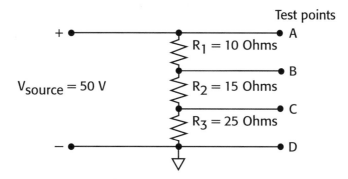

FIGURE 7.6 Moving the voltage reference point.

Whenever a voltage is specified at a point, it means that voltage is considered *with respect to ground*. In Figure 7.7, if the voltage at point A is required, it really means the voltage from point A *to ground*. Since ground in this circuit is point D, one needs to find the voltage from point A to point D. We specify this as V_{AD} for this circuit. Since you have already learned two methods of finding the voltage drops in a circuit, either method will do here. It is even simpler; the voltage across points A to D is the same as the applied voltage, E, or 50 volts.

What would be the voltage at point B? Since it is referenced to ground, point D, it is really V_{BD}. That represents the voltage drop across both resistors R_2 and R_3. Since we already know the values of all resistors and the applied voltage, the easiest method to find the voltage at point B is to use the voltage-divider formula.

$$V_B = V_{BD} = [(R_2 + R_3) \div R_T] \times E$$
$$= [(15\ \Omega + 25\ \Omega) \div 50\ \Omega] \times 50\ V$$
$$= (40\ \Omega \div 50\ \Omega) \times 50\ V$$
$$= 40\ V$$

Use the same method to find V_C.

$$V_C = V_{CD} = (R_3 \div R_T) \times E$$
$$= (25\ \Omega \div 50\ \Omega) \times 50\ V$$
$$= 25\ V$$

The voltage at point D is really the voltage from point D to itself, so it must be zero. Remember point D is the ground point for this circuit.

Be careful not to try to get these voltage values to add up to the applied voltage, since some of these voltages are across several resistors. The voltage at point A is actually dropped across all three resistors, and the voltage at point B is dropped across two resistors.

Moving the Ground Point

Since voltages specified at a point refer to ground, then moving the ground point should have some effect on how a voltage is calculated. Figure 7.7 is the same circuit as before, except that the ground reference has been relocated to point C. Find the voltages at each of the points for the circuit.

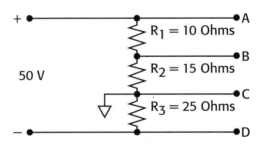

FIGURE 7.7 Moving the reference point with point C grounded.

$$V_A = V_{AC} = [(R_1 + R_2) \div R_T] \times E$$
$$= [(10\ \Omega + 15\ \Omega) \div 50\ \Omega] \times 50\ V$$
$$= 25\ V$$

$$V_B = V_{BC} = (R_2 \div R_T) \times E$$
$$= (15\ \Omega \div 50\ \Omega) \times 50\ V$$
$$= 15\ V$$

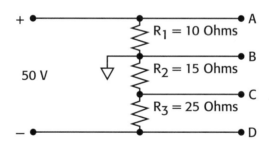

$$V_C = V_{CC} = 0V \quad \text{(Since point } C \text{ is ground).}$$
$$V_D = V_{DC} = (R_3 \div R_T) \times E$$
$$= (25\ \Omega \div 50\ \Omega) \times 50\ V$$
$$= -25\ V$$

This voltage is negative, since the reference point, C, is more positive than point D. It is important to recognize that the polarity of a voltage measured at a point in a circuit depends on the ground reference point used.

Try moving the ground in the circuit again. Refer to Figure 7.8.

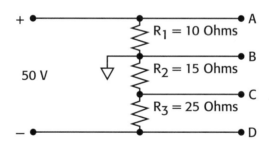

FIGURE 7.8 Moving the ground to point B.

$$V_A = V_{AB} = (R_1 \div R_T) \times E$$
$$= (10\ \Omega \div 50\ \Omega) \times 50\ V$$
$$= 7.77\ mA \times 43.5\ V$$

$$V_B = V_{BB} = 0\ V \quad \text{(since point } B \text{ is ground)}$$
$$V_V = V_{VB} = (R_2 \div R_T) \times E$$
$$= (15\ \Omega \div 50\ \Omega) \times E$$

$$= -15\ V \quad \text{(since point } C \text{ is negative with respect to ground)}$$
$$V_D = V_{DB} = (R_2 + R_3) \div R_T \times E$$
$$= [(15\ \Omega + 25\ \Omega) \div 50\ \Omega] \div 50\ V$$
$$= -40\ V \quad \text{(since point } D \text{ is negative with respect to ground)}$$

Example 7.9
Refer to Figure 7.9.
Solve for the following
values, and be sure to
indicate proper polarity:
V_A, V_B, V_C, V_D.

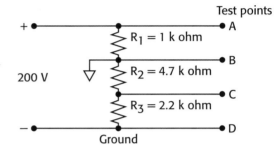

FIGURE 7.9   Self-test schematic diagram.

Solutions

Refer to Figure 7.9. Solve for the following values, and be sure to inidicate proper polarity:

$$V_A = 200\ V \qquad V_B = 175\ V \qquad V_C = 55.7\ V \qquad V_D = 0\ V$$

7.7   SUMMARY

- Current is the same value in all parts of a series circuit.
- The voltage dropped across a single resistor in a series circuit is equal to the resistor value times the current through it.
- The sum of the voltage drops in any series circuit must total the applied voltage.
- Power dissipated by a resistor in a series circuit can be found using the Ohm's law power formula.
- Total power in a series circuit can be found using Ohm's law. Total power dissipated by the circuit is equal to the sum of the power dissipated by each individual resistor.
- Multiple resistors can be used to form a voltage divider to provide any ratio of total applied voltage.

- Normally, voltages are specified to be from two points in a circuit. One of those points is usually the designated ground reference point. It is important to recognize that a voltage at a specified point must be referenced to some other point, and that by moving the reference point, the voltage measured will often be different.
- The series circuit is one of the basic circuits. A PC uses many such circuits. The PC power supply is the source; the motherboard, adapter cards, drives, LEDs, and so on are the loads. Considering a single one of these loads, the current flows from the power supply, through the load, and back to the power supply, in series.
- If any interruption occurs on a series circuit, the current drops to zero and none of the series loads will operate. A common example is a string of Christmas tree lights with each bulb wired in series. If any one bulb burns out, all lights will go out.
- Voltage measurements are made across (in parallel with) a device.
- Current travels through a device, and current measurements are made in series with a device.

7.8   KEY TERMS

Total resistance   Voltage across
Voltage-divider formula   Current through
With respect to ground

7.9   EXERCISES

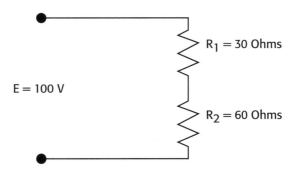

FIGURE 7.10   Exercises' schematic diagram.

1. Refer to Figure 7.10. Find I.

2. Refer to Figure 7.10. Find V_{R_1}.

3. Refer to Figure 7.10. Find V_{R_2}.

4. Refer to Figure 7.10. Find P_T.

5. Refer to Figure 7.10. Find P_{R_1}.

6. Refer to Figure 7.10. Find P_{R_2}.

7. A series circuit consists of two resistors. $R_1 = 1\ k\Omega$ and $R_2 = 3.3\ k\Omega$. The voltage applied across the two series resistors is 35 V. Find the value of V_{R_1}.

8. In the circuit in Question 7, find V_{R_2}.

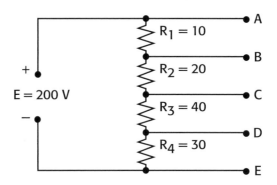

FIGURE 7.11   Schematic diagram for Questions 9–12.

9. Refer to Figure 7.11. What is the voltage at point A?

10. Refer to Figure 7.11. What is the voltage at point B?

11. Refer to Figure 7.11. What is the voltage at point C?

12. Refer to Figure 7.11. What is the voltage at point D?

8 ▪ Parallel Circuits

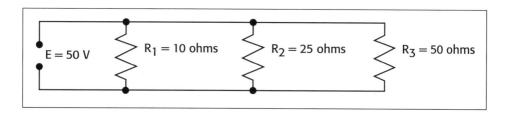

8.1   INTRODUCTION

Parallel circuits are the second type of simple electric circuit normally examined in a study of electronics. In Chapter 7, "Series Circuits," you learned that current in a series circuit is the same value in all parts of the circuit. You also learned that the voltage divides up across the different series components, and the sum of the voltage drops always equals the applied voltage, which is formally known as Kirchoff's Voltage Law for Series Circuits, or KVL for short.

In a parallel circuit, the *voltage* is common to all parallel components, and it is the *current* that divides among the different parallel paths. This chapter will introduce you to several styles of parallel circuits, and you will learn to find the values of voltage, current, and power for each component in the circuit.

8.2   THE TWO-RESISTOR CIRCUIT

Figure 8.1 shows a two-resistor parallel circuit. Resistors R_1 and R_2 are connected across the source voltage, E.

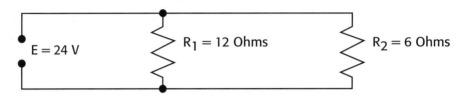

FIGURE 8.1   A two-resistor parallel circuit.

    The voltage across parallel circuit components is equal.

Key Ideas

8.2.1   Finding Current Values in Each Branch

It is frequently necessary and desired to find the individual branch currents in a parallel circuit. The next section shows how to do this.

Example 8.1

Each resistor has the same 24 *V* across it. To find the current through each resistor, simply use Ohm's law. Remember, in order to find current when the values of voltage and resistance are known, use Equation 3.2:

$$I = E \div R \qquad\qquad (3.2)$$

For example, to find the current through R_1,

$$I_{R_1} = 24\ V \div 12\ \Omega$$
$$= 2\ A$$

To find the current through R_2,

$$V_{R_2} = 24\ V \div 6\ \Omega$$
$$= 4\ A$$

To find the total circuit current, the two currents are added. The currents through each resistor are called *branch currents*. Refer to Figure 8.1. The total current flowing from the voltage source divides when it reaches the branch point. The current divides into that part that flows through R_1, termed I_{R_1}, and the remainder of current continues on to flow through resistor R_2, termed I_{R_2}. The two branch currents then recombine to form total current.

Key Ideas

The sum of the branch currents equals the total circuit current. This is formally known as Kirchoff's Current Law for parallel circuits, or KCL.

Check to see if KCL is satisfied for this circuit. Since the value of each branch current is known, what is needed is a method to find the total current. Another rule for parallel resistors is required.

Key Ideas

The equivalent resistance, R_{E_q}, is always less than the smallest value parallel resistor.

There are several equations to help solve this type of parallel resistor circuit, but the simplest to use in this case is Equation 8.1:

$$R_{E_q} = (R_1 R_2) \div (R_1 + R_2) \tag{8.1}$$

This formula is often called the *product-over-the-sum-formula*, since that is the process used to solve it. Substituting the resistor values in our circuit,

$$R_{E_q} = (12\ \Omega \times 6\ \Omega) \div (12\ \Omega + 6\Omega)$$
$$= 4\ \Omega$$

Then, using Ohm's law, find the total current:

$$I_{Total} = E \div R_T$$
$$= 24\ V \div 4\ \Omega$$
$$= 6\ A$$

Finally, compare this value to that obtained by adding the two branch currents.

$$I_{Total} = I_{R_1} + I_{R_2} \; ?$$
$$6\,A = 2\,A + 4\,A$$
$$6\,A = 6\,A$$

So, the sum of the individual branch currents does equal the total current.

8.2.2 Finding Power Values

The value of power dissipated by indivdual resistors is sometimes required. The next section shows how to do this.

Example 8.2

Find the value of power dissipated by each of the resistors in Figure 8.2. Since the resistance value is known, as well as the voltage and the current values, any one of the three Ohm's law formulas for power already learned could be used. If the current had not already been found, however, then a formula that does not require finding an additional value should be used. Since the voltage and resistance values are already known, use Equation 3.7

$$P = E^2 \div R \qquad\qquad (3.7)$$

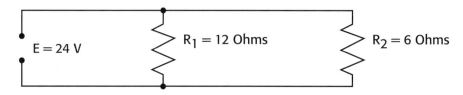

FIGURE 8.2 Two-resistor parallel circuit schematic diagram.

For resistor R_1:

$$P_{R_1} = E^2 \div R_1$$
$$= (24\ V)^2 \div 12\ \Omega$$
$$= 48\ W$$

For resistor R_1:

$$P_{R_2} = E^2 \div R_2$$
$$= (24\ V)^2 \div 6\ \Omega$$
$$= 96\ W$$

To find total power, P_T, use $P_T = E^2 \div R_T$. Since the value of R_T, 4 Ω is already known, it is necessary only to substitute this value:

$$P_T = (24\ V)^2 \div 4\Omega$$
$$= 44\ W$$

To check the answer, see if the sum of individual powers equals the total power.

$$P_T = P_{R_1} + P_{R_2}\ ?$$
$$144\ W = 96\ W + 48\ W$$
$$144\ W = 144\ W$$

Since the results check, it proves the math is correct.

The sum of individual power dissipations equals the total circuit power.

Key Ideas

8.3   THE MULTIPLE RESISTOR CIRCUIT

Figure 8.3 shows a three-resistor parallel circuit. The current value through each resistor can be found in the same manner as was used for the two-resistor circuit; by using Ohm's law for each resistor. If finding the total current is all that is needed, however, then a different formula should be used. Remember finding the total resistance, R_T, was a required step in order to use Equation 3.2 to find total current:

$$I_T = E \div R_T$$

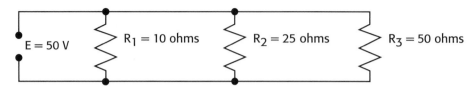

FIGURE 8.3   Three-resistor parallel circuit.

This circuit has three resistors, each with a different resistance value. The formula for total resistance used for the two-resistor circuit cannot be used here. A new formula (Equation 8.2) is required, usually referred to as the long formula:

$$R_T = 1 \div [(1 \div R_1) + (1 \div R_2) + (1 \div R_3 +) \ldots (1 \div R_N)] \tag{8.2}$$

The term R_N simply means that you continue to add each resistor in the parallel circuit, up through the highest number resistor. This is the same idea already learned for the total resistance formula for series circuits. Then substitute the values of the circuit into the formula,

$$R_T = 1 \div [(1 \div 10\ \Omega) + (1 \div 25\ \Omega) + (1 \div 50\ \Omega)]$$
$$= 1 \div [(0.10) + (0.04) + (0.02)]$$
$$= 1 \div 0.16$$
$$= 6.25\ \Omega$$

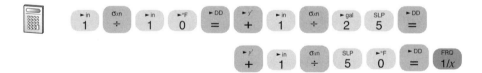

Now that the total resistance is known, it's easy to find the value of total current.

$$I_T = 50\ V \div 6.25\ \Omega$$
$$= 8\ A$$

If the current through each resistor needs to be found, use the same method as before:

$$I_{R_1} = E \div R_1$$
$$= 50 \ V \div 10 \ \Omega$$
$$= 5 \ A$$

$$I_{R_2} = E \div R_2$$
$$= 50 \ V \div 25 \ \Omega$$
$$= 2 \ A$$

$$I_{R_2} = E \div R_3$$
$$= 50 \ V \div 50 \ \Omega$$
$$= 1 \ A$$

In order to check the answers, see if the sum of the branch currents equals the total current.

$$I_T = I_{R_1} + I_{R_2} + I_{R_3}$$
$$8 \ A = 5 \ A + 2 \ A + 1 \ A$$
$$8 \ A = 8 \ A$$

8.3.1 Finding Power Values

The power dissipated by each resistor is found using the method already presented. Since the currents through each resistor have already been found, and the applied voltage and the value of each resistor were given, there are three equations, presented in Chapter Three, "Electrical Terms, Notations, and Prefixes," that could be used to find the power of each resistor.

$$P = IE \tag{3.5}$$

$$P = I^2 R \tag{3.6}$$

$$P = E^2 \div R \tag{3.7}$$

Example 8.3

Find the total and individual power dissipations for the circuit in Figure 8.4.

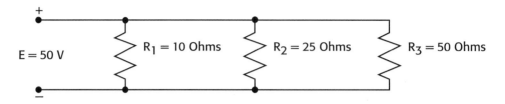

FIGURE 8.4 Parallel circuit.

Using Equation 3.5,

$$P_{R_1} = I_{R_1} \times E$$
$$= 50\ A \times 50\ V$$
$$= 250\ W$$

$$P_{R_2} = I_{R_2} \times E$$
$$= 2\ A \times 50\ V$$
$$= 100\ W$$

$$P_{R_3} = I_{R_3} \times E$$
$$= 1\ A \times 50\ V$$
$$= 50\ W$$

Using Equation 3.6,

$$P_{R_1} = (I_{R_1})^2(R_1)$$
$$= (5\ A)^2(10\ \Omega)$$
$$= 250\ W$$

$$P_{R_2} = (I_{R_2})^2(R_2)$$
$$= (2\ A)^2(25\ \Omega)$$
$$= 100\ W$$

$$P_{R_3} = (I_{R_3})^2(R_3)$$
$$= (1\ A)^2(50\ \Omega)$$
$$= 50\ W$$

Using Equation 3.7,

$$P_{R_1} = E^2 \div R_1$$
$$= (50\ V)^2 \div 10\ \Omega$$
$$= 250\ W$$

$$P_{R_2} = E^2 R_2$$
$$= (50\ V)^2 \div 25\ \Omega$$
$$= 100\ W$$

$$P_{R_3} = E^2 \div R_3$$
$$= (50\ V)^2 \div 50\ \Omega$$
$$= 50\ W$$

Any of the three equations will work to calculate the resistor currents. The choice usually depends on what other circuit values are already known or can most easily be found.

8.4   EQUAL-VALUE RESISTORS

There is a special case in which a circuit has several equal ohmic value resistors in parallel. The total circuit resistance could be found using either of the equations presented so far. There is an easier method, however. If the value of any one of the resistors (they are all equal) is used and this value is divided by the number of resistors, the total resistance value can be found.

For instance, if two 50 Ω resistors are parallel, simply divide 50 Ω by 2 to get $R_T = 25\ \Omega$. If there are four 400 Ω resistors in parallel, the total resistance would simply be 400 Ω/4 resistors, or 100 Ω.

Example 8.4

Figure 8.5 shows three resistors in parallel. Find the total resistance and the currents in the circuit.

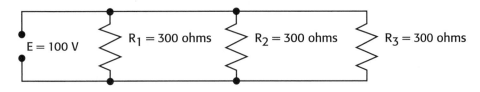

FIGURE 8.5 Three equal-value parallel resistors circuit.

First, use the special case Equation 8.3 to find the total current:

Example 8.5

$$R_T = (Value\ of\ one\ resistor) \div (\#\ of\ resistors) \qquad (8.3)$$
$$R_T = 300\ \Omega \div 3$$
$$= 100\ \Omega$$

So the total current is:

$$I_T = E \div R_T$$
$$= 100\ V) \div 100\ \Omega$$
$$= 1\ A$$

And the current through each resistor will be ⅓ of the total current:

$$I_{R_1} = I_{R_2} = I_{R_3} = I_T \div 3$$
$$= 1\ A \div 3$$
$$= 1/3\ A$$

333.3333333 *mA* or 333 *mA* rounded to three places.

Example 8.6

Refer to Figure 8.6. Find the value of total resistance, total current, and current through each resistor.

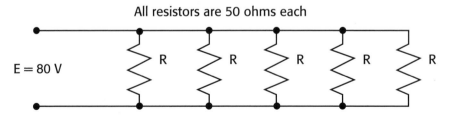

All resistors are 50 ohms each

E = 80 V

FIGURE 8.6   Five equal-value parallel resistors circuit.

$$R_T = 50 \ \Omega \div 5$$
$$= 10 \ \Omega$$

$$I_T = E \div R_T$$
$$= 80 \ V \div 10 \ \Omega$$
$$= 8 \ V$$

$$I_{R_1} = I_{R_2} = I_{R_3} = I_{R_4} = I_{R_5}$$
$$= I_T \div 5$$
$$= 8 \ A \div 5$$
$$= 1.6 \ A$$

8.4.1 Finding Power Values

The power dissipation values can be found using the same means as before. Since all values, voltage, current, and resistance for each component are known, there are three choices in this case. Since each resistor has the same ohmic value, simply compute the power for one resistor.

Example 8.7

The power may be calculated like this, using Equations 3.5–3.7:

$$P_{R_1} = P_{R_2} = P_{R_3} = P_{R_4} = P_{R_5} = P_T$$

1. $I_T E$
   $$= (1.6 \ A)(80 \ V)$$
   $$= 128 \ W$$

2. $(I_R)^2 R = (1.6 \ A)^2 (50 \ \Omega)$
   $$= 128 \ W$$

3. $E^2 \div R$
   $$= (80 \ V)^2 \div 50 \ \Omega$$
   $$= 128 \ W$$

Total power can be found by:

1. $P_T = IE$ (3.5)
   $$= (8 \ A)(80 \ V)$$
   $$= 640 \ W$$

2. $P_T = (I_T)^2 R_T$ (3.6)
    $= (8\ A)^2 (10\ \Omega)$
    $= 640\ W$

3. $P_T = E^2 \div R_T$ (3.7)
    $= 80\ V^2 \div 10\ \Omega$
    $= 640\ W$

Then check the results to see if the sum of the individual powers equals the total power.

$$P_T = P_{R_1} + P_{R_2} + P_{R_3} + P_{R_4} + P_{R_5}\ ?$$
$$640\ W = 128\ W + 128\ W + 128\ W + 128\ W + 128\ W$$
$$640\ W = 640\ W$$

The formula you use will depend on what values are known in the circuit, or which values can most easily be measured with test instruments. In practice, voltage readings are usually much more convenient to make than current measurements. This is because current measurements require opening the circuit, which is not easily done with soldered components. Desoldering, testing, and then resoldering a component requires specialized skills, and the removal and replacement nearly always leaves telltale tool marks on a circuit board. In addition, unless special grounded-tip soldering irons are used, the voltage present on the iron's tip can damage voltage-sensitive devices such as CMOS components.

8.5   PARALLEL CIRCUITS IN THE PC

There are many examples of parallel circuits in the PC. The next few sections detail some of the more common ones.

8.5.1   Bus Slots

The motherboard bus slots (PCI, ISA, et cetera) are connected in parallel, as that way they can share the same data, address, and power lines. That is the definition of a computer bus: a common connection between two or more circuits. All the drive motors share a common +12 VDC line. The drives also share a common +5 VDC line to power their logic boards.

8.5.2   Bypass Capacitors

There is another example of components in parallel in PCs. All the ICs on a motherboard or expansion card utilize small-value capacitors connected in parallel with the pin connected to the supply voltage line called V_{CC}. The other terminal of the cap (capacitor) goes to the ground bus. These capacitors help to stabilize the power line voltage at the chip. When other circuits in the PC switch on and off, something that happens all the time, the slight resistance of the wires connecting boards and chips to the power supply is enough to drop a meaningful amount of voltage. The variation in power rail (the main voltage supply lines) voltage to the chips could cause loss of data or a system crash. So these small bypass caps serve to shunt any voltage spikes to ground and help stabilize the supply voltage applied to each chip. Any short-term voltage drop will also be absorbed by the capacitors. This is an effective second stage of voltage regulation and is commonly used for all ICs. More detail on capacitors is given in Chapter 13, "Inductors and Capacitors." Refer to Figure 8.7 for a schematic diagram of an IC with a V_{CC} line bypass capacitor.

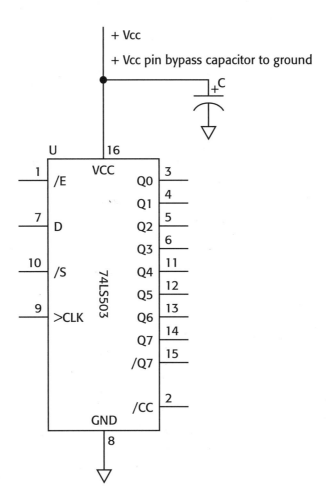

FIGURE 8.7 IC with bypass capacitor connected from the V_{CC} pin to ground.

8.6   SUMMARY

- The value of voltage across parallel circuit components is equal, as is obvious by inspection.
- The sum of the branch currents equals the total circuit current. This is known as Kirchoff's Current Law for parallel circuits, or KCL for short.
- The equivalent resistance is always less than the smallest value parallel resistor.
- The sum of the individual power dissipations equals the total circuit power.
- To find the equivalent resistance of two unequal resistors, use the "product-over-the-sum" equation.
- To find the equivalent resistance of three or more unequal value resistors, use the "long equation."
- To find the equivalent resistance of two or more equal-value parallel resistors, use the "special case equation."
- When you add expansion cards to the motherboard bus, you are adding electrical loads. These additional loads are in parallel with each other as far as the power supply is concerned. They offer decreased circuit resistance and increased current draw. They consume more power. Keep this in mind when sizing a PC power supply. Additional capacity to add expansion cards in the future should be taken into consideration in order to avoid power supply overload and the resulting subsystem problems. Refer to Figure 8.8.

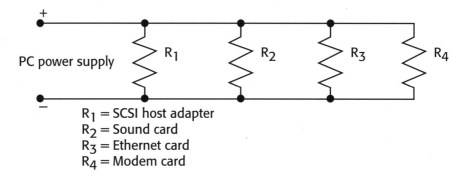

R_1 = SCSI host adapter
R_2 = Sound card
R_3 = Ethernet card
R_4 = Modem card

FIGURE 8.8   Adding additional expansion cards increases the power required from the power supply.

- The drives in a PC share the +12 VDC power rail from the supply in order to run their motors. They share the common + 5 VDC power rail from the supply to run their logic boards. These are parallel circuits; all the + 12 VDC equipment is connected in parallel and all the + 5 VDC equipment is connected in parallel.
- Integrated circuit chips mounted on motherboards and expansion cards commonly use small-value capacitors in parallel with the chips' positive power rail line to ground, in order to stabilize the voltage to the chip, and bypass short transient voltage spikes.

8.7 KEY TERMS

Branch current   Current-divider   Kirchoff's Current Law (KCL)
Parallel   Current junction

8.8 EXERCISES

1. Refer to Figure 8.9. Find the total circuit resistance.

2. Find the value of total current in Figure 8.9.

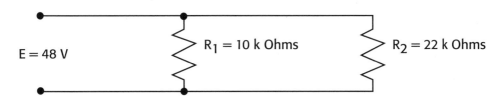

$E = 48\ V$ $R_1 = 10\ k\ Ohms$ $R_2 = 22\ k\ Ohms$

FIGURE 8.9 Two-resistor parallel circuit for Exercises 1–10.

3. Find the value of current through R_1 in Figure 8.9.

4. Find the value of current through R_2 in Figure 8.9.

5. Refer to Figure 8.9. Find the total resistance.

6. Find the value of power dissipated by R_1 in Figure 8.9.

7. Find the value of power dissipated by R_2 in Figure 8.9.

8. Refer to Figure 8.10. What is the total resistance of the circuit?

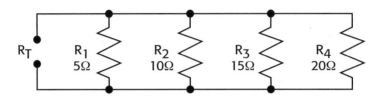

FIGURE 8.10   Circuit schematic for Question 8.

9. A parallel circuit consists of the following resistors:

   $R_1 = 100\ \Omega$, $R_2 = 150\ \Omega$, $R_3 = 200\ \Omega$, $R_4 = 300\ \Omega$, $R_5 = 1\ k\Omega$. What is the total resistance of this circuit?

10. Three 900 Ω, 1 W resistors are connected in parallel. What is the total resistance of the circuit?

11. If 10 V were applied across the circuit in Question 10, how much total power would be dissipated?

12. Refer to Question 11. What value of power would be dissipated by a single resistor in the circuit?

13. Based on your knowledge of resistors as presented so far in this text, would you say the circuit in Questions 13–15 is safe? Why or why not?

14. A parallel string of 6 light bulbs is designed to operate on 12 VDC. The total power consumed by the string is 24 W. How much current flows through each bulb? How much power is consumed by each light bulb?

15. A circuit has three light bulbs connected to the voltage source. If one or more lights burn out, the others will remain lit. What type of circuit is this?

16. In a parallel circuit, the total resistance is always_____ than the smallest resistance value in the circuit.
    a. higher
    b. lower
    c. the same value as

17. What is common to all components in a parallel circuit?
    a. Voltage
    b. Current
    c. Power
    d. Resistance
    e. None of the above

18. You build a PC and install the following expansion cards into the PC motherboard bus slots:

    A video card that draws 4 *A* @ 5 *V*
    A sound card that draws 3 *A* @ 5 *V*
    An Ethernet card that draws 1.5 *A* @ 5 *V*

    Draw a schematic diagram for the circuit and calculate the total current drawn from the PC's 5-volt power supply section by the expansion cards.

19. Calculate the resistance each card exhibits to the 5-volt supply.

20. Calculate the total resistance the expansion cards "look like" to the 5-volt supply.

21. Calculate the total power drawn by the expansion cards from the 5-volt supply.

22. A PC power supply's fan runs on 12 *V* @ 500 *mA*, and an additional case fan runs on 12 *V* @ 750 *mA*. You want to add another larger case fan to help cool the system. The new fan runs on 12 *V* and draws 1 *A*. Calculate the total power required to run all the fans.

23. Why are small capacitors connected from an IC's power supply voltage pin to ground?

9 Series-Parallel Circuits

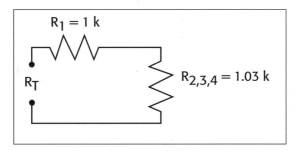

9.1 INTRODUCTION

A *series-parallel* circuit has elements of both a series and a parallel circuit. In this chapter you will learn to examine various series-parallel circuits, and to solve for values of resistance, current, voltage, and power in those circuits. Although this type of circuit is the most complex type presented so far, you will discover that by using the rules already learned for series circuits and parallel circuits, it is an easy matter to solve for these values.

The use of a convenient table will help to summarize the values found in these circuits. It lists values found for resistance (R), current (I), voltage drops (V), and power (P). This table can be known as an *RIVP summary* table. It is a valuable aid to summarizing and understanding the circuit operation of series-parallel circuits.

9.2 FINDING TOTAL RESISTANCE

Usually, solving a series-parallel circuit for the value of total circuit resistance is a required step for solving for other values in the circuit, such as total current. The next section shows how to do this.

Example 9.1

Figure 9.1 shows a simple series-parallel circuit. The current must flow through the entire circuit. It is best to tackle the "most difficult part" of the circuit first; that is the parallel portion of the circuit.

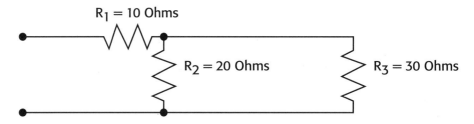

FIGURE 9.1 Series-parallel circuit schematic diagram.

Solution

Resistor R_1 is in series with the parallel combination of resistors R_2 and R_3. First, determine the equivalent resistance of the two parallel resistors. Use the "product-over-the-sum" formula for parallel resistors to find the equivalent resistance like this:

$$R_{23} = (R_2 R_3) \div (R_2 + R_3)$$
$$= (20\ \Omega \times 30\ \Omega) \div (20\ \Omega + 30\ \Omega)$$
$$= 12\ \Omega$$

It is always helpful to draw an equivalent showing the value found for the two parallel resistors. An *equivalent circuit* shows a simplified, but electrically equivalent, version of a circuit. It is a method to aid in the solution of a given circuit problem. Figure 9.2 shows the original circuit redrawn to indicate the equivalent resistance of the two parallel resistors.

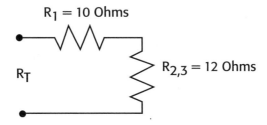

FIGURE 9.2 Equivalent parallel portion circuit schematic diagram.

Notice that a series circuit has been drawn in the equivalent circuit. One series resistor represents R_1, and the second resistor represents the equivalent value of R_2 in parallel with R_3. Next, find the total circuit resistance by following the rules for series circuits. This equivalent circuit can be thought of as simply a two-resistor series circuit. Use Equation 9.1 to solve for R_T.

$$R_T = R_1 + R_{2,3} \tag{9.1}$$
$$= 10\ \Omega + 12\ \Omega$$
$$= 22\ \Omega$$

Now draw an equivalent circuit, using the value just found for R_T (Figure 9.3).

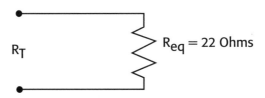

FIGURE 9.3 Equivalent parallel circuit schematic diagram.

Example 9.2

Figure 9.4 shows another series-parallel circuit, this one having three parallel resistors. Solve for equivalent circuit resistance in a similar manner as before, only this time using the formula for three different value resistors.

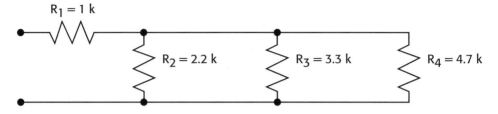

FIGURE 9.4   Series-parallel circuit.

Solution

Solve for the parallel equivalent resistance:

$$R_{2,3,4} = 1 \div [(1 \div R_2) + (1 \div R_3) + (1 \div R_4)]$$
$$= 1 \div [(1 \div 22 \ k\Omega) + (1 \div 3.3 \ k\Omega) + 1 \div 4.7 \ k\Omega)]$$
$$= 1,031 \ \Omega$$

or 1.03 $k\Omega$ in standard format for this text.

The equivalent circuit will reduce to a simple series circuit (see Figure 9.5).

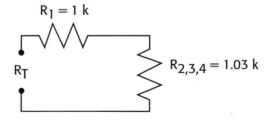

FIGURE 9.5   Series-parallel equivalent circuit schematic.

If the two resistance values of the circuit are added to obtain a single value of total resistance, the final equivalent circuit can be drawn (Figure 9.6). This is the method used in series circuits.

$$R_T = R_1 + R_{2,3,4} \qquad\qquad (9.2)$$
$$= 1\ k\Omega + 1.03\ k\Omega$$
$$= 2.03\ k\Omega$$

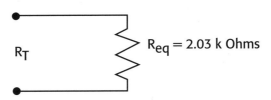

R_T R_{eq} = 2.03 k Ohms

FIGURE 9.6   Series-parallel circuit schematic diagram.

Example 9.3

Next, we will examine a more complex circuit, one with both multiple series as well as multiple parallel circuit elements. Refer to Figure 9.7.

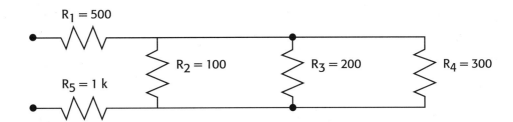

$R_1 = 500$

$R_5 = 1\ k$

$R_2 = 100$ $R_3 = 200$ $R_4 = 300$

FIGURE 9.7   Series-parallel circuit schematic diagram.

Solution

First, solve for the parallel element equivalent resistance:

$$R_{2,3,4} = 1 \div [(1 \div R_2) + (1 \div R_3) + (1 \div R_4)$$
$$= 1 \div [(1 \div 100\ \Omega) + (1 \div 200\ \Omega) + (1 \div 300\ \Omega)]$$
$$= 54.5\ \Omega$$

Next, draw the first equivalent circuit showing the parallel equivalent resistance value. See Figure 9.8.

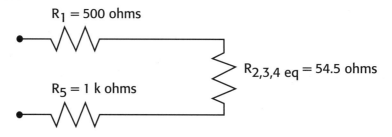

FIGURE 9.8 First equivalent circuit.

Next, add the two series resistors, R_1 and R_5, to get the equivalent resistance of those elements. Then draw another equivalent circuit based on the results. See Figure 9.9.

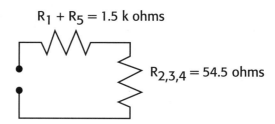

FIGURE 9.9 Equivalent circuit schematic diagram.

$$R_1 + R_5 = 500 \ \Omega + 1 \div k\Omega = 1.5 \ k\Omega$$

Finally, add the two remaining equivalent resistance values to get the total resistance value (see Figure 9.10).

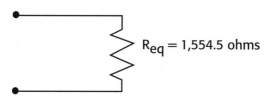

FIGURE 9.10 Equivalent resistance for series-parallel circuit.

$$R_T = 1.5 \ k\Omega + 54.5 \ \Omega$$
$$= 1{,}554 \ \Omega \ (\text{or } 1.55 \ k\Omega \text{ to 3 significant figures})$$

9.3   FINDING TOTAL CURRENT

Usually, the next step in solving a series-parallel circuit is finding total circuit current. This section shows how to do this.

Example 9.4

Find the total resistance and total current for Figure 9.11 using Ohm's law:

Solution

See the following.

$$I_T = E \div R_T$$
$$= 22 \ V \div 22 \ \Omega$$
$$= 1 \ A$$

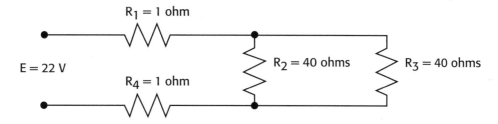

FIGURE 9.11   Schematic diagram.

Example 9.5

Use a similar method to find the total current for Figure 9.12.

Solution

$$I_T = E \div R_{Eq}$$
$$= 12 \ V \div 2.03 \ k\Omega$$
$$= 5.91 \ mA$$

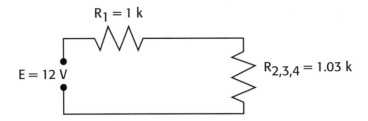

FIGURE 9.12   Schematic diagram.

Example 9.6

Find the total current in the circuit of Figure 9.13. This is another circuit for which total circuit resistance was already found.

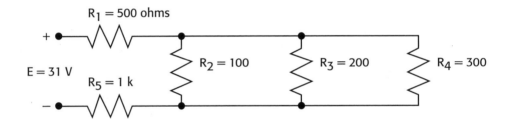

FIGURE 9.13   Schematic diagram.

Solution

It is easiest to use the equivalent circuit that shows all resistances in the circuit as a single value equivalent resistance. Refer to Figure 9.14.

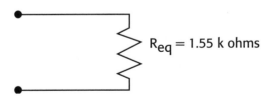

FIGURE 9.14   Schematic diagram.

$$I_T = R \div R_T$$
$$= 31\ V \div 1.55\ k\Omega$$
$$= 20\ mA$$

Key Ideas

Use the proper equivalent circuit in order to simplify the calculation for total current. It is not always easy to look at the original circuit and "see" the steps required to find the total current. Using equivalent circuits makes the task much easier. Be sure to actually draw any needed equivalent circuits on your scratch paper, and refer to them when solving the problem.

9.4   FINDING VOLTAGE DROPS

To find the voltage drops in a series-parallel circuit, any of the formulas already learned, such as Ohm's law, the voltage-divider formula, and the current-dividing formula, can be used.

Example 9.7

Find the voltage drops for the circuit in Figure 9.15. The best schematic diagram to use in this case is the first equivalent circuit drawn, Figure 9.16. Since there are two series resistances shown in the diagram, it is easy to use the voltage-divider formula to solve for the drops. Refer to Figure 9.17.

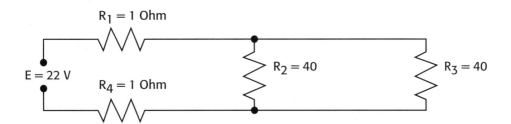

FIGURE 9.15   Schematic diagram.

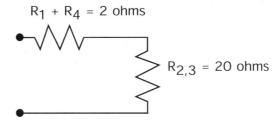

FIGURE 9.16   Equivalent circuit schematic diagram.

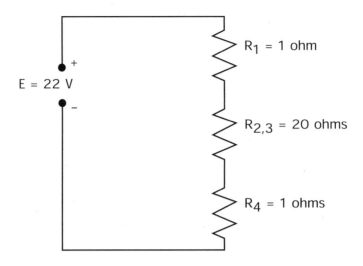

FIGURE 9.17 Voltage-divider schematic diagram.

Solution

Calculate the voltage drops this way:

$$V_{R1} = (R_1 \div R_T) \times E$$
$$= (1\ \Omega \div 22\ \Omega) \times 22\ V$$
$$= 1\ V$$

Since R_4 has the same value as R_1, and has the same value of current through it, it will also drop the same amount of voltage, 1 *V*.

Since there are two parallel resistors, R_2 and R_3, there will be a single voltage drop across those resistors, which will be:

$$V_{R2,3} = (R_{2,3} \div R_T) \times E$$
$$= (20\ \Omega \div 22\ \Omega) \times 22\ V$$
$$= 20\ V$$

Example 9.8

Refer to Figure 9.18. This circuit is equivalent to the circuit in Figure 9.4. It is the best one to use to find the voltage drops, since the voltage-divider circuit should be easy to recognize.

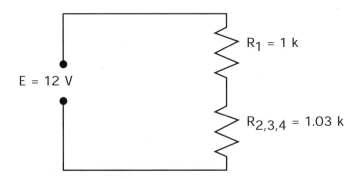

FIGURE 9.18 Schematic diagram.

Solution

Using the voltage-divider formula (see Equation 9.3),

$$V_{R_1} = [(R_1) \div (R_1 + R_{2,3,4})] \times E \qquad (9.3)$$
$$= [(1\ k\Omega) \div (1\ k\Omega + 1.03\ k\Omega)] \times 12\ V$$
$$= 5.91\ V$$

This voltage is the same across all three parallel resistors, and is equal to:

$$V_{R_{2,3,4}} = [(R_{2,3,4}) \div (R_1 + R_{2,3,4})] \times E$$
$$= [(1.03\ k\Omega) \div (1\ k\Omega + 1.03\ k\Omega)] \times 12\ V$$
$$= 6.09\ V$$

Check

Check to make sure that the sum of the two voltage drops equals the applied voltage.

$$E = V_{R1} + V_{R_{2,3,4}}?$$
$$= 5.91\ V + 6.09\ V$$
$$12\ V = 12\ V$$

9.5 FINDING POWER VALUES

The value of power dissipated by the resistances in a series-parallel circuit can be found in the same manner as before using Ohm's Law power formulas. A few examples will help illustrate the idea.

Example 9.9

Refer to Figure 9.19. This is one of the circuits for which the values of resistance, current, and voltage drops have already been determined.

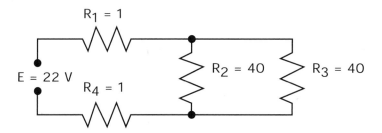

FIGURE 9.19   Schematic diagram.

Solution

First, list the given values, as well as those already found:

$$E = 22 \ V, \ I_T = 1 \ A, \ V_{R_1} = 1 \ V, \ V_{R_{2,3}} = 20 \ V$$
$$R_T = 22 \ \Omega, \ R_1 = 1 \ \Omega, \ R_2 = 20 \ \Omega, \ R_3 = 20 \ \Omega, \ R_4 = 1 \ \Omega$$

The Ohm's law power equations give:

$$P_T = IE$$
$$= 1 \ A \times 22 \ V$$
$$= 22 \ W$$

$$P_{R_1} = IR_1$$
$$= 1 \ A \times 1 \ \Omega$$
$$= 1 \ W$$

Since the branch currents that flow through R_2 and R_3 have not yet been calculated, it would be easier to use the following method to solve for P_{R_2} and P_{R_3}. See Equation 9.4.

$$P = V^2 \div R \tag{9.4}$$

$$P_{R_2} = V_{R_{2,3}})^2 \div R_2$$
$$= (20 \ V)^2 \div 40 \ \Omega$$
$$= 10 \ W$$

$$P_{R_3} = V_{R_3})^2 \div R_3$$
$$= (20 \ V)^2 \div 40 \ \Omega$$
$$= 10 \ W$$

Check

To check, just add the individual power values to see if the total equals the total power value found earlier.

$$P_T = P_{R_1} + P_{R_2} + P_{R_3} + P_{R_4}?$$
$$22 \ W = 1 \ W + 10 \ W + 10 \ W + 1 \ W$$
$$22 \ W = 22 \ W$$

Example 9.10

As an alternative method for finding the power dissipated by R_2 and R_3, one can first find the amount of current that flows through each resistor, and then use $P_{R_2} = (I_{R_2})^2 R_2$ and $P_{R_3} = (I_{R_3})^2 R_3$ to solve for power. The resulting values found should match those found using the first method. Refer to Figure 9.20.

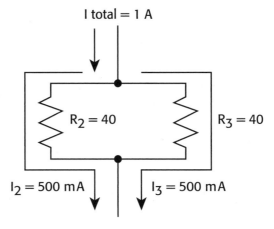

FIGURE 9.20 Schematic diagram.

Solution

First, find each of the branch currents (see Equation 9.5):

$$I_{R_2} = [(R_3) \div (R_2 + R_3)] \times I_T \tag{9.5}$$
$$= [(40 \ \Omega) \div (40 \ \Omega - 40 \ \Omega)] \times 1 \ A$$
$$= 0.5 \ A \text{ or } 500 \ mA \text{ in the preferred form}$$

$$I_{R_3} = [(R_2) \div (R_2 + R_3)] \times I_T$$
$$= [(40 \ \Omega) \div (40 \ \Omega - 40 \ \Omega)] \times 1 \ A$$
$$= 0.5 \ A \text{ or } 500 \ mA \text{ in the preferred form.}$$

Check

As a check, make sure the two branch currents equal the total current value:

$$I_{R_2} + I_{R_3} = I_T?$$
$$50 \ mA + 500 \ mA = 1000 \ mA$$
$$1000 \ mA = 1000 \ mA \ (1000 \ mA = 1 \ A)$$

Then, use the current values just found to find the power dissipations (see Equation 9.6):

$$P_{R_2} + I_{R_2} = V_{R_2} \tag{9.6}$$
$$= 500 \ mA \times 20 \ V$$
$$= 10 \ W$$

$$P_{R_3} + I_{R_3} = V_{R_3}$$
$$= 500 \ mA \times 20 \ V$$
$$= 10 \ W$$

These answers match those found using the first method, and are just as accurate. The method you choose depends on the values already known, or that are easiest to find, for a given circuit.

Example 9.11

Find the power dissipations in a circuit for which the values of resistance, current, and voltage have already been determined. Refer to Figure 9.21.

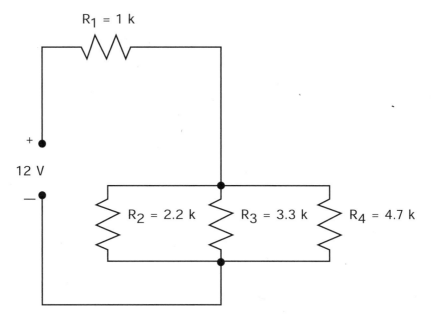

FIGURE 9.21 Schematic diagram for example 9.11.

Solution

First, list all of the known values.

$$E = 12 \ V, I_T = 5.91 \ mA, R_1 = 1 \ k\Omega, R_2 = 2.2 \ k\Omega,$$
$$R_3 = 3.3 \ k\Omega, R_4 = 4.7 \ k\Omega, R_T = 2.03 \ k\Omega$$

Solving for P_T, gives:

$$I_T = I_E$$
$$= 1 \ mA \times 12 \ V$$
$$= 12 \ mW$$

Notice that R_1 is in series with the rest of the circuit. Since all the current must flow through R_1 now, find the power it dissipates by using the value of I_T. See Equation 9.7.

$$P_{R_1} = I_T R_1 \tag{9.7}$$
$$= (5.91 \; mA)^2 (1 \; k\Omega)$$
$$= 34.9 \; mW$$

It is easiest to use the common voltage across the parallel elements to solve for power, as this does not require finding individual resistor currents. See Equation 9.8.

$$P_{R_2} = (V_{R_2})^2 \div R_2 \tag{9.8}$$
$$= (6.09 \; V)^2 \div 22 \; k\Omega$$
$$= 16.9 \; mW$$

$$P_{R_3} = (V_{R_3})^2 \div R_3$$
$$= (6.09 \; V)^2 \div 3.3 \; k\Omega$$
$$= 11.2 \; mW$$

$$P_{R_3} = (V_{R_3})^2 \div R_4$$
$$= (6.09 \; V)^2 \div 4.7 \; k\Omega$$
$$= 7.89 \; mW$$

Check

See if the sum of the individual power values equals the total power.

$$P_T = P_{R_1} + P_{R_2} + P_{R_3} + P_{R_4} \; ?$$
$$70.9 \; mW = 34.9 \; mW + 16.9 \; mW + 11.2 \; mW + 7.89 \; mW$$
$$70.9 \; mW = 70.9 \; mW$$

9.6   USE OF THE RIVP TABLE

When a circuit becomes complex, with many parts, it is very convenient to have a summary chart or table to record found values. The last example in this chapter is of a complex series-parallel circuit. It has several series elements and several parallel elements. It's a good example with which to introduce the use of the RIVP table. The *RIVP table* lists all values found for a circuit: resistances (R), currents (I), voltage drops (V), and power values (P) are all listed as they are determined. Refer to the circuit in Figure 9.22.

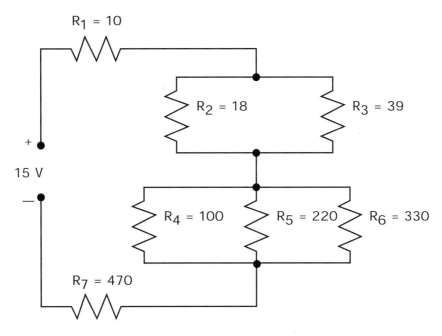

FIGURE 9.22 RIVP table schematic diagram.

The given values for the circuit are listed on the schematic diagram. Solve for all currents, voltage drops, and power dissipations. Because keeping track of all these values becomes difficult, it is easier to use the RIVP table to log each value as it is determined. This will then give a very clear summary chart of all the circuit values. The table can then be used to conveniently look up certain values in order to solve for others. Refer to Table 9.1 to see an RIVP table.

9.6.1 Given Values

Begin by filling-in all *given* values—those supplied or listed in the problem—in the RIVP table before actually starting to solve for the values in the circuit. Record the values for all resistors and the applied voltage in the table, since those are the only known values in the circuit.

TABLE 9.1 RIVP Summary Table

Component(s)	R	I	V	P
R1				
R2				
R3				
R4				
R5				
R6				
R7				
R2, R3				
R4, R5, R6				
Total Value				

9.6.2 Total Equivalent Resistance

Next, solve for total equivalent resistance, R_T. As presented earlier, it is best to solve for values in the parallel portions of a series-parallel circuit first. First find the equivalent resistance values for the two parallel portions of Figure 9.23. Values from the portion with the two resistors in parallel, R_2 and R_3, are determined first.

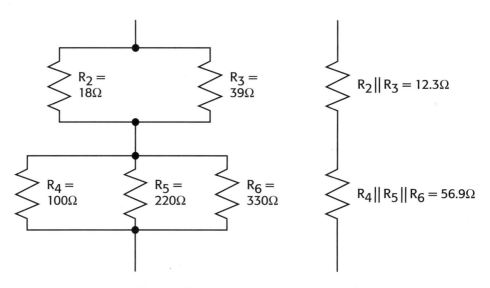

FIGURE 9.23 Circuit schematic diagram.

$$R_{2,3} = (R_2 R_3) \div (R_2 + R_3) \tag{8.1}$$
$$= (18\ \Omega \times 39\ \Omega) \div (18\ \Omega + 39\ \Omega)$$
$$= 123\ \Omega$$

Enter this value into the RIVP table at the intersection of the R column and row $R_{2,3}$. Refer to Table 9.2.

TABLE 9.2 Partially Completed RIVP Summary Table

Component(s)	R	I	V	P
R1	10 Ω			
R2	18 Ω			
R3	39 Ω			
R4	100 Ω			
R5	220 Ω			
R6	330 Ω			
R7	470 Ω			
R2, R3	12.3 Ω			
R4, R5, R6	56.9 Ω			
Total Value	549.2 Ω			

Next, solve for the equivalent resistance of the three parallel resistors, $R_{4,5,6}$. Use the formula introduced earlier for finding the value of several different value parallel resistors:

$$R_{4,5,6} = 1 \div [(1 \div R_4) + (1 \div R_5) + (1 \div R_6)]$$
$$= 1 \div [(1 \div 100\ \Omega) + (1 \div 220\ \Omega) + (1 \div 330\ \Omega)]$$
$$= 56.9\ \Omega$$

Record this value into the RIVP table. Now draw an equivalent circuit using the values found for the parallel portions of the circuit. Notice that the result is a series circuit. Figure 9.24 shows the first equivalent circuit, based on the values found so far.

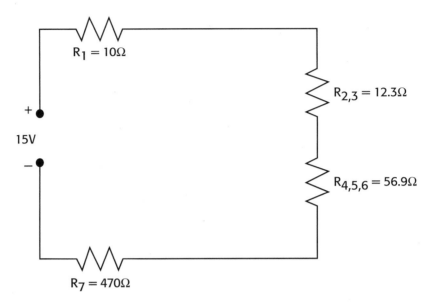

FIGURE 9.24 First equivalent circuit.

Simply add the different resistor values from the equivalent circuit in order to find total resistance. This method was already presented in Chapter 7, "Series Circuits."

$$R_T = R_1 + R_{2,3} + R_{4,5,6} + R_7$$
$$= 10\ \Omega + 12.3\ \Omega + 56.9\ \Omega + 470\ \Omega$$
$$= 549.2\ \Omega$$

Record this value for R_T in the RIVP table.

9.6.3 Finding Current Values

Next, solve for either currents or voltage drops. Arbitrarily choose to solve for the currents first. Since the applied voltage and the total resistance are known, it is easy to solve for total current. Using Ohm's law,

$$I_T = E \div R_T$$
$$= 15\ V \div 549.2\ \Omega$$
$$= 27.3\ mA$$

Record this value in the table.

Now go around the circuit and solve for each current. It is best to use a systematic approach, so start with R_1, and generally proceed around the circuit in a clockwise fashion. The current through R_1 is the same as the total current, since R_1 is in series with the rest of the circuit. Record this current value in the correct place in the table. You should see that resistor R_7 is in series with the rest of the circuit. Although this resistor is out of order, it is easy to see the current value, as it is the same as for R_1. So fill out the same current value for it in the RIVP table.

Next, find the value for the currents through R_2 and R_3. For these two resistors, it is best to use the current-dividing formulas learned earlier.

$$I_{R_2} = [R_3 \div (R_2 + R_3)] \times I_T \qquad (9.5)$$
$$= [39\ \Omega \div (18\ \Omega + 39\ \Omega)] \times 27.3\ mA$$
$$= (0.684105)(27.3\ mA)$$
$$= 18.7\ mA$$

Use the same method to find the current through R_3.

$$I_{R_3} = [R_2 \div (R_2 + R_3)] \times I_T \qquad (9.5)$$
$$= [18\ \Omega \div (18\ \Omega + 39\ \Omega)] \times 27.3\ mA$$
$$= (0.3157894)(27.3\ mA)$$
$$= 8.62\ mA$$

When solving for current through the three parallel resistors, R_4, R_5, and R_6, it is easier to use the Ohm's law formula. Remember, since they are in parallel, all three resistors have the same voltage drop across them. Parallel components always have the same voltage drop across them. The equivalent resistance of these three resistors has already been found. Simply use the RIVP table to find this equivalent value of 56.9 Ω. Since all the circuit current must flow through this equivalent resistance value, use Ohm's law to solve for the voltage drop:

$$V_{R_{3,5,6}} = I_T R_{4,5,6}$$
$$= (27.3\ mA)(56.9\ \Omega)$$
$$= 1.55\ V$$

This is the value of voltage across each of the three parallel resistors. Fill out the table for voltage drops for R_4, R_5, and R_6 as well as $R_{4,5,6}$. Then, using Ohm's law, find the individual current values:

$$I_{R_4} = V_{R_{4,5,6}} \div R_4$$
$$= 1.55 \ V \div 100 \ \Omega$$
$$= 15.5 \ mA$$

$$I_{R_5} = V_{R_{4,5,6}} \div R_5$$
$$= 1.55 \ V \div 220 \ \Omega$$
$$= 7.05 \ mA$$

$$I_{R_6} = V_{R_{4,5,6}} \div R_6$$
$$= 1.55 \ V \div 330 \ \Omega$$
$$= 4.7 \ mA$$

Check

An easy way to check the calculations so far is to add up the individual branch currents through the resistors and see if they equal the total current. In other words, does the circuit satisfy KCL?

$$I_{R_4} + I_{R_5} + I_{R_6} = I_T ?$$
$$15.5 \ mA + 7.05 \ mA + 4.7 \ mA = 27.3 \ mA$$
$$27.25 \ mA = 27.3 \ mA$$
(rounded off to three significant digits, or 27 rounded to two significant digits)

Remember that the answers for the individual currents were rounded off slightly; the answer does agree pretty well with the total current value. In fact, when 27.25 *mA* is rounded off to three significant digits, it rounds to 27.3 *mA*. Common sense should tell you that when solving for branch currents, a branch current can never be greater than the total circuit current. So in this circuit, KCL is satisfied.

9.6.4 Finding Voltage Drops

The current value for R_7 was already found, so the currents are finished. Continue to find the remaining voltage drops. The next one to find is V_{R_1}.

$$V_{R_1} = I_T R_1$$
$$= (27.3 \ mA)(10.0 \ \Omega)$$
$$= 273 mV$$

Record this value into the table, and find the next voltage drop. Use the equivalent resistance of R_2 and R_3 to find the drop across those two parallel resistors:

$$V_{R_{2,3}} = R_{2,3}I_T$$
$$= (12.3 \ \Omega)(27.3 \ mA)$$
$$= 336 \ mV$$

Record these voltage drop values in the table. Since R_2 and R_3 are in parallel, they share the same voltage drop across them. Fill in the same value under E and across from R_2, R_3 as well as $R_{2,3}$.

The voltage drop across the three parallel resistors, R_4, R_5, and R_6, was found previously. From the RIVP table, it is 1.55 V. So then, find the last voltage drop, V_{R_7}. R_7 is in series with the rest of the circuit, so the total circuit current must flow through it.

$$V_{R_7} = I_T R_7$$
$$= (27.3 \ mA)(470 \ \Omega)$$
$$= 12.8 \ V$$

Record this value in the RIVP table. Notice that the smallest voltage drops are across the smallest resistances, and the greatest voltage drops are across the largest resistances. Most of the voltage is dropped across the 470 Ω resistor. This is because voltage drop is directly proportional to resistance, and the 470 Ω resistor is the largest value resistor in the circuit.

Check

Check on the values found for voltage drop by adding up the voltage drops across *equivalent resistance* values, and see if the result matches the applied voltage. Remember to use only one value for voltage drop across parallel components. It is the same value, not two or three voltages that appear across any number of parallel components.

$$V_{R_1} + V_{R_{2,3}} + V_{R_{4,5,6}} + V_{R_7} = E?$$
$$= 273 \ mV + 33.6 \ mV + 1.55 \ V + 12.8 \ V = 15.0 \ V$$
$$14.959 \ V \cong 15.0 \ V$$

The values are close, but not exact. After considering that the answers were rounded off slightly for each individual voltage drop answer, the results match pretty well. Be careful when doing this check that you do not wind up showing a voltage drop that exceeds the applied voltage. Such a result is a sure indication of a math error, and you should continue to recheck your calculations up to that point until you get close agreement with the correct answer.

9.6.5   Finding Power Values

The RIVP table has enough values now to solve for the power dissipation in all resistors. There is a choice of formulas to use, since at least two values for each of the resistors are known. When you have a choice of formulas, pick one that does not involve squaring a number. The accuracy is improved, since each of our values in the table has already been rounded. Squaring will only increase the error.

For R_1, the value can be found without squaring, by using:

$$P_{R_1} = I_T V_{R_1}$$
$$= (27.3\ mA)(273\ mV)$$
$$= 7.45\ mW$$

For R_2, find power dissipation in a similar manner:

$$P_{R_2} = I_{R_2} V_{R_2}$$
$$= (18.8\ mA)(33.6\ mV)\ \text{(values used from RIVP table)}$$
$$= 6.32\ mW$$

The power dissipated by R_3 is found in the same way:

$$P_{R_3} = (8.62\ mA)(33.6\ mV)$$
$$= 2.90\ mW$$

There is a place in the summary table to record the value of power dissipated by both R_2 and R_3 together. Power is additive, so add the values of dissipation for both resistors:

$$P_{R_2} + P_{R_3} = 6.28\ mW + 2.90\ mW$$
$$= 9.18\ mW$$

Finding the power dissipation of R_4 gives:

$$P_{R_4} = I_{R_4} V_{R_4}$$
$$= (15.5\ mA)(1.55\ V)$$
$$= 24.0\ mW$$

Record these values in the table.

Next, solve for the power dissipated by R_5:

$$P_{R_5} = I_{R_5}V_{R_5}$$
$$= (7.05 \ mA)(1.55 \ V)$$
$$= 10.9 \ mW$$

Power dissipated by R_6 is:

$$P_{R_6} = I_{R_6}V_{R_6}$$
$$= (4.7 \ mA)(1.55 \ V)$$
$$= 7.29 \ mW$$

The summary table has a place to record the combined power dissipation of R_4, R_5, and R_6. After adding the individual power values, the result is:

$$P_{R_4} + P_{R_5} + P_{R_6} = 24.0 \ mW + 10.9 \ mW + 7.29 \ mW$$
$$= 42.2 \ mA$$

Then find the power dissipated by the last resistor, R_7.

$$P_{R_7} = I_T V_{R_7}$$
$$= (27.3 \ mA)(12.8 \ V)$$
$$= 349 \ mW$$

Finally, find the total power dissipated by the circuit (see Equation 9.9).

$$R_T = I_T E \hspace{4cm} (9.9)$$
$$= (27.3 \ mA)(15.0V)$$
$$= 409.5 \ mW$$

Check
Check to see if the individual power dissipation values add up to equal the total power dissipated.

$$P_{R_1} + P_{R_2} + P_{R_3} + P_{R_4} + P_{R_5} + P_{R_6} + P_{R_7} = P_T ?$$
$$7.45 \ mW + 6.28 \ mW + 2.90 \ mW + 24.0 \ mW + 10.9 \ mW + 7.29 \ mW + 349 \ mW = $$
$$409.5 \ mW$$
$$407.82 \ mW = 409.5 \ mW$$

The answer is close, but does not agree exactly. That's OK because the values for each power value were rounded off, and those were calculated using rounded current and voltage values. So, there is pretty good agreement with the value found for total power; it does prove the calculation method is sound. Remember what was presented in Chapter 6, "Resistors." Commonly used resistors have tolerances of plus or minus ten or twenty percent each. The values found in the aforementioned circuit would most certainly be more than accurate enough for all but the most critical of real circuits. RIVP Table 9.3 has all the given and solved values listed.

TABLE 9.3 Completed RIVP Table

Component(s)	R	I	V	P
R1	10 ΩΩ	27.3 mA	273 mV	7.45 mW
R2	18 Ω	18.7 mA	336 mV	6.28 mW
R3	39 Ω	8.62 mA	336 mV	2.90 mW
R4	100 Ω	15.5 mA	1.55 V	24.0 mW
R5	220 Ω	7.05 mA	1.55 V	10.0 mW
R6	330 ΩΩ	4.7 mA	1.55 V	7.29 mW
R7	470 Ω	27.3 mA	12.8 V	349 mW
R2, R3	12.3 Ω	27.3 mA	336 mV	9.18 mW
R4, R5, R6	56.9 Ω	27.3 mA	1.55 V	42.3 mW
Total Value	549.2 Ω	27.3 mA	15 V	409.5 mW

9.7 PC POWER SUPPLY OUTPUT CALCULATIONS

By this point you are probably already thinking about PCs and the power requirements put on the power supply. A properly sized supply is critical to ensure continued long-term operation of the PC. Besides the AC input voltage and current ratings, modern PC power supplies are rated both in terms of total power capability as well as current ratings per voltage section.

As an example, the power supply from a popular brand of PC case/supplies is listed as having the following output capabilities:

+ 3.3 VDC @ 14 A

+ 5 VDC @ 30 A

+12 VDC @ 8 A

−12 VDC @ 0.5 A

−5 VDC @ 0.5 A

+5 VDC standby @ 1.5 A

In addition, the + 3.3 VDC and the + 5 VDC sections together are specified as being able to provide a maximum of 150 watts. The supply is advertised and listed on the sticker as being a "300 Watt" supply. Here's how the numbers actually work out.

It is possible to multiply each section to find the wattage for that section, and then add up the wattages. Remember, however, that the sticker lists a limitation of 150 W maximum for the + 3.3 VDC and the + 5 VDC sections combined. So calculate the other sections for their power capability and add the resulting wattage to 150 W.

The +12 VDC section can provide 12 V × 8 A or 96 watts

The -12 VDC section can provide 12 V × 0.5 A or 6 watts

The − 5 VDC section can provide 5 V × 0.5 A or 2.5 watts

The + 5 Volts standby section can provide 5 V × 1.5 A or 7.5 watts

The total for these sections is 112 watts.

Add to that the 150 watts for the combined + 3.3 VDC and + 5 VDC sections and it gives a grand total of 262 watts, rather less than the advertised 300 watts on the sticker!

These calculations bring up a couple of interesting points. First, you can't really put too much trust in the accuracy of most PC power supply sticker ratings. Second, you should learn to calculate your PC's actual power requirements. An improperly sized PC power supply can cause all sorts of problems. Random lockups and reboots are a common for a too-small PC power supply. More on power supplies will be presented in Chapter 14, "Power Supplies."

9.8   SUMMARY

- Solve for values in the parallel portions of a series-parallel circuit first. These are often considered to be the "most difficult" parts of the circuit.
- Use equivalent circuits to aid in the solution of circuit problems.

■ A series-parallel circuit can be simplified to a series circuit, for which values can be easily determined.

■ Summarize important circuit values in an RIVP table.

■ Ohm's law still applies.

■ Most of the circuits in a PC are the series-parallel type.

9.9 KEY TERMS

RIVP table
Equivalent circuit
Series-parallel

9.10 EXERCISES

Questions 1–13 are about the following circuit:

A circuit consists of one resistor in series with two resistors that are connected in parallel. The series resistor is R_1, and the parallel resistors are R_2 and R_3. R_1 has a resistance of 20 Ω, R_2 has 50 Ω, and R_3 has 100 Ω. There is a voltage of 106.6 volts applied to the circuit.

1. What is the total circuit resistance?

2. What is the total current value?

3. What is the voltage drop across the parallel pair of resistors?

4. What is the voltage drop across R_1?

5. What is the current through R_2?

6. What is the current through R_3?

7. What is the total power dissipated?

8. How much power is dissipated by R_1?

9. How much power is dissipated by R_2?

10. How much power is dissipated by R_3?

11. Make your own RIVP table. Fill in all the data as you answered for Questions 1–10.

12. Using the RIVP table you just made, what observation can you make about the voltage drop values as related to the resistor sizes?

13. Using the data from your RIVP table, what observation can you make concerning the power dissipations of resistors in the circuit?

    The remaining question is concerned with the circuit diagrammed in Figure 9.25.

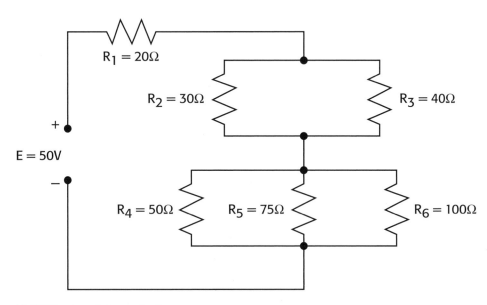

FIGURE 9.25 Schematic diagram.

14. Copy the circuit diagram into your notebook. Fill out RIVP Table 9.1, using all the given values for the circuit. Do not solve for any values yet.

10 Magnetism and Electromagnetism

10.1 INTRODUCTION

This chapter will introduce you to the fields of magnetism and electromagnetism. Although magnets were known thousands of years ago, exactly how magnetism works is still being investigated today in an attempt to better understand this invisible force. This chapter will present several magnetic theories, principles, and devices.

10.2 THE DISCOVERY OF MAGNETISM

Thousands of years ago, people discovered a special type of iron ore, first called *lodestone* (which meant "leading stone"), that could attract small metal objects containing iron. The ancient Chinese shaped a piece of lodestone into the shape of a ladle, or small cup, which when freely pivoted on a smooth surface, would always tend to point north. The device was used to "lead the way north," and gave rise to the name "leading stone." Today we call the iron ore *magnetite*, and the modern version of the Chinese pointing device is called a *magnetic compass*, or simply a compass, as shown in Figure 10.1.

FIGURE 10.1   A magnetic compass.

10.3   MAGNETIC POLES

A compass consists of a small, lightweight, freely pivoting magnet, suspended on a near-frictionless bearing. It works because of the principles of magnetic attraction and repulsion. Recall the laws governing electrically charged objects. In a similar way, opposite ends of a piece of magnetic material have oppositely behaving properties. The ends of a magnet are called its *poles*. A magnet has a *north pole* and a *south pole*. Opposite magnetic poles will attract, and like poles will push apart. You can easily demonstrate this for yourself by trying to force the north poles or the south poles of two magnets together. You will find it very hard to do. You will feel an invisible force that pushes the two magnets apart. If you try the same experiment with the opposite poles of two magnets, that is a north pole of one magnet and the south pole of another, you will find the magnets are pulled together with a considerable force, as shown in Figure 10.2.

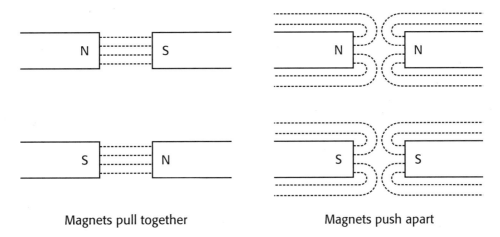

Magnets pull together   Magnets push apart

FIGURE 10.2   Magnetic poles showing attraction by unlike poles and repulsion by like poles.

10.4   THE EARTH'S MAGNETIC FIELD

A compass is able to show direction because its magnetized pointing needle reacts to the magnetic field emanating from, and surrounding, the earth. About 400 years ago, scientists discovered the magnetic field of the earth. They could not explain its reason for existence, but they theorized it must be due to some property of the earth itself. Today, scientists believe the center of the earth contains a large, molten mass of iron-bearing material. This molten material is in slow, constant motion, due to the earth's rotation, tidal forces (gravity) of the moon and oceans, as well as convection movements caused by the great heat of the earth's core. *Convection* is the tendency for warmer objects to rise and cooler objects to sink. This creates convection currents of material flowing around in circles. Take a look at soup boiling in a pan on a stove, and you will understand the concept. The modern scientific view is that this slow movement of the core material generates the magnetic field that we can observe on the surface of the earth. Even though the core material moves extremely slowly, it involves the movement of a huge amount of material over a very large area of the earth, and is able to sustain the magnetic field we observe on the surface of the earth. One way this field can be observed is by seeing how it affects a magnetic compass needle. See Figure 10.3.

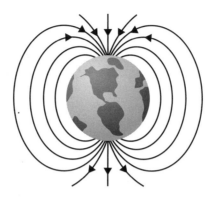

Imaginary field lines
leaving the South Pole
and entering the North
magnetic pole.

FIGURE 10.3 The earth's magnetic field.

10.5 TYPES OF MAGNETS

There are different types of magnets. Some are naturally occurring, while others are man-made. Some are permanent and some are temporary. The next sections present these different types of magnets.

10.5.1 Natural and Artificial Magnets

Magnetite is an example of a *natural magnet.* A second form of magnet can be created by rubbing a piece of soft iron with a piece of magnetite. An *artificial* (man-made) *magnet* is created in this way. A third form of magnet involves using a coil of wire carrying electric current, wrapped around a piece of iron-alloy metal. This is how an *electromagnet* is formed. The electric current flowing through the wire creates a magnetic field surrounding the wire, which also passes through the material in the center of the coil, and magnetizes the iron-alloy material placed inside the coil.

10.5.2 Permanent and Temporary Magnets

Some materials are able to stay magnetized for long periods of time without losing much of their magnetic properties. Magnets made from these materials are called *permanent magnets.* Other magnets are designed to be magnetized for short periods of time, and later lose their magnetism; they are known as *temporary magnets.*

Retentivity refers to a magnet's capability of holding its magnetism over time. A permanent magnet has good retentivity, and a temporary magnet has poor retentivity. Retentivity is determined by the atomic structure of the elements in the material used to make the magnet.

A device called a *keeper* is made from material that is easy for magnetic flux lines to pass through. It is placed across the poles of a magnet not in use, in order to keep the magnetism from decreasing over time. Figure 10.4 shows a powerful magnet used in a hard drive read/write head actuator mechanism.

FIGURE 10.4   A hard drive R/W head actuator magnet. It is the semicircular piece shown in the lower left part of the drive.

10.6   THEORIES OF MAGNETISM

Several different magnetic theories have been put forth though the years in an attempt to understand and explain magnetism. Two of these theories are described next.

10.6.1   The Molecular Theory

One early theory of magnetism was the *molecular theory*. It held that molecules of a magnetic material could become aligned in a particular direction. If a piece of magnetic material had most or all of its molecules aligned with the same orientation, the material was magnetized. It is possible to demagnetize magnetic materials as well. The molecular theory held that if enough force was applied to the magnet, in a form such as heat or mechanical energy, the molecules would be randomly moved about into new, disoriented positions. An example of this idea is the observation that if a magnet was repeatedly struck with a hammer, it would lose much of its magnetism. A modern example of this principle can be seen on the warning labels on computer floppy disks. The labels advise users to keep the disks away from heat sources such as direct sunlight, since heat will tend to demagnetize the disk, and all data on the disk would be lost. Of course, excessive heat also tends to warp the plastic shell around the disk material, rendering it unusable. Floppies contain extremely small areas that are magnetized in a special pattern to represent numbers. Disks are organized or divided up by magnetic markers into tracks (concentric circles) and sectors (pie-shaped wedges).

10.6.2   The Domain Theory

The more modern theory of magnetism describes small areas of a material that have a particular magnetic orientation as magnetic *domains*. Domains are believed to behave as they do because of the direction of *spin of electrons* in the atoms of the magnetic material itself. Scientists have discovered that in addition to orbiting around the nucleus of an atom, electrons also spin. As an electron spins on its axis, rather as the way the Earth spins, it is a charge in motion. A change in motion of a charged particle always creates a magnetic field around it. The direction of the field, or what we call its *magnetic polarity*, is dependent upon the direction of spin of the electrons. A magnetic domain is thought to be a very small area of a material in which all or most of the electrons in the atoms have the same spin direction. The area is not necessarily the size of an individual molecule.

10.7   MAGNETIC MATERIALS

Materials are classified into one of three magnetic categories: ferromagnetic, paramagnetic, and diamagnetic. *Ferromagnetic* materials are those usually made from iron or iron-alloys. An *alloy* is a mixture of two or more metals. Ferromagnetic materials tend to become highly magnetized in the same direction as the magnetic field they are exposed to. These are the kinds of metals used to make most permanent magnets and devices that are designed to be operated with magnetism. Examples of ferromagnetic materials are soft iron, many iron alloys, nickel, cobalt, and some alloys of the latter two. Some special alloys of these metals have many times the magnetic force of magnetite. Some examples of these special alloys are Permalloy®, supermalloy, alnico, and cunife. *Alnico* is an alloy of aluminum, nickel, and cobalt, and has been used for more than 60 years for loudspeaker magnets and other devices. *Cunife* is an alloy of copper, nickel, and iron (ferrite). Some of these modern magnetic alloys of iron are used to make magnets for DC motors, loudspeakers, and dynamic microphones, as well as the read/write head actuators for hard drives, to list just a few examples.

Battery-powered drills and other DC-powered, portable electric hand-tools are made possible largely by the powerful motors that use these special alloys. The success of these small, powerful motors is due to the combined technological improvements made by the sciences of metallurgy, mechanics, and electronics.

Paramagnetic materials acquire only a small amount of magnetism when they are placed in a strong magnetic field. These types of materials don't make good permanent magnets. Examples include certain alloys of aluminum, beryllium, cobalt, magnesium, manganese, and nickel.

Diamagnetic materials tend to become slightly magnetized in the opposite polarity of the magnetic field they are placed in. Most people consider materials in this category to be nonmagnetic. Examples are: bismuth, paraffin wax, wood, and silver. They are not used to make magnets.

10.8   USES OF MAGNETISM

The earliest known use of magnetism by man was that of the compass, for direction finding. Later, as technology evolved, new devices using magnets and electromagnets came about. Early uses included telegraph sounders and electric motors. Modern examples include loudspeakers, microphones, tape recorders, magnetic storage devices for computers, and the control of electron beams used to form the pictures on TVs, oscilloscopes, and radar screens. The small powerful motors used in portable tools, and small, radio-controlled hobby vehicles, such as cars, boats, and planes, are some other examples. Recall the electromagnetic relay, presented previously. It also operates using the principle of magnetism to remotely control one electrical circuit with another, without any direct electrical connection between the two circuits.

10.8.1   Medical Uses for Magnetism

In the past, many people believed that magnets could be used to treat, and even cure diseases such as arthritis. Today, modern medicine does not believe there is any special power of magnetism to cure disease. But modern science does use magnetism to help diagnose medical problems that could not be discovered by previously available methods. MRI stands for *magnetic resonance imaging.* It uses a very powerful magnetic field to "see" inside a person's body, better than X-rays or other methods can, with less risk to the patient. It can often reveal more detail about soft tissue such as the brain.

10.8.2   Burglar Alarms

A magnetic proximity switch can be made to close or open an electric circuit if it is placed in a magnetic field. One popular use of such a switch is in burglar alarms for windows or doors. The switch consists of two parts: a small, magnetically operated switch, using two thin metal "reeds," and a small magnet. The reed switch is mounted on the stationary part of the door or window—the doorjamb, for instance. The magnet is mounted on the moving part of the closed door or window, immediately next to the switch. The magnetic field from the magnet penetrates the reed switch, inducing each part of the switch to act as a magnet. The two parts are drawn together due to the induced magnetic force, and the switch is closed. The closed switch completes an electric circuit to a sensor in the burglar alarm. If the door or window is opened, as during a burglary, the magnet is moved away from the switch, removing the magnetic field from the switch, causing it to open, which trips the sensor thus sounding an alarm.

10.8.3   Magnetic Navigation

The interaction of a compass and the earth's magnetic field is the basis for magnetic direction-finding. Ship's captains have used magnetic compasses for navigation for hundreds of years. The magnetic north direction is not constant for all locations. In addition, it is in constant motion. To help compensate for this fact, sailors use books that list the latest position and strength of the magnetic field at various locations on the earth. This listing must be constantly updated in order to be accurate. Some scientists now believe that the earth's magnetic field has changed widely in the past. They also believe they now have solid evidence that the north and south magnetic poles of the earth have actually reversed position; in fact, many times. This magnetic pole reversal could possibly account for some changes in climate, animal and plant life extinctions, and other otherwise-unexplained events.

Even thirty years ago the role magnetism plays in the animal world was widely ignored or disbelieved. Today, however, it is widely known that many animals have a keen sense of the earth's magnetic field. Some scientists believe these creatures use the earth's magnetic field to help them know the time of day, the time of ocean tides, the direction to their home, and the season. NASA even did magnetic experiments on people a few years ago, but stopped research when they discovered that the abnormal magnetic fields used in the experiment seemed to induce the signs of stress in test animals, and affected measurable changes in the human subjects.

The idea seemed to be that the earth has a distinctive "magnetic signature," and animals have grown used to it over time. If the animals are exposed to a magnetic field having significantly different strengths, or if the fields changed at different rates than the normal earth field, the animals were upset, or could not find their way home.

Military use of magnetism also includes the detection of underwater submarines by sensing the disruption of the earth's magnetic field, called a *magnetic anomaly*, as a submarine passes through the ocean. Since a ship such as a submarine is composed of a large mass of ferromagnetic metal (steel), it tends to disrupt the magnetic flux lines as it passes through an area in the sea. These anomalies can be sensed by detectors carried aboard aircraft flying over the oceans. Some newer spy satellites also have this sensing capability.

10.8.4 Magnetism in the PC

Floppy disks, hard drives, and all the motors for the drives and fans inside a PC operate using magnetism and electromagnetism. Data is written and read from magnetic media on floppy and hard drives, Iomega® Zip® drives, et cetera. The small loudspeaker that sounds beep codes uses a permanent magnet. Hard drives use permanent magnets and a structure known as a *voice coil actuator*, which uses an electromagnet to generate a magnetic field that works with the permanent magnet to move the read/write heads in and out above the disk surface. A CRT-based computer monitor uses electromagnetism to direct a single stream of electrons rapidly enough across the inside of the tube face to simulate pictures and text. The drive motors all use electromagnetism to spin the disks. All the wireless components use electromagnetism to create the radio links between various units. Refer to Figure 10.5 for a picture of a hard disk read/write head assembly.

FIGURE 10.5 Hard disk read/write head assembly.

10.9   ELECTROMAGNETISM

There is a special relationship between electric current and magnetism that cannot be separated. This relationship has two parts. First, whenever current flows through a conductor, there is a magnetic field that surrounds the conductor. The strength of this magnetic field depends on the amount of current flow. The greater the current flow, the more powerful the magnetic field. The magnetic polarity of the field depends on the direction of current flow.

10.9.1   Motors

This first part of the two relationships between current and magnetic fields is the basis of operation of electric motors. A motor operates because of the interaction of two magnetic fields. The rotating motion of the motor, called *torque*, is developed through the interaction of the two magnetic fields. Remember, magnetic fields will either attract or repel, depending on their respective polarities. One of the magnetic fields in an electric motor is produced by what is called the *field magnet*. The field can be produced by either a permanent magnet inside the motor, or it can be produced by a coil consisting of many turns of wire. When current flows through the coils of wire, a very powerful magnetic field is produced. The coil is called the *field coil* for this reason. Another magnetic field is produced by a separate coil, able to rotate within the field coil's magnetic influence, called the *armature*. The magnetic attraction and repulsion set up between the two fields in the motor cause the motor shaft to turn at high speed. We can use the motor shaft to do work for us, such as operating a fan, washing machine, starting a car's engine, and so on. This same motor action is used in many analog or moving-needle gauges and meters. See Figure 10.6 for a picture of a magnetic field surrounding a current-carrying conductor.

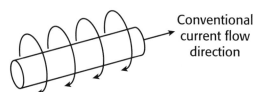

Conventional
current flow
direction

FIGURE 10.6   A magnetic field surrounds a current-carrying conductor.

10.9.2 Generators

The second part of the electromagnetic relationship is responsible for the operation of electric generators. An English physicist and chemist named Michael Faraday, 1791–1867, discovered this principle. Whenever there is relative motion between a magnetic field and a conductor, there will be a voltage produced in the conductor. Relative motion means simply that of the two elements, the magnet and the conductor, one is in motion, and the other is stationary. It also means that both could be moving, but at different speeds or directions. The amount of the voltage created in the conductor in this way depends on four factors:

The strength of the magnetic field: The stronger the field, the greater the induced voltage.

The speed of interaction: The faster the speed, the greater the induced voltage.

The number of conductors immersed in the magnetic field: The greater the number of conductors in the field, the greater the induced voltage.

The angle between the field and the conductors: Maximum voltage is induced when the field and conductors are perpendicular to each other; minimum voltage is induced when the field and conductors are parallel to each other.

Generators are made with coils of wire with many turns in order to create a great amount of *induced voltage*. Induced voltages will be treated in greater detail in Chapter 13, "Inductors and Capacitors."

An electric generator is constructed in essentially the same way as an electric motor. The main difference is that in a generator, the rotating mechanical force (torque) is applied to the input shaft in order to turn the armature. The interaction of the magnetic fields produces a reaction. The result is the production of induced voltage in the other generator coil. The resulting voltage is connected to external power circuits.

10.10 MAGNETIC SHIELDING

Although it is impossible to actually stop magnetic flux field lines, it is possible to redirect them to flow around something. Special metal alloys are used to construct covers for magnetically sensitive equipment such as the *cathode-ray tubes* (CRT) used in older oscilloscopes and some types of television monitors. The pictures on these displays would be distorted by the interaction of stray magnetic fields, and so must be magnetically shielded from these effects. Many pieces of specialized medical monitoring equipment are protected in this way. The material used to form the shields is very permeable to magnetic flux. *Permeability* is a measure of how easily something can pass through a material. Magnetic permeability refers to how easily magnetic flux lines can pass through a material. So, if the sensitive equipment is enclosed in very magnetically permeable material, the magnetic flux lines will be deflected and pass through only the permeable material, rather than taking the less-permeable path through the protected piece of equipment. Most computer speakers incorporate magnetic shielding to prevent distortion of the picture on the video monitor's cathode ray tube (CRT). The picture produced by CRT-based monitors (and TV sets as well), as discussed earlier, is formed by a stream of electrons moving to strike the phosphor coating on the inside of the picture tube. The electron stream has an inherent magnetic field, so it can be moved or deflected by an external magnetic field such as that produced by the speakers. One form of magnetic shielding material is known as μ-metal (mu metal) and was at one time commonly used to surround old CRT oscilloscope tubes.

Another common PC application of magnetic shielding is the use of a metal shield surrounding all the input and output motherboard connectors on an ATX-style PC. Without such a shield, there would be much greater likelihood of interference both to the PC from outside electromagnetic signals as well as interference to other devices from the PC's internal clock and control signals, which are really radio frequency (RF) signals. Figure 10.7 shows the use of a highly-permeable magnetic shield, also known as an *I/O shield*, since it covers the open area around the input and output connectors.

FIGURE 10.7   Electro-magnetic I/O shield on the rear panel of a PC.

10.11   COMMON ELECTROMAGNETIC CONCERNS WITH PCs AND RELATED EQUIPMENT

- Floppy and hard drives are magnetic devices. Care must be taken when working with them not to subject them to strong, close-by magnetic fields, or loss of data can occur. Most hard drives, because they have metal housings, are much better shielded than floppies, since floppies have only plastic, a nonmagnet material, for protection. Floppy disks are quite susceptible to outside magnetic fields. A technician should exercise care when using magnetized tools such as screwdrivers, especially the motorized versions, which contain powerful magnets in the motors. Do not store floppy disks in the immediate vicinity of a CRT-based monitor or a power supply. Most CRT-based monitors include a coil of wire mounted just inside the case and surrounding the picture tube that is energized for a second or two each time the monitor is first turned on. This *degaussing coil* is used to neutralize the magnetic charge on a metal screen used to steer the electron beam to its intended targets, the small phosphor dots coating the inside of the tube face. When this screen, called a *shadow mask*, becomes magnetized over time, the result is known as misconvergence, meaning the electron stream does not focus or converge properly anymore and the image is polluted. The degaussing coil is a powerful electromagnet that can cause data loss in magnetic media, such as floppy disks, if they are left close to the monitor. Handheld degaussing coils used to be standard tools of the TV serviceman. Refer to Figure 10.8 for a picture of a CRT and degaussing coil.

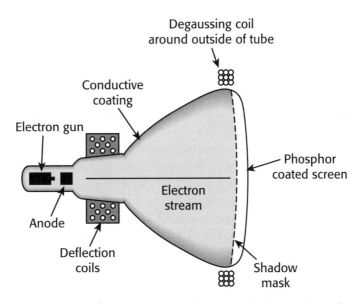

FIGURE 10.8 A CRT showing the location of a degaussing coil.

■ Removing metal from the case of a computer defeats the magnetic shield, which increases the chance for interference both to and by the PC. Popular case "mod kits" and cases sold with clear plastic replacing part or all of the normal metal case may look interesting and flashy, but are a very bad idea from an electromagnetic interference standpoint.

■ The proper motherboard I/O metal shield should always be used to ensure an "RF-tight" enclosure. This means preventing the high-frequency clock and control signals inside the PC from radiating outside the case.

■ Operating a computer in a strong nearby RF environment, such as in a commercial or amateur radio station, requires strict reliance on good electromagnetic shielding techniques and good electrical grounding as well.

10.12   SUMMARY

■ Magnetite is a naturally occurring iron ore that is strongly magnetic.

■ A magnetic compass works by the interaction of a small, moveable piece of magnetized material suspended on a near-frictionless bearing, with the earth's magnetic field.

■ A magnet has two poles, a north and a south pole. Like poles repel and unlike poles attract each other.

■ Artificial magnets are man-made and can be made to be either permanent or temporary. A common form of temporary magnet is called an electromagnet.

■ Permanent magnets are made from special metal alloys and are used in motors, generators, loudspeakers, hard drives, floppy drives, and solenoids, among many other uses.

■ Retentivity is a relative rating of a material's ability to hold onto a magnetic field after the magnetizing force is removed from it.

■ The domain theory is a modern magnetic theory that holds that small areas of a material can become magnetized in the same polarity due to most or all the electrons spinning in the same direction.

■ The molecular theory of magnetism holds that materials become magnetized when a material's molecules are aligned in the same direction.

■ Ferromagnetic materials can become strongly magnetized with the same magnetic polarity as the external magnetizing force.

■ Diamagnetic materials can become weakly magnetized with the opposite polarity as the external magnetizing force.

■ Paramagnetic materials can become weakly magnetized with the same polarity as the external magnetizing force.

■ A field magnet is one used to establish a magnetic field in a generator or motor, which will interact with the magnetic field developed by the armature. In loudspeakers the "armature" is called the voice coil. A similar naming convention is used to describe the mechanism that moves the read/write heads in modern hard drives (voice-coil actuators).

- A voltage is induced in a conductor whenever there is relative motion between the conductor and a magnetic field.
- A dipole magnetic field is one with two magnetic poles, north and south.
- Permanent magnets and electromagnets are used in the motors that spin fans and drives.
- CRT-based computer monitors use electromagnets in the form of deflection coils in the yoke to move the electron stream back and forth to draw the image on a phosphor-coated screen.
- Special highly permeable material can be used to redirect magnetic fields in order to shield sensitive parts of equipment.
- CRT monitors use powerful electromagnets called degaussing coils to provide a pulse of AC current when the monitor is first turned on. This AC current randomizes the magnetic field on the shadow mask. Keep magnetic media away from the immediate vicinity of a CRT-based monitor to avoid data loss.
- The radiation levels of electromagnetic fields from a CRT-based monitor are addressed by two ratings: ELF and VLF. ELF refers to *extremely low frequency*; the vertical refresh or synchronization rate of the monitor falls in this range. VLF refers to *very low frequency*; a monitor's horizontal refresh rate falls in this range. Watch for CRT-based monitors that meet the low emission standards of the TCO (a safety standard).

10.13   KEY TERMS

Magnetite	Retentivity	Induced voltage
Magnetic compass	Domain theory	Dipole field
Poles	Ferromagnetic	Magnetic shielding
Natural magnet	Paramagnetic	Degaussing coil
Artificial magnet	Diamagnetic	Shadow mask
Permanent magnet	Field magnet	ELF
Temporary magnet	Field coil	VLF
Electromagnet	Armature	

10.14   EXERCISES

1. A magnet consisting of an ore that is found buried in the ground is called a (an):
   a. natural magnet
   b. artificial magnet
   c. electromagnet

2. A magnet made by man is called:
   a. natural
   b. artificial
   c. synthesized

3. A magnet that tends to stay magnetized for a long time is called:
   a. permanent
   b. temporary
   c. read/write

4. A magnet that quickly loses its magnetism is called:
   a. permanent
   b. temporary
   c. lossy

5. List at least three devices that operate using magnetism.
   1.
   2.
   3.

6. List the two essential concepts that describe *electromagnetism*.
   1.
   2.

7. What was the earliest known use of magnetism by man?

8. List a modern medical use for magnetism.

9. What is the principle of magnetic attraction and repulsion?

10. What is a *magnetic domain*?

11. What scientist is given credit for discovering the relationship between electricity and magnetism?

12. How does an electromagnetic relay make use of magnetism?

13. How does a magnetic switch help prevent burglary?

14. The magnetic field produced by the earth is:
    a. constant
    b. changing

15. Some scientists believe the magnetic poles of the earth have reversed many times in the past.
    a. true
    b. false

16. A permanent magnet will eventually lose some or all of its magnetism unless a keeper is kept across its magnetic poles while the magnet is not in use.
    a. True
    b. False

17. The greatest care should be exercised when using a power screwdriver around what?
    a. Floppy disks
    b. Hard drives
    c. LCD monitors
    d. Computer cases
    e. Keyboards

18. Which of the following use electromagnetism to store data?
    a. Floppy disks
    b. Hard disks
    c. Zip disks
    d. Punch cards
    e. Both a, b, and c

19. The motors used in a PC's drives and fans use which type of magnets?
    a. Permanent
    b. Electromagnet
    c. Both a and b

20. Electromagnetism is used to create the image on which type of display?
    a. LCD
    b. CRT

11 Inductors and Capacitors

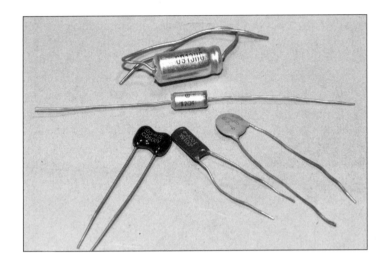

11.1 INTRODUCTION

This chapter introduces two new electrical components, inductors and capacitors. They both store energy but in different ways. Both are used in many different types of circuits. A PC technician should understand their uses and names.

11.2 INDUCTORS

Like an electromagnet, introduced in the last chapter, there is another basic device that makes use of the special relationship between electric current and magnetism. It is called an *inductor*. Inductors are used to do many different jobs in electronic circuits, but the principle of operation is the same in all cases: the capability of limiting the rate of current change in a circuit.

Key Ideas

Inductors limit changes in current.

Inductors are usually described in two ways: how they act in a circuit, and how they are constructed. It is the inherent property of inductors to oppose a change in the amount of current flowing through them. This is what makes them so useful in electronic circuits. They accomplish the opposition to current change by temporarily storing energy in a magnetic field surrounding the inductor.

Inductors are usually thought of as being coils of wire, and in many circuits they take that form. All conductors, at or near room temperature, tend to act somewhat inductively, even if the conductor length is quite short; so not all inductors are large and bulky coils of wire. One type of inductor used on a miniature circuit on a printed circuit board resembles a small figure made of concentric copper traces—a very thin conducting spiral of metal deposited on the insulating board.

All current-carrying conductors have a magnetic field that surrounds them; the greater the current, the stronger this magnetic field. Inductors are designed to have a specific amount of inductance, and inductance depends on the capability of temporarily storing energy from the circuit in a magnetic field. For packaging and economic reasons it is desirable to have the greatest amount of energy storage possible for a given amount of space. This not only reduces the overall inductor size, but also the size of the equipment in which the inductor is installed. One of the simplest ways of increasing the magnetic field effect around a conductor is to form it into a loop. The magnetic field will be intensified inside the loop. If many loops are formed in the conductor, the energy storage capability as well as the inductance value will increase greatly. This is why many large-value inductors are made with many coils, or turns of wire; Figure 11.1 shows this idea.

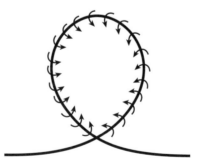

FIGURE 11.1 Conductor loops increase the concentration of a magnetic field.

11.3   INDUCTANCE

Inductance is the property of an inductor that opposes a change in current. It can also be defined as the capability to store energy in an electromagnetic field. When the value of current in a circuit increases, some of the increased energy from the circuit, as a result of the increasing current, is temporarily stored in a magnetic field around the inductor. As the current continues to increase, the magnetic field proportionally increases around the inductor. Once the circuit current value stabilizes, the magnetic field around the inductor also stabilizes. If the circuit current decreases, the magnetic field will shrink, and in doing so, will transfer the temporarily stored energy of the magnetic field back into the circuit.

11.4   TYPES OF INDUCTORS

Inductors can be classified into groups by intended circuit use and by rate of current change applied to them. Inductors are designed to handle different values of current, and are constructed differently. Depending on the rate of current change (frequency) over which the inductor must operate, different construction methods are used as well.

11.4.1   Frequency Range

Frequency can be thought of as the number of repetitions per unit of time of a voltage or current in a circuit. If something happens many times per unit of time, it has a high rate of change, or high frequency. Things that happen a few times per unit of time are said to have a low rate of change, or a low frequency. The standard unit of time used for expressing frequency is the second. So, frequency is the number of times something repeats per second. The unit used to denote frequency is the hertz (Hz) named for the German physicist, Heinrich Hertz (1857–1894).

Low-frequency Inductors

When the frequency of current change is low, about 1 to 50,000 times per second (1 Hz–50 kHz), inductors are most often constructed using coils of wire wrapped around a large metal form called a *core*. Inductor cores of this type are made of special types of metal that have a very high permeability. Remember from Chapter 10, "Magnetism and Electromagnetism," that permeability refers to the tendency of a material to concentrate lines of magnetic flux. The permeable core gives the inductor much greater inductance, or capability of storing energy in a magnetic field, than if the inductor had no core. Cores are made from special steel alloys, or mixtures of metals, that provide many times the permeability of air. The number of times greater than air a material concentrates flux lines is known as *relative perme-ability*, and is represented by the symbol μ (the Greek letter "mu").

Laminated Iron Cores

Most low-frequency inductors (often called power inductors) use cores made up of many thin strips made from special metal alloys. The strips are each coated with var-nish and oven-dried before being stacked together, sandwich style. The stack of metal core strips is either clamped or bolted together to form a solid assembly. This provides a large magnetic mass, which greatly increases the inductance of the coil, but will not provide a large electrical path for current to flow in the core material itself.

Inductor Losses

The undesired currents flowing in the core dissipate power, since according to the Ohm's law power formula, $P = I^2 R$. This power dissipation results from cur-rents (I) flowing in the core, called *eddy currents*, through the imperfect conductors (R)—the steel plates, which have a fair amount of electrical resistance. The thin laminations help decrease power losses (P) by limiting the eddy current power loss. The smaller current pathways provided by the thin laminations will not conduct as much current as a solid metal core. Figure 11.2 shows the schematic symbol and a picture of a power inductor.

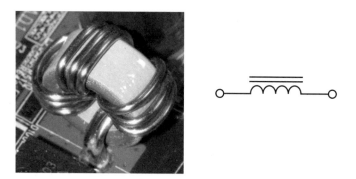

FIGURE 11.2   Schematic symbol and a picture of a power inductor.

Intermediate-frequency (IF) Inductors

As the rate of current change (frequency) increases, the losses in an inductor in-crease greatly. Part of the increase is due to greater core losses due to increased eddy currents. To combat these power losses, inductor core metal is ground up and mixed with a nonconducting material, usually a ceramic, plus some glue called a *binder*. The material is heated and pressed into the desired shape, usually a rod, or a doughnut shape called a *torus*. Wire is then wrapped around the core material to form the inductor. By making the individual pieces of core material very tiny, and insulating them from each other, the eddy current pathways are even further reduced, which tends to limit the power lost due to these currents. New designs for power inductors and transformers (refer to Chapter 13, "Transformers") use toroidal cores, which offer better shielding than laminated iron cores.

Hysterysis Loss

Another way in which a core can rob power from an inductor circuit is called *hysterysis*. Hysterysis can be thought of as a sort of "electrical inertia." A good example of mechanical inertia is push-starting an automobile. It takes a great amount of energy to get the car moving and it takes a great amount of energy to stop the car, compared with the energy required to just keep the car moving. A similar thing happens in an electronic circuit containing inductance. It is impossible to magnetize or demagnetize a magnetic material instantaneously. When a magnetic material such as an inductor core is repeatedly magnetized, first in one direction and then in the opposite direction, the core cannot instantly give up the magnetism and reverse magnetic polarity. That is exactly what it is forced to try to do, however, when the inductor is placed in a circuit using alternating current. AC first causes current to flow one direction in a circuit, and then the direction of flow is reversed. This process repeats over and over, at some fixed frequency.

Hysterysis is not normally a big problem at very low frequencies, since the speed at which the core material has to give up its magnetism is relatively long. As the frequency rises, the time required to reverse magnetic polarity becomes shorter. The core materials used for the power inductors simply cannot reverse polarities fast enough, and power is lost due to the greater energy required to reverse core polarity. Remember the example of trying to stop the moving car. To combat this tendency, core designers use different mixes, called *alloys*, of materials for higher-frequency cores. These alloys consist of iron or iron alloys plus the ceramic materials and binder. These materials are referred to as powdered iron and ferrite. The engineering required combines the science of metallurgy as much as electronics. Inductors made in this manner are known as intermediate-frequency (IF) inductors. Figure 11.3 shows the schematic symbol and picture of an intermediate-frequency inductor.

High-frequency Inductors

If the frequency is high enough, over several megahertz, even the core materials used at intermediate frequencies have too much power loss, so inductors used for this frequency range do not use any core at all. They are simply self-supporting coils of wire, or they may use small strips of good insulating material, such as glass,

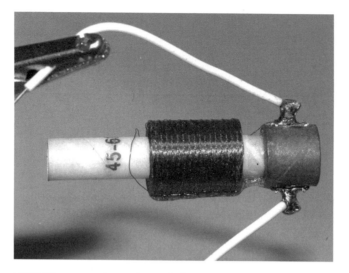

FIGURE 11.3   Schematic symbol and photo of an IF inductor.

ceramic, or plastic, to support the coil. They are known as air core inductors. Figure 11.4 shows the schematic symbol and picture of an air core inductor.

FIGURE 11.4   Schematic symbol and photo of an air core inductor.

Summary of Inductor Losses

Inductors can suffer from several different types of power losses, including:

> **Resistive losses due to the wire used in the winding:**   This is known as copper loss, and follows the Ohm's law formula $P = I^2R$, where P = power loss, I = winding current value, and R = the resistance of the winding conductor.

Eddy current losses resulting from currents flowing in the core material:
Laminations are used at low frequencies, powered metal cores are used at
intermediate frequencies, and no cores are used at very high frequencies.

Hysterysis losses: These occur due to the "inertia" of the magnetic core material, which prevents the core from being instantly magnetized and then demagnetized. Various metal alloys are used to combat this effect.

Direct radiation loss: Modern inductors use toroidal designs, which tend to
keep the electromagnetic field from escaping the core.

Increased wire resistance due to high-frequency effects: These will be explained in Chapter 12, "Alternating Current Circuits."

11.4.2 Inductor Core Styles

Inductors are also named by core construction style. The low-frequency inductors
that use the laminated cores are known as *shell core* types. They are built up using a
series of metal strips called "*E*" and "*I*" pieces. Newer, low-frequency inductors are
wound on toroidal cores. Intermediate-frequency inductors are made with either a
rod core or a toroidal core. Recent advances in the state of the art in inductor manufacturing allow efficient designs to be built using the doughnut, or toroid core
style. In addition to offering lower core losses, these inductors also tend to lose less
of their stored magnetic field due to direct energy radiation. A magnetic field in
motion tends to radiate energy away from the source, in a manner similar to how
radio or TV signals are radiated from a transmitting antenna. This amounts to a
loss of energy from the circuit, which is undesirable. Toroidal inductors are considered self shielding, since they tend to more completely confine the magnetic
field within the core material itself. It is now possible to construct tiny inductors,
called *microchip inductors*, for use with integrated circuits. Figure 11.5 shows several
inductor core styles.

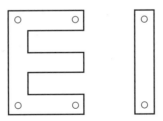

Toroid core

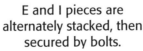

E and I pieces are
alternately stacked, then
secured by bolts.

FIGURE 11.5 Inductor *"E"* and *"I"* core pieces and a toroid core.

11.5   INDUCTOR CIRCUITS WITH CHANGING CURRENT

When current through an inductor changes, there is a reaction due to the inductance. The following sections describe what happens during inductor current changes.

11.5.1   Increasing Current

When current in a circuit attempts to increase, as when the switch is closed, a magnetic field is created around the conductor carrying the current. This magnetic field expands outward, beginning in the center of the conductor. Remember, whenever there is relative motion between a magnetic field and a conductor, there will be a voltage created in the conductor, called an *induced voltage*. This induced voltage is always created opposite to the original circuit voltage. It is commonly called a back-voltage or *counter electromotive force*, and is abbreviated *CEMF*. If the conductor is part of a complete circuit, then the induced voltage will cause a current, again opposite to the original circuit current. This current will tend to oppose the instantaneous current rise in the circuit. The opposition is greatest the instant the current first attempts to increase, and the effect dies out over time. Figure 11.6 shows an illustration of induced voltage.

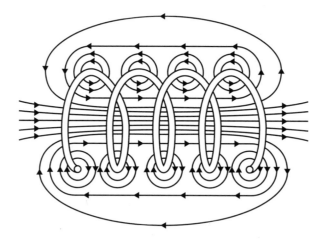

FIGURE 11.6   Voltage induced in a conductor.

The mathematical relationship that predicts the length of time required for an inductive circuit to reach a constant or *steady-state* condition is called the circuit *time constant*. The time constant for an inductive circuit is:

$$t = L \div R \tag{11.1}$$

where t = time in seconds
        L = inductance value in henries
        R = resistance in ohms

During a single time constant, the circuit current will increase only by 63.2%. As long as there is a driving force that is attempting to increase current, the increase will be limited to 63.2% per time-constant period. We can use a handy chart to solve any time-constant problem. The chart is called the *universal time-constant chart*. It shows two curves, one increasing and one decreasing, that can be used to predict the value of current in an inductive circuit. Figure 11.7 shows a universal time-constant chart, or "set of curves."

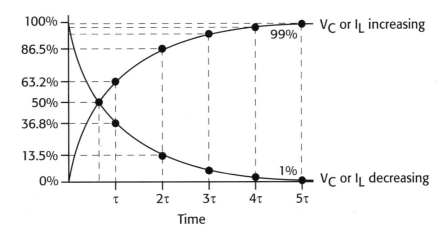

FIGURE 11.7   Universal time-constant chart.

The time interval over which the value of current is increasing, but before a steady-state condition is reached, is known as the *transient* time. Transient simply means a value is changing, in this case, circuit current though the inductor is changing. The time-constant chart is used to predict the value of current during transient times.

The Practical Time Constant

Often, it is desirable to know how long it takes the current in a circuit to reach 50 percent of its maximum or minimum value. That length of time is known as the *practical time constant, t_p* and is equal to 70 percent of the circuit time constant.

Example 11.1

Find the practical time constant of a series circuit with a 1 H inductor and 200 ohms resistance.

Solution

First, compute the circuit time constant, using $t = L \div R$

$$t = 1H \div 200\Omega$$

$$t = 5 \; ms$$

Next, find the practical time constant,

$$t_p = t \times 0.7$$

$$= 3.5 \; ms$$

Solving Increasing-current Problems

In many cases, it is necessary to predict the value of circuit current some specific amount of time after a switch is closed in an inductive circuit. The method of solution is fairly easy. The steps required are:

1. Compute the time constant of the circuit, using given circuit values and

$$t = L \div R$$

2. Divide the time in question by the circuit time-constant time.
3. Compute the total circuit current value using Ohm's law.

$$I = E \div R$$

4. Use the universal time-constant chart, using the number of time constants found in Step 2, to find the percentage of total current in the circuit.
5. Multiply the percentage of current found in Step 4 by the total current value computed. This is the value of current flowing at the time required.

Example 11.2

Refer to Figure 11.8. Find the circuit time constant and find the value of current flowing in the circuit one time constant after the switch is closed.

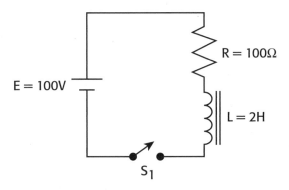

FIGURE 11.8 Time-constant schematic circuit.

Solution

1. Compute the circuit time constant.

$$t = L \div R$$
$$= 2\,H \div 100\,\Omega$$
$$= 2\,ms$$

2. Find the total circuit current, using Ohm's law.

$$I_T = E \div R$$
$$= 100\,V \div 100\,\Omega$$
$$= 1\,A$$

3. Look up percentage of total current using the universal time-constant chart. In this case, one time constant will correspond to a value of 63.2% of total circuit current (1 A).
4. Multiply total circuit current by the percentage value found using the chart. In this case,

$$63.2\% \times 1\,A$$
$$= 632\,mA$$

Example 11.3

Find the values of current at the other time constants. The current value, after two time constants, or 40 ms (milliseconds) for this circuit, reaches 86% of total current, or:

$$86\% \times 1\,A$$
$$= 860\,mA$$

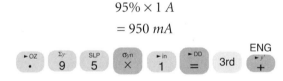

Solution

After three time constants, or 60 ms for this circuit, there will be:

$$95\% \times 1\,A$$
$$= 950\,mA$$

After four time constants, or 80 ms for this circuit, we will have:

$$98\% \times 1\,A$$
$$= 980\,mA$$

After five time constants, or 100 ms for this circuit, there will be:

$$99\% \times 1\,A$$
$$= 990\,mA$$

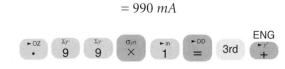

After each additional time constant, the circuit current will increase by 63.2% of the *difference* between its present value and the maximum possible value. For example, after six time constants, the current will increase an additional 63.2% of the difference between 99% and 100% of 1 A. The current value will never reach 1 A. To simplify things, it is common practice when dealing with most everyday electrical circuits to agree that the circuit current has reached its maximum value after five time constants. It is considered close enough to round off this way. For this example circuit, this means that after five time constants, or 100 ms, the current is agreed to be equal to the maximum value of 1 A. This convention will be used throughout the rest of this book.

Five time constants is the time required for a circuit to reach a steady-state condition.

Key Ideas

Example 11.4

Refer to Figure 11.9. Find the value of current flowing in the circuit after 70 μs (microseconds).

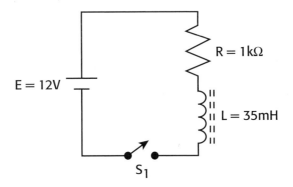

FIGURE 11.9 Example 11.4 circuit schematic diagram.

Solution

1. The circuit time constant is

$$t = L \div R$$
$$= 35 \ mH \div 1 \ k\Omega$$
$$= 35 \ \mu s$$

2. Divide the time in question by the time-constant time.

$$70 \ \mu s \div 35 \ \mu s$$
$$= 2 \text{ time constants}$$

3. Find total circuit current, using Ohm's law. In this case:

$$I_T = E \div R$$
$$= 12 \ V \div 1 \ k\Omega$$
$$= 12 \ mA$$

4. Use the chart to find the corresponding percentage of total current. In this case,

$$2 \text{ time constants} = 86\% \text{ of } I_T.$$

5. Multiply the percentage found in Step 4 by the value of total current:

$$I = 0.86 \times 12 \ mA$$
$$= 10.3 \ mA$$

11.5.2 Inductive Circuits with Decreasing Current

There are two possible ways current can decrease: by turning off the power, or by lowering the voltage applied. When a switch opens and the current in the circuit tries to stop, it does not actually stop immediately. The rapid rate of current change going toward zero causes the magnetic field that was established around the inductor to rapidly collapse back through the conductor in the winding. This induces a very high CEMF. The induced voltage is enough to cause an arc across the switch contacts as they move apart. The arc maintains current flow for a short time after the circuit is open. The arc is stopped when the contacts become too far apart to sustain the arc anymore. Contact arcing in switches used to control inductive loads is a serious problem that ultimately destroys the switch after many on-off cycles. It is beyond the scope of this text to calculate the current in an arc. The problem involving lowering the applied voltage, however, can be easily solved. See Example 11.5.

Example 11.5

Refer to Figure 11.10. Assume the applied voltage, E, was 10 V, and the voltage had been applied for greater than five time constants. So the circuit current has reached steady-state value. Now if the applied voltage is instantly reduced to 5 V, what will happen to the current?

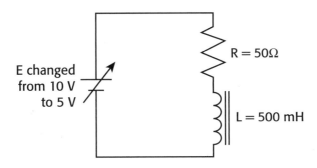

FIGURE 11.10 Example 11.5 circuit schematic diagram.

Solution

As soon as the voltage is reduced, some of the magnetic field collapses back through the inductor winding. This induces a CEMF, which tries to maintain the same value of original current flowing. As the magnetic field collapses, it eventually stops, since there is a new value of current established by the source voltage and circuit resistance. What happens is a gradual decline in current to the new value.

First find the circuit time constant:

$$t = L \div R$$
$$= 500 \; mH \div 50 \; k\Omega$$
$$= 10 \; \mu s$$

Next, find the original current value, using Ohm's law:

$$I_1 = E_1 \div R$$
$$= 10\ V \div 50\ \Omega$$
$$= 200\ mA$$

Then, find the new current value,

$$I_2 = E_2 \div R$$
$$= 5\ V \div 50\ \Omega$$
$$= 100\ mA$$

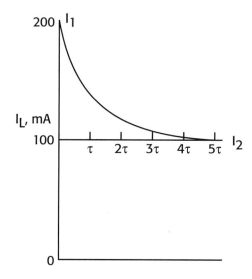

Since it takes five time constants for the current to reach its new value, it takes $5 \times 10\ ms = 50\ ms$ for the current value to change. Figure 11.11 shows a graph of the current.

FIGURE 11.11 Example 11.6 current graph.

11.6 SERIES AND PARALLEL INDUCTORS

The total inductance value of inductors wired in series are generally calculated by adding the individual inductance values. Inductors can be connected so that their windings go in the same direction. In this case, they are considered to be in a *series-aiding connection*. They could also be wired in opposite winding directions. Such a connection would be considered *series-opposing*.

For a simple series-aiding connection, the formula for total inductance is Equation 11.2:

$$L_T = L_1 + L_2 \tag{11.2}$$

Example 11.6

Refer to Figure 11.12. The two inductors are wired in a series-aiding configuration. Find the equivalent inductance value.

L_1 L_2

FIGURE 11.12   Two series-aiding inductors circuit schematic diagram.

Solution

The total value of inductance is equal to:

$$L_T = L_1 + L_2$$
$$= 1\,H + 1H$$
$$= 2\,H$$

Mutual Inductance

The previous example assumes that the two inductors are physically placed far enough apart so that none of the magnetic field created around the first inductor will cut through the second inductor. If any of the magnetic flux lines do cut through the second inductor, there is said to exist some *mutual inductance* between the two inductors. This must be taken into account in the computation of equivalent inductance value. We will define mutual inductance as L_M. It is difficult to figure the value of mutual inductance, and it is beyond the scope of this text to do so. For our purposes, the value of mutual inductance will be given in problems. Coils with mutual inductance are said to be *inductively coupled*. Another method that can be used to minimize the mutual inductance between inductors that must be placed close together in a circuit is to mount them perpendicular to each other. More on mutually coupled inductors will be presented in Chapter 13.

Equation 11.3 is used when two inductors are wired in series, with mutual inductance, is:

$$L_T = L_1 + L_2 + 2(L_M) \tag{11.3}$$

Example 11.7

Assume the two inductors used in this example are moved close enough together so they now have 0.5 H mutual inductance. Find the total equivalent inductance value.

Solution

$$L_T = L_1 + L_2 + 2(L_M)$$
$$= 1\ H + 1\ H + 2(0.5\ H)$$
$$= 3\ H$$

Series-opposing Inductors

If the inductors are wired so that their windings are in opposite directions, the connection is considered to be series opposing. Equation 11.4 is used to find total inductance with series-opposing inductors is:

$$L_T = L_1 + L_2 - 2(L_M) \tag{11.4}$$

11.7 INDUCTOR RATINGS

Inductors are rated using several important values. These include:

The value of inductance: Large power inductors may be in henries, IF inductors are commonly in millihenries (*mH*), and RF inductors are commonly rated in microhenries (*μH*).

The DC current rating: This depends on the gauge and length of wire used to manufacture the inductor. Large inductors can have a significant value of DC resistance, often several hundred ohms.

Frequency range: An inductor should never be used at a frequency range other than that for which it was designed, as it will either not provide the required inductance value, or will overheat due to excessive power losses, and be damaged.

Test voltage: An inductor should never be operated in a circuit in which this voltage value will be exceeded. Failure to observe the rated working voltage may cause arcing between coils of wire or from the wire to the core. This will cause insulation breakdown and possibly destroy the inductor.

11.8 PRACTICAL INDUCTOR CIRCUITS

The next sections show some practical uses for inductors, including their typical uses in a PC.

11.8.1 Power Filter Inductors

One of the most common applications for large power inductors, in the past, was in AC-to-DC power supplies. Chapter 12 describes AC, or alternating current, in more detail. For the purpose of this chapter, it is enough to say that AC simply means that current flows first in one direction in a circuit, and then reverses direction. This pattern is repeated many times each second. As explained in Section 11.4, this repetition of the current reversing is called frequency. The common frequency of AC in the United States is 60 times per second. Electronic equipment, however, predominantly requires direct current to properly operate. So an AC-to-DC power supply is used to provide the required DC from the commercially available AC line current. The job of the *filter inductor,* also called a *filter choke,* is to greatly reduce the variations in current supplied to the DC equipment by the DC power supply. It accomplishes this because it temporarily stores and then returns energy to the circuit. Without the filter inductor, the reversing property of AC causes variations called *ripple* in the DC output from the supply. Power filter inductors greatly reduce this AC ripple. Figure 11.13 shows the use of a filter inductor in an AC-to-DC power supply. Modern DC power supplies do not normally use large inductors for filtering. Instead they use smaller, more reliable and much cheaper integrated circuits to accomplish the ripple reduction.

FIGURE 11.13 DC power supply filter inductor.

11.8.2 Other Inductor Uses

Other uses for inductors are presented next.

Tuned Circuits

Inductors are still used in some radio, television, and other communication circuits to perform frequency-selective functions. A radio is tuned by adjusting components in a frequency-sensitive circuit, or circuits optimized for a particular frequency. In the past, virtually all these circuits used individual inductors, plus other components such as capacitors to select the operating frequency. Most modern versions of these circuits are now built using integrated circuit (IC) technology.

Limiting the Rate of Current Change

Large-scale industrial use of inductors includes limiting the rate of current change through solid-state components that would otherwise be damaged or destroyed. Many loudspeaker systems still employ inductors as part of the system that channels the proper musical tones (frequencies) to the respective speakers. Inductors are used in these *crossover* systems to prevent high-frequency (treble) notes from entering the low-frequency speaker (woofer). Since the large mass of a woofer's cone prevents it from vibrating at high frequencies, the high-frequency energy applied to a woofer magnetically saturates and heats its voice coil, thus lowering the efficiency of the speaker. Figure 11.14 shows a crossover circuit. More on crossover systems will be presented in Chapter 12.

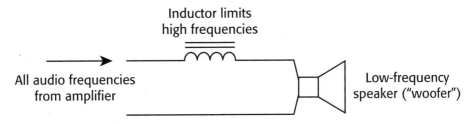

FIGURE 11.14 Inductor used in a speaker crossover system to suppress high frequencies from getting to the woofer.

Automotive Applications

The "coil" in a car's electrical ignition system is really a large autotransformer, a single tapped coil of wire with a laminated-iron core. A modern automobile operates on 12.6 volts DC. The "coil," operating at a low DC voltage, can provide anywhere from 10,000 to 40,000 volts, depending on the model and age of car, to operate the spark plugs. Modern car engines use a high-energy ignition system (HEI) that supplies about 40 *kV*, resulting in cleaner-burning engines with less tailpipe pollution. Spark plugs are used to ignite the gasoline/air mixture in the cylinders inside the engine block. This process of very high-voltage generation is accomplished by the *points* (a momentary contact switch within the distributor) turning off the current through the coil. The instant the current in the coil stops, a large magnetic field around the coil collapses, and the rapid cutting of the hundreds of turns of coil wire by the field induces the short, high-voltage pulse that is fed to the appropriate spark plug to ignite the fuel mixture. Modern car engines use a coil at each spark plug to greatly reduce high voltage loss that was common using the old-fashioned ignition wires, which were up to four feet long. Newer vehicles continue to use a common distributor—a form of high-voltage rotary switch. Figure 11.15 shows a picture of an automobile coil.

Current Smoothing

Another use of inductors in automotives is filtering the DC power used by a car stereo. The car's alternator, used to provide electrical energy for keeping the DC battery charged, is actually an AC generator. It produces a constantly varying current, and this ripple must be filtered out to provide a smooth, constant direct current. Most car radios contain a small DC filter inductor inside their case that prevents the radio from producing hum in the speakers. Figure 11.16 shows a DC filter schematic for a car radio.

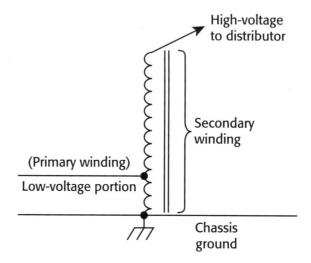

FIGURE 11.15   Automobile coil " schematic.

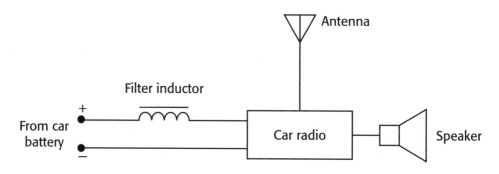

FIGURE 11.16   Car radio DC filter inductor circuit.

PC Applications

Inductors are an integral part of the switching power supplies used in both a PC's main power supply as well as on the motherboard. For example, the +12 VDC supplied to the motherboard via the new 4-pin cable on ATX +12 PC power supplies actually runs a second power supply on the motherboard. This second switching supply converts the +12 VDC down to + 3.3 VDC to run the modern processors. The inductors are used to smooth out the current. Refer to Figure 11.17 for a picture of a PC power supply inductor, and Figure 11.18 for an inductor used in a PC motherboard switching power supply.

FIGURE 11.17 An inductor used in a PC's main power supply.

FIGURE 11.18 An inductor used on a motherboard power supply.

11.9 SAFETY CONSIDERATIONS FOR INDUCTORS

Inductors should never be operated at voltages exceeding their design value, called its *working voltage*. This parameter is usually printed on the side of large power inductors, or listed on the manufacturer's spec sheet. Since inductors can produce very high CEMF the instant current in a circuit is interrupted, always exercise extreme care when working around inductor circuits. Exceeding an inductor's rated current value can cause the winding to overheat, and possibly fail open. If an inductor is operated beyond its design current value, the core will become saturated. *Saturation* means the core cannot accept any more magnetic flux. An increase in current, and therefore flux beyond the saturation point, causes problems. One such problem is the creation of interference to other circuits in the form of radiated electromagnetic energy at new frequencies. The core is said to become nonlinear, and the inductor can cause interfering frequencies called spurious harmonics (spurs) to be created, which can cause problems to other devices operating at radio frequencies. These new frequencies are the result of the core's saturation, and they can travel a fair distance by direct radiation to affect other circuits. Some types of low-frequency inductors have an air gap in the core to prevent saturation. Figure 11.19 shows an inductor core with an air gap. Interestingly, the read/write heads on hard drives use an electromagnet, essentially an inductor, with an air gap. The gap allows the magnetic flux to "leak out" enough to affect the recording medium (the disk's magnetic surface coating).

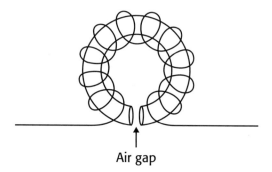

Air gap

FIGURE 11.19   An inductor air gap.

Sometimes electronic technicians and hobbyists must build an inductor to meet a certain design value. They may not be able to obtain the required unit readily available or "off the shelf." This situation often occurs in prototype design. As you will recall, a prototype is the first circuit built to test its functionality before full-scale production begins. Inductor cores are manufactured with a specific permeability value. Different materials are used for different frequency ranges. Using the wrong core type to construct an inductor will result in the wrong inductance value. Different core alloys, called *mixes*, are used, depending on the required frequency and power level.

PLATE 1 (FIGURE 1.14) A basic PC repair tool kit.

PLATE 2 (FIGURE 12.26) A bank of six power-factor-compensating capacitors mounted on an electric utility power pole cross beam.

PLATE 3 (FIGURE 10.5) Hard disk read/write head assembly.

PLATE 4 (FIGURE 4.6) Motherboard SPST jumpers.

PLATE 5 (FIGURE 1.3) A PC motherboard.

PLATE 6 (FIGURE 4.23) Geothermal electric power generating sites. Photo courtesy of Pacific Gas and Electric Company.

PLATE 7 (FIGURE 5.4) Back-probing the power supply connector in order to measure power supply voltages.

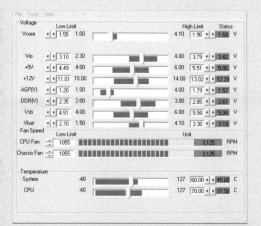

PLATE 9 (FIGURE 14.13) A hardware monitoring program that is invoked from the desktop.

PLATE 8 (FIGURE 10.4) A hard drive R/W head actuator magnet. It is the semicircular piece shown in the lower left part of the drive.

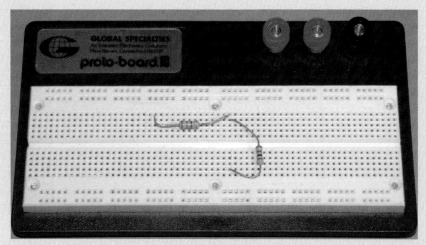

PLATE 10 (FIGURE 4.45) A breadboard circuit.

PLATE 12 (FIGURE 4.20) A light emitting diode (LED).

PLATE 11 (FIGURE 11.34) Integrated circuit bypass capacitors are the small rectangles labeled "c" with a number.

PLATE 13 (FIGURE 3.3) Microprocessor heat sinks.

11.10   STRAY INDUCTANCE

As explained earlier, all conductors have inductance. In low-frequency circuits, such as DC and most audio circuits, this inductance is not a serious problem, and for the most part, can be ignored. However, as the frequency of signals increases, the physical length of connecting wires in circuits, and even on circuit boards becomes critical. As you probably know, computers operate at very high speeds. They calculate using rapidly occurring signals. These are high-frequency signals, which attempt to cause current to change rapidly. Inductance acts to limit the rate of current change in a circuit. That is the reason that integrated circuits are constantly being made smaller—not only to pack more devices into a given area, but also to lessen the effects of lead inductance between individual IC chips. Circuits that operate very fast need to have all their components as close together as possible to limit signal degradation due to stray lead inductance. As an example, assume a computer signal is supposed to consist of square waves. If lead inductance is too great, the square waves will be severely distorted, as shown in Figure 11.20.

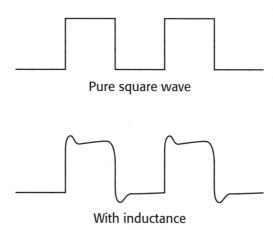

Pure square wave

With inductance

FIGURE 11.20   Square wave signal degradation due to lead inductance.

11.11   CAPACITORS

Capacitors (*caps*, for short) work by temporarily storing energy in an electrostatic field (a field of charges at rest). Like inductors, they are used in many different types of circuits to perform many different functions.

Capacitors store energy in an electrostatic field.

Key Ideas

11.11.1 A Brief History of Capacitors

Capacitors, in one form or another, have been around for a long time. The next section presents some of the early forms of capacitors.

The Leiden Jar

In the mid-1700s, a scientist placed thin metal foil on both the outside and inside of a glass bottle. He used a cork, (a good insulator) with a metal rod through the middle of it. A small metal chain inside of the bottle connected the metal rod to the inside foil. The top of the metal rod on the outside of the glass bottle had a metal ball attached to it. The ball on top and the outside metal foil were electrodes. The glass functioned as a good insulator, separating the inner and outer electrodes. In 1745 Pieter van Musschenbroek, a physicist and mathematician in the city of Leiden, Netherlands, invented a cheap and convenient source of electric sparks. He was able to deposit an electrical charge on the foil plates of the jar device, and later he could cause a spark to jump across the connection between the ball and the outside foil by placing a conductor across the two points. People did not know the nature of electricity at that time. Early experiments such as the one described here led some people to believe that electricity was an invisible fluid. By charging the bottle with this "fluid" they thought that somehow they had managed to "condense" some of the electric fluid in the bottle. The device, now commonly known as a Leiden jar, was an early form of capacitor. In those times people referred to the device as a condenser, since they believed that the jar was somehow magically able to "condense" the invisible electric "fluid" from the air. That is the origin of the term condenser for a capacitor. Sadly, many people still use the term "juice" for electricity today. Automobile mechanics still refer to the small capacitor inside a car engine's distributor housing as a "condenser." Old ideas are not easily abandoned. See Figure 11.21 for a picture of a Leiden jar.

Capacitors are devices used to temporarily store an electrical charge. They do this by storing energy in an electrostatic field between two electrically conducting *plates*. The plates are usually made of metals such as aluminum, brass, or copper. Certain types of capacitors use conductive chemical solutions, called *electrolytes*. All capacitors consist of at least two conducting plates, plus an insulator between the plates called the *dielectric* material. Typical dielectric materials used in capacitors include waxed paper, plastic, glass, ceramics, air, oil, and special high-tech materials such as Mylar®, a type of plastic film.

A relative rating of how well a given material supports an electric field, compared to air, is known as the *relative dielectric constant*, and is represented by the

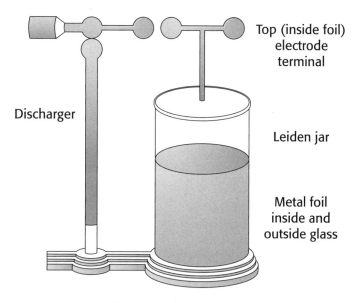

FIGURE 11.21   A Leiden jar, an early form of capacitor.

symbol ε (the Greek letter "epsilon"). Air is considered to have a relative dielectric constant of 1. Some materials have dielectric constants many times that of air.

11.11.2   Capacitance

Capacitance can be defined several ways. First, capacitance is usually thought of as the capability of storing a charge on the plates of the capacitor. However, the modern view of precisely how capacitors store a charge focuses on the dielectric material, rather than the plates. It is thought that the electron orbits around the nucleus of the atoms within the dielectric material in a capacitor are influenced by the electric field existing between the plates. The electrons, which have a negative electric charge, are pushed away from the negative capacitor plate, and are attracted to the positive plate. It is believed that the energy storage occurs due to this change in the electrons' orbit. The energy is stored in the electrons' displacement from their usual paths around the atomic nuclei in the dielectric material. Figure 11.22 illustrates this idea. When a capacitor is discharged, the electrons' orbits return to normal.

The second method of describing capacitors is by their action in a circuit. All capacitors tend to oppose a change in voltage across their terminals. This means it is impossible to instantly change the voltage across a capacitor. Capacitors are often

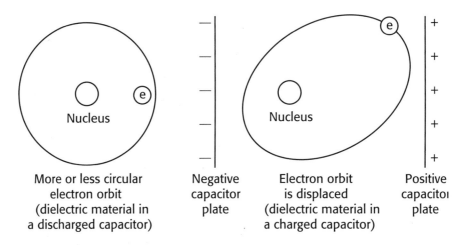

FIGURE 11.22   Electron orbit displacement in a charged capacitor's dielectric material.

used in circuits that make use of this capability. They are also used to form filter circuits that can discriminate between different rates of voltage change (frequencies).

Finally, capacitance is defined mathematically as charge quantity per potential difference, or coulombs/volt. A capacitance of 1 farad is equal to 1 coulomb per volt.

Capacitors oppose a change in voltage.

Key Ideas

Factors Affecting Capacitance

The specific value of capacitance depends on several things:

Plate area: Larger plate area means more capacitance.

Plate spacing: The closer the plates, the greater the electric field effect between them, therefore the greater the capacitance.

Dielectric material: Different materials used between the plates are able to support different electric field strengths.

Temperature: Some capacitors gain capacitance as the temperature rises, which is known as *positive temperature coefficient*. The oppposite effect is known as a *negative temperature coefficient*.

Capacitance can be defined by Equation 11.5:

$$C = kA \div t \qquad (11.5)$$

where:

C equals coulombs/volt.

k equals the relative dielectric constant of the material.

A equals the area of overlay of the plates.

t equals the thickness of the dielectric material.

11.12   CAPACITOR TYPES AND DESIGNATIONS

Capacitors are grouped into those that can be varied over a range of capacitance, and those that have a fixed value of capacitance. They are further designated as to shape, frequency range, and intended circuit use. The types of material used to construct the plates and the dielectric is also used to designate capacitors. The next sections present some of these categories.

11.12.1   Fixed and Variable Types

The two general categories into which capacitors are placed are *fixed* and *variable*. Fixed capacitors have a single value of capacitance, which cannot be changed or adjusted. Variable capacitors can be adjusted to change their capacitance over some limited range.

In addition to the general categories of fixed and variable, capacitors are further described by combinations of how they are used in a circuit, how they are constructed, and in some cases, even by how they look.

11.12.2   Named by Circuit Function

Some of the names given to capacitors according to their action or intended use in an electronic circuit include tuning, trimmer, filter, coupling, compensating, crossover, neutralizing, energy storage, standoff, feed-through, resonating, padder, and transmitting. A very common example of a tuning capacitor is that used to tune a radio receiver. The tuning, which means selecting the desired radio signal, is accomplished by adjusting the capacitance value of an air-dielectric variable capacitor. This changes the frequency or station received by the radio. A very small variable capacitor used to carefully fine-tune a circuit is called a *trimmer* capacitor. Trimmer capacitors are often used with a much larger capacitor, such as the one used to tune the radio. They are usually adjusted only at the factory, in a process known as *alignment*.

11.12.3   Named by Shape

Some examples of capacitor shapes include can, tubular, disk, chip, doorknob, bathtub, and piston. Figure 11.23 shows examples of various capacitor shapes.

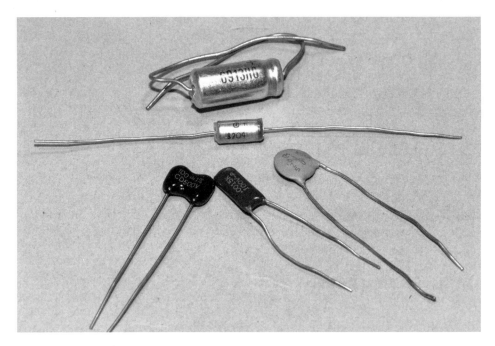

FIGURE 11.23   Capacitor outline shapes.

11.12.4   Named by Dielectric and/or Plate Materials

Capacitors are also given names that reflect the materials used to construct the plates and dielectric insulating material. Names such as: ceramic, plastic, paper, Mylar, polystyrene, (a special type of plastic), silver-mica, air, vacuum, oil, glass, and electrolytic (discussed next), refer to some of the capacitor types in use today.

11.12.5   Electrolytic and Nonelectrolytic Capacitors

A final method of classifying capacitors, into two types, is by the general material used in the dielectric. A nonelectrolytic capacitor uses the types of materials listed in the preceding section: glass, paper, plastic, ceramic, Teflon, vacuum, and so on. These types of capacitors can be connected in an electric circuit without regard to polarity, with no risk of damage. There is no positive or negative terminal of these capacitors. Another designation for this type is *electrostatic* capacitor.

Electrolytic capacitors are made differently. They contain a paste solution of chemical-soaked gauze material sandwiched between two plates made from aluminum or some other conducting metal. During manufacture, a DC voltage is applied across the plates, in a process called *forming*. When an electrolytic capacitor is formed, the chemical solution in the gauze chemically reacts with the plate material. The surface of the plates becomes etched, which increases the effective surface area. This causes the capacitance value to increase greatly. During the forming process, a very thin oxide layer forms on one of the plates. The oxide is a very good insulator, and since this layer is so thin, it decreases the plate spacing. This close plate spacing effect further increases the capacitance of the electrolytic capacitor. The overall effect of the electrolytic manufacturing process creates a unit that has many times the capacitance of a similar sized, nonelectrolytic capacitor. The two main types of electrolytic capacitors are aluminum and tantalum. These are the metals used to form the insulating oxide layer.

11.12.6   Leakage Current

This increased capacitance has some drawbacks. The electrolytic solution is not a perfect insulator, and there can be a significant amount of current between the plates, called *leakage* current. The leakage current is many times greater than for a nonelectrolytic capacitor. Electrolytic capacitors are also more expensive. Their capacitance value can decrease with use, age, and excess heat. The electrolyte dries out, causing loss of capacitance. The large electrolytic capacitors commonly used in the power supplies of radios and stereos often lose much of their capacitance as they age. The function of these capacitors is to smooth out, or filter the variations in the DC voltages of the power supply, called ripple voltage. They are termed *filter* capacitors for this reason. When they age, the decreased value of capacitance becomes less effective in *attenuating* (opposing or holding down) the variations in the DC applied to the amplifiers (tubes or transistors). This reduced attenuation allows the ripple to be amplified by the tubes or transistors in the radio, and we hear the resulting *hum* in the loudspeaker.

Because the electrolytic capacitors are formed with a definite voltage polarity applied to each terminal, these units must always be used in a circuit with proper respect to the polarity of connections. Failure to observe proper polarity will result in catastrophic failure of the capacitor. Electrolytic capacitors must be used within their specified voltage and temperature range, as well. Electrolytic capacitors used improperly tend to explode as the capacitor generates gas pressure within, which cannot easily escape the sealed container. Always use extreme caution when working around electrolytic capacitors.

11.13   THE CAPACITOR CHARGE CIRCUIT

Refer to Figure 11.24. Assume that both capacitor plates have an equal number of free electrons. There is no overall potential difference existing between the plates.

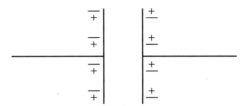

FIGURE 11.24 Charges on capacitor plates.

The dielectric material between the plates serves as an insulator, and for all practical purposes, we will assume there can be no current through the dielectric. When the switch is closed, excess electrons are forced out of the negative source terminal, and they flow onto the capacitor plate connected to it. These electrons exhibit an electric field that repels electrons on the opposite capacitor plate. Those electrons are also attracted to the positive source terminal, and so flow to the source. For each electron that flows onto the plate connected to the negative source terminal, there will be one electron repelled from the opposite plate. The overall effect is to cause a shift of electrons in the circuit, called *displacement current*. Remember, no current flowed *through* the capacitor, yet current did flow in the capacitor circuit.

While the electrons flow in the circuit, a voltage builds up across the plates of the capacitor due to the increasing electric field between plates. At some point in time, electron flow stops. That point is reached when the voltage across the capacitor plates is exactly equal to the source voltage. The capacitor is said to be charged at this point. The switch may be opened, and a voltage will be seen to exist across the capacitor equal to the source voltage, even though the source is no longer connected. The capacitor will store this potential difference across its plates for some time. A perfect capacitor would store the voltage forever. However, real capacitors don't have perfect dielectric material, and some small amount of leakage current will flow. Nonelectrolytic capacitors, as explained earlier, have very little leakage current, and many capacitors can store a voltage for days, or weeks. A technician should be aware of this since a very real shock hazard exists even with equipment that has been turned off an unplugged for some time.

11.13.1 The Charge Curve

Plotting the current that flows in the capacitor circuit as the capacitor charges results in a graph like the one shown in Figure 11.25.

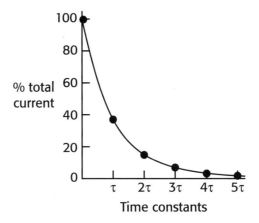

FIGURE 11.25   Capacitor charge current graph.

The current flow is at maximum the instant the switch is closed, connecting the source to the capacitor. As current flows over time, the rate decreases, until at full charge, the capacitor will no longer accept any further electrons. This action is similar to the charging of a cell or battery, although the internal action differs.

The voltage existing across the capacitor can be plotted over time. Figure 11.26 shows that as the current flow decreases, the voltage increases until the capacitor voltage equals the source voltage.

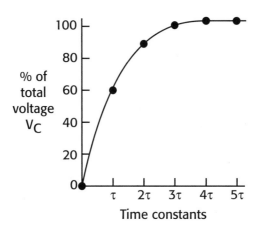

FIGURE 11.26   Capacitive charge voltage graph.

11.13.2   Capacitive Time Constant

Figure 11.27 shows the curves for current and voltage drawn on the same scale. These curves should look familiar. These are another example of the universal time-constant curve set. Notice that the X-axis is calibrated for time constants. The capacitor obeys the same characteristics, as do inductors. With capacitors, the main interest is the voltage across them, whereas with inductors the main concern is with current through them.

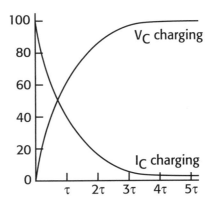

FIGURE 11.27   Capacitor voltage versus current.

11.13.3   RC Charge Problems

Circuits containing a resistor and a capacitor are called *resistor-capacitor* or *RC* circuits. This section presents examples of RC circuits.

Example 11.8

Refer to Figure 11.28. What happens in this circuit when the switch is closed? Assume the capacitor was completely discharged initially, so it had 0 volts across its plates. When the switch is closed, current flows in the circuit, and voltage builds up across the capacitor. Now find what voltage exists at the various time-constant points.

Example 11.9

Refer to Figure 11.28. Find the capacitor voltage at each of five time-constant points, starting from t_0. Assume the capacitor is discharged initially.

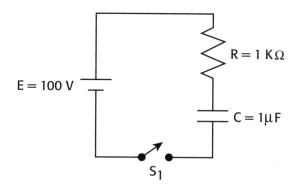

FIGURE 11.28 Circuit for Example 11.8 and 11.9.

Solution

1. As with any time-constant problem, begin by computing the circuit time constant.

2. So, in 1 ms, the capacitor voltage V_c will have risen to 63.2% of the source voltage, or

$$V_c = 100 \ V \times 0.632$$
$$= 63.2 \ V_c$$

3. At the two-time-constant point, the voltage V_c will have risen to 86.5% of the source voltage, or:

$$V_c = 100 \ V \times 0.865$$
$$= 86.5 \ V$$

4. After three time constants, or 3 ms, the capacitor voltage will be:

$$V_c = 100 \ V \times 0.95$$
$$= 95 \ V$$

5. After five time constants have passed, capacitor voltage will be

$$V_c = 100 \ V \times 1.00$$
$$= 100 \ V$$

At this point, the capacitor is fully charged.

Example 11.10

Refer to Figure 11.29. Compute the circuit time constant, and find the voltage across the capacitor, V_C, after 4.65 ms. How much time will it take for the capacitor to charge to 48 volts?

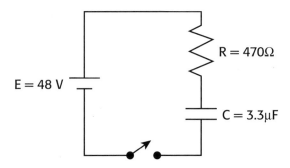

FIGURE 11.29   Example 11.10 circuit.

Solution

1. The circuit time constant is found with Equation 11.6:

$$t = RC \qquad\qquad (11.6)$$
$$= 470 \ \Omega \times 3.3 \ \mu F$$
$$= 1.55 \ ms$$

2. Find the number of time constants in 4.65 ms.

$$4.65 \ ms \div 1.55 \ ms = 3 \ TCs$$

3. Use the time-constant chart to find the percent of maximum voltage across the capacitor at that point.

$$3 \ TCs = 95\% \ \text{maximum voltage}$$

4. Multiply the percentage found in the chart by the source voltage.

$$48 \ V \times 0.95 = 45.6 \ V$$

5. It will take five time constants for V_c to reach 48 volts.

$$5 \ TCs = 5 \times 1.55 \ ms$$
$$= 7.75 \ ms$$

Key Ideas

Electrolytic capacitors have a considerable amount of leakage current. They may never charge to equal the source voltage because of this. The amount of leakage current generally increases with the age of the capacitor. Eventually, the leakage current can become sufficient to cause the capacitor to fail, short out, and possibly explode.

11.14 CAPACITIVE CIRCUIT WITH DECREASING CURRENT

The following examples describe the discharge action of capacitors.

Example 11.11

Refer to Figure 11.30. This is the same circuit as Figure 11.29 with the addition of a separate resistor path. This circuit can be used to demonstrate capacitor discharge action. Assume that the switch was in the position shown in Figure 11.30 (A) for longer than five time constants. Then the switch was moved to the position shown in Figure 11.30 (B). The second resistor, R_2, will now provide a path for current to flow out of the capacitor. It will take five time constants for the capacitor to fully discharge.

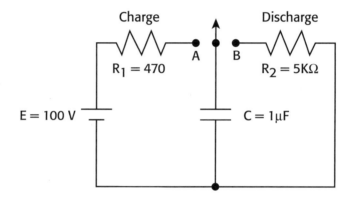

FIGURE 11.30 Example 11.11 circuit.

Example 11.12

Compute the discharge time constant of the circuit, and find the value of V_c at each of the five time-constant points. Assume that $V_c = 100\ V$ initially.

Solution

1. The time constant of the discharge path is found with Equation 11.7:

$$TC = R_2 C \qquad (11.7)$$
$$= 5\ k\Omega \times 1\ \mu F$$
$$= 5\ ms$$

2. After one time constant (5 ms), the voltage on the capacitor is 36.8% of the initial capacitor voltage, or:

$$V_C = 100\ V \times 0.368$$
$$= 36.8\ V$$

3. After two time constants (10 ms), the capacitor voltage has decreased to 13.5% of the initial capacitor voltage, or:

$$V_C = 100\ V \times 0.135$$
$$= 13.5\ V$$

4. After three time constants (15 ms), the capacitor voltage has decreased to 5% of the initial capacitor voltage, or:

$$V_C = 100\ V \times 0.05$$
$$= 5\ V$$

5. After four time constants, the capacitor voltage has decreased to only 2% of the initial voltage, or:

$$V_C = 100\ V \times 0.02$$
$$= 2\ V$$

6. After five time constants, the capacitor has completely discharged to zero volts. It has taken:

$$5TCs = 5 \times 5 \ ms$$
$$= 25 \ ms$$

11.15   SERIES AND PARALLEL CAPACITORS

Capacitors are often used in series or parallel connection. Series capacitors are treated similarly to resistors in parallel.

11.15.1   Series Capacitors

Overall values of capacitors in series are solved for similarly to resistors in parallel. The overall value is found by using the product-over-the-sum equation.

Example 11.13

Refer to Figure 11.31. Find the equivalent value of the series capacitors.

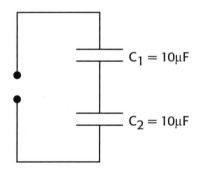

FIGURE 11.31   Example 11.13 circuit.

Solution

The formula for finding the equivalent value of two series capacitors is:

$$C_T = (C_1 C_2) \div (C_1 + C_2) \tag{11.8}$$

For this circuit:

$$C_T = (10 \ \mu F \times 10 \ \mu F) \div (10 \ \mu F + 10 \ \mu F)$$
$$= 5 \ \mu F$$

A shortcut can be used in this case, since all capacitor values are in microfarads.

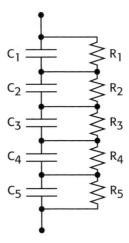

(Remember, the answer is in microfarads.)

Since in this case the capacitor values are the same, the voltage will be dropped equally across each capacitor. If capacitors rated for 10 WV (working voltage—see Section 11.16 for more information) are used, then the parallel connection would allow 20 volts across it with no damage.

When unequal capacitance value capacitors are used, however, the smallest capacitance unit will drop the largest voltage. Since capacitors, especially electrolytics, tend to age at different rates, they will tend to all have a slightly different capacitance value. Often voltage-equalizing resistors are connected across series-wired units to help evenly distribute the voltage. Refer to Figure 11.32.

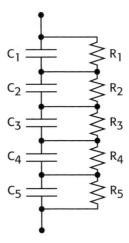

FIGURE 11.32 Voltage equalizing resistors across series capacitors.

Series capacitor circuits are usually employed to raise the working voltage. For example, a given circuit requires 10 μF of capacitance at a working voltage of 1,000 volts. Capacitors of this μF capacitance range are difficult to find at working voltage ratings in excess of 350 volts or so. One possible solution would be to use 10 series-wired capacitors each having a rating of 100 μF at 150 *WV*. Such a series connection would yield an equivalent value of 100 $\mu F \div$ 10 units = 10 μF at a working voltage value of 150 $V \times$ 10 units = 1,500 V. It is good engineering practice to allow the extra margin of working voltages on such a high-voltage circuit. It would also be advisable to use voltage-equalizing resistors of approximately 470 $k\Omega$ each across each capacitor. The choice of resistance value is a tradeoff between a lower value, which would cause excessive current drain, and a higher value, which would have little equalizing effect. Since electrolytic capacitors often lose capacitance value at different rates as they age, using the parallel-connected voltage-dropping resistors helps counteract the differential voltage drops that would normally occur on nonprotected series-wired electrolytics.

Key Ideas

Series capacitor connections are used to provide higher working voltage values at lower capacitance value than a single unit. Parallel capacitor connections are used to provide increased capacitance, with no increase in working voltage, compared to a single unit.

11.15.2 Parallel Capacitors

Values of capacitors wired in parallel are solved for similarly to resistors in series. That is, the total capacitance value is the sum of the individual capacitance values.

Example 11.14

Find the equivalent capacitance value of the parallel capacitors in Figure 11.33.

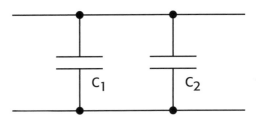

FIGURE 11.33 Two parallel capacitors.

Solution

Equation 11.9 is used to find parallel equivalent capacitance for two capacitors:

$$C_T = C_1 + C_2 \tag{11.9}$$
$$= 10\ \mu F + 10\ \mu F$$
$$= 20\ \mu F$$

Key Ideas

Since the capacitors are connected directly across the source, the voltage rating for the parallel connection is that of the source voltage. Capacitors connected in parallel are often used to increase the overall capacitance value. The working voltage rating is the same as for a single capacitor.

11.16 CAPACITOR RATINGS

Capacitors are rated according to several factors. The primary value is capacitance, rated in farads. A *farad* is the basic unit of capacitance. It is derived from the name of an English scientist, Michael Faraday. One farad is a very large unit of capacitance. Common metric subunits of capacitance are the *microfarad*, (μF), equal to 10^{-6}, the *nanofarad*, (ηF), equal to 10^{-9}, and the *picofarad* (ρF), equal to 10^{-12}. Capacitors of 1 F or more are used as backup batteries for some motherboards, and as "voltage stiffening" capacitors when connected in parallel across the DC power leads of high power car stereo amplifiers.

The second most important rating is voltage. There are several voltage values used. The most common is called the *working voltage*, abbreviated *WV*. This is a DC rating, so it is common to see the letters WVDC or a similar designation used. A capacitor should never be operated in a circuit in which this voltage value is exceeded. Some capacitors also carry a voltage value called *test voltage*. Test voltage should not be considered to be the safe working voltage value.

Capacitors are also designed to work over specific temperature ranges. Electrolytic capacitors, as previously explained, are the most sensitive to temperature, and often have their working temperature ranges printed on their side.

Size and shape are also important, especially in today's modern small circuits and systems. The higher the working voltage and the higher the capacitance, the larger the capacitor package. Technicians should keep in mind that it is always a tradeoff between these factors when deciding on using a particular unit. Replacement capacitors must fit into the same space, and have the same lead arrangement as the original unit. A capacitor designed with axial leads will not be the proper one for replacing a radial-lead unit designed for printed circuit board use. As presented earlier in the text, components that have one lead coming out of the component body at each end, on the same center-line axis, are called *axial-lead* components. Those that have leads coming out of one end, but all on the same radius, are called *radial-lead* components. This is the type of lead arrangement used for electrolytic capacitors intended for vertical mounting on circuit boards in order to save board space.

11.17   PRACTICAL CAPACITOR CIRCUITS

Capacitors are used to accomplish a variety of things. Capacitors can be used to block the flow of direct current, even while allowing the flow of AC in the same circuit. Actually, capacitors do not allow current to flow through them, but in AC circuits, current flows onto and off of the plates, back and forth at the frequency of the AC. This is called *displacement current*. Capacitors act as short-duration cells or batteries, and are used in some circuits to provide a temporary power backup for low-current circuits such as those used in computers. Recent advances in manufacturing techniques have allowed the construction of 1-farad electrolytic units, which is an extremely large amount of capacitance. Such a device can be considered a storage cell for backup purposes in computers. Because they can store so much energy, careful handling is required, for a short placed across their terminals can cause burns.

11.17.1   Power Filter Capacitors

Capacitors are commonly used in both linear and switching power supplies that convert AC to DC. Switching supplies are used in the main PC power supply as well as secondary power supplies on the PC motherboard, such as the one used to convert +12 VDC down to 3.3 VDC or 2.8 VDC and 1.8 VDC used by modern microprocessors. When used in this manner, they are commonly termed "filter capacitors," since they act to filter out the remnants of the AC, called ripple voltage, and pass the DC voltage. Refer to Figure 11.34. You should notice the use of the capacitor as a ripple reducer. The capacitor acts to maintain a constant voltage across its terminals. So when the input waveform decreases, the capacitor discharges into the circuit to make up the difference. More on the use of capacitors to reduce ripple in power supplies is presented in Chapter 14, "Power Supplies."

11.17.2   Other Capacitor Uses

As you will see in Chapter 16, "Communications," capacitors are used in circuits to provide frequency selectivity. Such uses are common in communication circuits. Capacitors are also used to provide a desired action on a waveform. Because they

FIGURE 11.34   Integrated circuit bypass capacitors are the small rectangles labeled "c" with a number.

serve to limit the rate of voltage change, they can be used for waveshaping. Capacitors are commonly used to reduce voltage transient spikes occurring on the voltage supply lines to integrated circuit chips. A small value capacitor bridging the voltage supply pin to the circuit board ground bus is used to bypass or short to ground these spikes, which could potentially corrupt data.

PC RAM

The most common memory types used in PCs is based on a type of transistor that operates in similar fashion to a very small value capacitor. DRAM and SDRAM use a family of transistors called complimentary metal oxide semiconductor (CMOS). The small capacitor exists between the gate structure and a conducting channel in this type transistor. This type of memory is very dense—a large quantity can be made in a relatively small volume—and relatively inexpensive, accounting for its widespread use. It does however, require a dedicated circuit to continually pulse the transistors with voltage to "refresh" the small capacitors; otherwise the contents of RAM will be lost. This refreshing happens very often, as the CMOS transistors can hold their digital state for only a few nanoseconds. The job of memory refreshing is the highest priority of the PC so it rightly gets the highest priority interrupt or IRQ.

11.17.3   Stray Capacitance

Just as the physical layout of wires has a certain amount of undesired inductance, as presented earlier in the chapter, undesired capacitance between conductors is also present. Depending on the frequency of circuit operation, this *stray capacitance* may be quite important. A strange thing happens when the operation of an inductor at very high frequencies is analyzed. At some frequency, an inductor will act in a predominantly capacitive rather than an inductive manner. That results from capacitance that exists between adjacent turns of the winding. This inter-winding capacitance will render the inductor useless beyond a certain frequency, depending on the physical construction of the inductor. Refer to Chapter 12 for more on frequency.

All high-frequency circuits are constructed to be very rigid, as any movement of components will result in a slightly different value of stray capacitance as the components shift positions. It is common to see heavy-gauge metal housings used as well as many screws holding the sides of enclosures together, to maintain rigidity.

11.18   SAFETY CONSIDERATIONS FOR CAPACITORS

The two main worries with using capacitors are voltage and temperature. Exceeding either of these will shorten the life of the capacitor. Electrolytic capacitors have the added concern of polarity. If connected backwards in a circuit, they tend to explode. If connected across a voltage that exceeds the rated working voltage of the capacitor, the capacitor can explode as well. Never attempt to replace a given capacitor with another having a lower working voltage or temperature rating. High-voltage capacitors can store a lethal charge for weeks. It is normal practice to discharge these types many times before touching them. It is also advisable to wire across the capacitor terminals after having thoroughly discharged the capacitor. This will help ensure that no voltage will build up on the capacitors. If you visit surplus electronic stores or electronics flea markets, you may notice some large, high-voltage capacitors. They often will have a wire connected across their terminals for safety.

Capacitors, especially electrolytics, don't like heat. Always allow plenty of air circulation around electrolytic capacitors. This is a major concern with the large numbers of electrolytic capacitors used on a typical PC motherboard as well as the main power supply. Always do a visual check for blistered or ruptured electrolytic caps on a motherboard.

11.19 SUMMARY

- Inductors oppose a change in current.
- Inductors produce a counter voltage or CEMF when current through them changes. The effect is to limit the rate of current change.
- Inductors are constructed based on the frequency range they must operate over, and the amount of current and voltage they must handle.
- To determine the total inductance in a circuit with inductors wired in series, add the total inductor values.
- Values of inductors wired in parallel are solved for in the same way as are resistors wired in parallel, using the "product-over-the-sum formula".
- Inductors are used to limit current change in power supplies.
- Inductors are used with capacitors to form tuned circuits.
- Inductors are used in stereo speaker systems to limit the high frequencies (treble) directed to the woofer. They are referred to as speaker crossover components when used like this.
- Inductors are used in a PC's main power supply as well as on smaller supplies on the motherboard, for the purpose of smoothing out current pulses from the supply.
- The main ratings for an inductor are its inductance value, the frequency range it is intended to operate over, the maximum current, the DC resistance, and the working voltage.
- Always exercise care when working around energized inductive circuits, as they can produce a very high voltage. This principle is used in automotive spark systems to produce over 40 kV.
- The henry is the unit of inductance.
- A capacitor consists of two or more electrodes in the form of plates that can be made of metal, metal film, or even a conductive solution, and an insulating dielectric material between the plates. The charge is stored on the plates and places electrical stress on the dielectric material. One view of how capacitors store a charge is that the electron orbits of the dielectric material become offset

from normal. When the capacitor discharges, the electron orbits return to their former positions.

■ An electrolyte is a chemical conductive solution used in electrolytic capacitors between metal foil plates to increase the capacitance by forming a very thin insulating layer on one of the plates.

■ A dielectric is a material used between capacitor plates as an insulator that supports electric field flux.

■ To charge a capacitor means the action of developing a voltage across the plates of a capacitor. Electrons flow onto the negative plate and away from the positive plate.

■ To discharge a capacitor means the action of removing the voltage across the plates of a capacitor. Electrons flow off the negative plate and onto the positive plate until the number of electrons on the two plates is equalized.

■ The capacitive RC time constant is the time required for the voltage on a capacitor to change by 63.2% of the difference between its present value and some ultimate value that is normally either the source voltage value or zero.

■ The farad is the basic unit of capacitance, named after the English scientist, Michael Faraday. One farad is equal to one coulomb of charge developing one volt of potential on a capacitor.

■ Capacitors should never be connected to a voltage source that exceeds the capacitor's working voltage.

■ Electrolytic capacitors are polarized and must be connected properly in a circuit, or else they can explode. Special types of nonpolarized electrolytic capacitors are made using a special internal connection consisting of two polarized electrolytic capacitors. This nonpolarized type is commonly used for speaker crossover and AC electric motor-starting uses.

■ To determine the capacitance value of capacitors wired in series, add the values of the individual capacitors. The voltage appearing across each capacitor must be determined using the series-capacitor voltage formula, so as to not exceed the capacitor's working voltage rating. In general, the smallest capacitance value capacitor in series will drop the highest voltage.

■ Capacitors are used to perform all sorts of electronic jobs such as voltage storage, voltage stabilization, bypassing noise, and as part of tuned circuits.

■ Capacitors can be found in nearly all types of power supplies and on all the circuit boards in a modern PC, and network hub, switch, and router.

■ Capacitors store energy in an electrostatic field in the dielectric material.

■ Electrolytic capacitors need airflow around them to avoid failure caused by overheating.

11.20   KEY TERMS

Inductors
Inductance
Frequency
Core
Eddy current
Hysterysis
Self-induction
Induced voltage

CEMF
Steady-state
Time constant
Practical time constant
Mutual inductance
Transient
Saturation
Plates

Electrolyte
Dielectric
Dielectric constant
Charge
Discharge
RC Time constant
Farad
Electrolytic

11.21   EXERCISES

1. The capability of opposing a change in current is called:

2. The action of inducing a voltage in a conductor when there is relative motion between the conductor and a magnetic field is called:

3. The number of times a voltage or current varies, or repeats during a certain time interval is called:

4. What is the name given to the material placed inside an inductor in order to increase the concentration of flux lines?

5. Undesired currents that flow through an inductor core are known as:

6. A resistance to changing magnetic flux, or inertia, is known as:

7. When a conductor has a changing current flowing through it, a voltage is induced in the conductor. This is known as:

8. The voltage created as a result of relative motion between a conductor and a magnetic field is called a (an):

9. The voltage created in an inductor as a result of changing current through the conductor is known as what?

10. If a current or voltage is constant, or unchanging, it is considered:

11. The amount of time required for a current to increase or decrease by 63.2% in a circuit is called the circuit's _____.

12. When a current or voltage is changing, it is considered to be:

13. When a magnetic core is conducting the maximum amount of magnetic flux it can sustain, it is considered to be:

14. Refer to Figure 11.35. What is the time constant of the circuit?

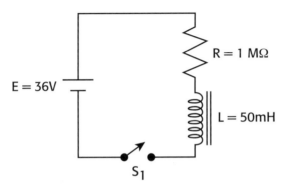

FIGURE 11.35   Time-constant circuit for Question 14.

15. List at least two uses for inductors in a PC.

16. A capacitor consists of two or more _____ and a _____.

17. Capacitors are designed to allow a large amount of current to flow through them.
   a. true
   b. false

18. Capacitors tend to oppose a change in _____.

19. Capacitors store energy in a (an) _____.

20. In a charging circuit, how many time constants are required before a capacitor will reach full voltage?

21. In a discharging circuit, how many time constants are required for a capacitor to fully discharge?

22. A circuit consists of a 24 V source, a switch, a 200 ohm resistor, and a 5 μf capacitor in series. Assume the switch is open, and the capacitor voltage is 0 volts. How long (in seconds), after closing the switch will the voltage on the capacitor be 15.17 volts?

12 ⋮ Alternating Current Circuits

12.1 INTRODUCTION

This chapter presents alternating current (AC) circuits. Alternating current is the type of electric current delivered to our homes and work places. It is the type that runs the power supply inside our computers, operates our home appliances such as radios and TVs, dries our clothes in electric dryers, provides heat from electric heaters, and so on. Alternating current, or AC for short, first flows one direction in a circuit and then reverses, flows the opposite direction, and repeats this action over and over in what are termed *cycles*. The most common type of AC is in the form of a sine wave, as shown in Figure 12.1.

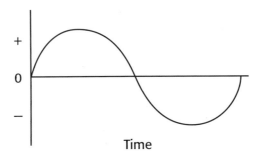

FIGURE 12.1   An AC sine wave.

12.2   AMPLITUDE, PERIOD, FREQUENCY, AND WAVELENGTH

Amplitude, period, frequency, and wavelength are terms commonly used to describe AC waveforms. The next sections describe them in more detail.

12.2.1   Amplitude

The *amplitude* of a waveform is its magnitude or height as seen on a graph or display such as that produced by an oscilloscope. AC waveforms are most easily measured for amplitude in terms of peak-to-peak value; at least it is usually easiest for students to understand this value. Other methods of expressing the amplitude of an AC waveform include just the positive peak, just the negative peak, the average value, and the effective value. Figure 12.2 shows these various values.

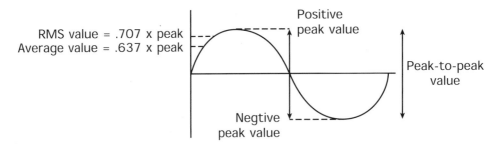

FIGURE 12.2   Various values are used to express the amplitude of an AC waveform.

RMS Values

You should understand that the most common form of expressing AC current, voltage, or power is in *RMS* values. RMS stands for *root mean squared*. It is not necessary to understand the math of how it is derived, but you should know that RMS is also known as the *effective value*, since this is the same value of DC that will cause an equal degree of heat in a resistor.

When people the world over express the value of AC current, voltage, or power, and they do not add an additional designation after the value, it is universally assumed the person means RMS values. For instance, when any scrupulous manufacturer specifies how much power a stereo amplifier will produce, it is in terms of RMS power, per channel. Specifications for PC power supplies are also in terms of RMS watts output. When we speak of the "120 VAC line," the figure 120 V is in RMS volts, not peak, peak-to-peak, or average value.

12.2.2 Period

The length of time for an AC cycle to occur or repeat is known as the cycle's time period or simply the *period*, measured in seconds. Refer to Figure 12.3.

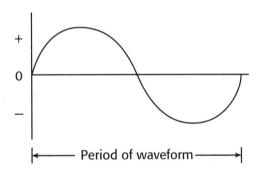

FIGURE 12.3 The period of a waveform is the time it takes to repeat.

12.2.3 Frequency

How many times in one second an AC waveform repeats is called its *frequency*. For example, in the United States and Canada, the standard power line frequency is 60 repetitions per second. Cycles per second are formally represented by the unit hertz

(Hz). A waveform's time period and frequency can be related mathematically by two equations, 12.1 and 12.2.

$$t = 1 \div f \tag{12.1}$$

Where t is the time period of the waveform in seconds and f is the frequency of the waveform in Hz.

$$f = 1 \div t \tag{12.2}$$

Again, t is in seconds and f is in Hz.

12.2.4   Wavelength

The length of the wave as it travels along a conductor in a circuit is known as the *wavelength*, measured in meters. The equation commonly used for wavelength is:

$$\lambda \cong C \div f \tag{12.3}$$

Where λ (the Greek letter lambda) represents the wavelength in meters, C is the speed of light, roughly 300,000,000 meters per second, and f represents frequency, in Hz. This is an approximation, used for many years to calculate wavelength. A closer approximation of the speed of light is 299,792,458 m/s. For our work, the number 300,000,000 is close enough.

Low-frequency waves have a long wavelength and high-frequency waves have a short wavelength. You have probably heard the term "shortwave" applied to a type of radio signals called high frequency (HF), commonly used for long-distance radio transmission.

Example 12.1

Calculate the wavelength of a waveform with a frequency of 3 MHz.

Solution

$$\lambda = C \div f$$
$$= (300 \times 10^6 \, m/s) \div (3 \times 10^6 \, Hz)$$
$$= 300 \div 3$$
$$= 100m$$

Example 12.2

Calculate the wavelength of a radio signal with a frequency of 50 MHz.

Solution

$$\lambda = C \div f$$
$$= (300 \times 10^6 \, m / s) \div (50 \times 10^6 \, Hz)$$
$$= 6m$$

The Federal Communications Commission (FCC) assigns and regulates the electromagnetic spectrum in the USA. The particular frequency range or *band* of frequencies from 50–54 MHz is allocated for use by certain license class holders of amateur radio (ham) operators. The "hams" commonly refer to this range of frequencies as the "six meter band," which is really a description of the approximate wavelength used, not the frequencies.

Here's one more example.

Example 12.3

Calculate the wavelength of the commonly used frequency of commercial power in the USA, 60 Hz.

Solution

$$\lambda = C \div f$$
$$= (300 \times 10^6 \, m / s) \div (60 \, Hz)$$
$$= 5 \times 10^6 \, m$$
$$= 5,000,000m$$

which is 5,000 km or roughly 3,000 miles.

12.3   AC WAVEFORMS

Although the sine wave is the most common form of an AC waveform, there are other forms as well. An AC sine wave can be distorted due to the addition of other frequency components, noise, circuit imbalances, or passage through a nonlinear device such as a diode. Also, DC and AC can both be present in a conductor together. Figure 12.4 shows some other AC waveforms. More on diodes and the rectifying process is presented in Chapter 14, "Power Supplies." For now it is enough to understand that *rectifying* means changing an AC waveform into a pulsating DC waveform.

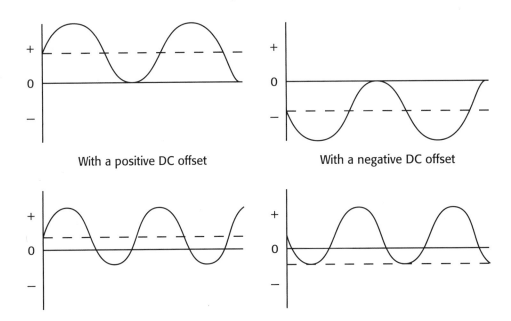

FIGURE 12.4   Other types of AC waveforms.

12.4   REACTANCE AND IMPEDANCE

In addition to resistors, inductors and capacitors offer opposition to AC. The opposition to AC current offered by inductors and capacitors is known as *reactance*, represented by the symbol X, and is measured in units of ohms. The net or overall opposition to AC in a circuit is known as *impedance*, represented by the symbol Z, and also measured in units of ohms. Impedance has three main components: resistance, inductive reactance, and capacitive reactance.

12.4.1   Inductive Reactance

When AC flows through an inductor, the inductor opposes the change in current and takes time to stabilize. This relationship is called the circuit's inductive time constant, as presented in Chapter 11, "Inductors and Capacitors." When the current is constantly changing, as in an AC sine wave, the result is called *inductive reactance*, which is represented by the symbol X_L, and is measured in units of ohms (Ω). The equation for determining inductive reactance involves the value of the inductor, L, in henries, and the frequency, f in Hz. Equation 12.4 shows this relationship.

$$X_L = 2\pi f L \qquad (12.4)$$

Figure 12.5 shows this relationship graphically. Notice that inductive reactance is directly proportional to both the frequency and the value of inductor.

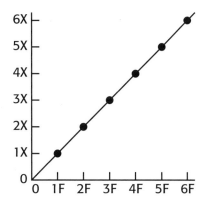

FIGURE 12.5 A graph of inductive reactance over frequency.

Example 12.4

Calculate the inductive reactance of a 1-H inductor with an AC current of 1 kHz flowing through it.

Solution

Use equation 12.4 and insert the given values.

$$X_L = (2\pi)(1kHz)(1H)$$
$$= (2\pi)(1 \times 10^3 Hz)(1H)$$
$$= 6,283.1$$
$$= 6.3 \text{ k}\Omega$$

Example 12.5

Calculate the inductive reactance that a 35 mH inductor offers to an AC current at a frequency of 1 MHz.

Solution

Use equation 12.4 and substitute the values given.

$$X_L = (2\pi)(1MHz)(5mH)$$
$$= (2\pi)(1\times10^6)(35\times10^{-3})$$
$$= 219,911$$
$$= 220\text{ k}$$

12.4.2 Capacitive Reactance

Capacitors react to a change in voltage across their terminals that is produced when AC is applied to them. The opposition to AC offered by a capacitor is known as *capacitive reactance*, X_C, measured in units of ohms. Equation 12.5 shows the relationship between the value of a capacitor (C) in farads (F), the frequency (f) of AC applied in hertz, and the resulting reactance to the AC in ohms.

$$X_C = 1 \div (2\pi fC) \tag{12.5}$$

Figure 12.6 shows this relationship graphically. Notice that capacitive reactance is *inversely* proportional to the frequency and capacitor value.

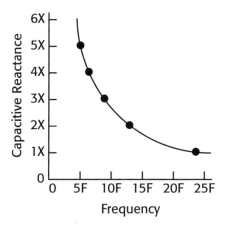

FIGURE 12.6 Capacitive reactance plotted against frequency.

Example 12.6

Calculate the capacitive reactance to a 60 Hz sine wave offered by a 1000μF capacitor.

Solution

Use equation 12.5 and substitute the given values.

$$
\begin{aligned}
X_C &= 1 \div ((2\pi)(60\,Hz)(1000\mu f)) \\
&= 1 \div ((2\pi)(60)(1000 \times 10^{-6})) \\
&= 1 / 0.3769911 \\
&= 2.7\,\Omega
\end{aligned}
$$

Example 12.7

Calculate the opposition to a 40 kHz AC signal offered by a 50ρF capacitor.

Solution

Use Equation 12.5 and substitute the given values.

$$
\begin{aligned}
X_C &= 1 \div (2\pi)(40\,kHZ)(50\rho f) \\
&= 1 \div (2\pi)(40 \times 10^{3})(50 \times 10^{-12}) \\
&= 1 \div 0.000012566 \\
&= 79577\,\Omega \\
&= 79.6\,k\Omega
\end{aligned}
$$

12.4.3 Impedance

Impedance is the net or total opposition to AC in a real circuit. The effects of resistance, inductive reactance, and capacitive reactance all figure into the equation. Equation 12.6 is used to solve for impedance when the value of resistance and both inductive and capacitive reactance values are known.

$$Z = \sqrt{(X_L - X_C)^2 + R^2} \tag{12.6}$$

Example 12.8

Calculate the impedance offered by a series circuit consisting of a 1-k resistor, a 100pF capacitor and a 25 mH inductor. Assume that a 100-kHz signal is applied to the circuit.

Solution

First find the reactance values. To solve for the inductive reactance, use equation 12.4 and substitute the given values.

$$
\begin{aligned}
X_L &= 2\pi(25mh)(100kHz) \\
&= 2\pi(25 \times 10^{-3})(100 \times 10^{3}) \\
&= 2\pi(2500) \\
&= 15707.9\Omega \\
&= 15.7k\Omega
\end{aligned}
$$

Next, solve for the capacitive reactance. Use Equation 12.5 and substitute the given values.

$$
\begin{aligned}
X_C &= 1/[(2\pi)(100kHz)(100\rho f)] \\
&= 1/[(2\pi)(100 \times 10^{3})(100 \times 10^{-12})] \\
&= 1/(2\pi \times .00001) \\
&= 1 \div (.000062832) \\
&= 15915.49 \\
&= 15.9k\Omega
\end{aligned}
$$

Finally, use equation 12.6 and substitute the values given.

$$
\begin{aligned}
Z &= \sqrt{(15.9k\Omega - 15.7k\Omega)^2 + 1k\Omega^2} \\
Z &= \sqrt{.2k\Omega^2 + 1k\Omega^2} \\
&= \sqrt{40000 + 1000000)} \\
&= \sqrt{1040000)} \\
Z &= 1019.8\Omega \\
Z &= 10.2k\Omega
\end{aligned}
$$

12.5   OHM'S LAW FOR AC CIRCUITS

Now that you know about reactance and impedance, you should know these other useful equations:

$$V = IX_L \tag{12.7}$$

$$V = IX_C \tag{12.8}$$

$$V = IZ \tag{12.9}$$

$$I = V \div X_L \tag{12.10}$$

$$I = V \div X_C \tag{12.11}$$

$$I = V \div Z \tag{12.12}$$

$$P = I^2 X_L \tag{12.13}$$

$$P = I^2 X_C \tag{12.14}$$

$$P = I^2 Z \tag{12.15}$$

12.6   PHASORS

To make a graphical representation of the relative magnitudes and phase relation-ships among resistance, inductive reactance, capacitive reactance, and impedance, several methods can be used. Usually the easiest for students to grasp is the rectan-gular method, which will now be presented.

Resistance is located on the X-axis and reactance is plotted on the Y-axis. Ca-pacitive reactance, X_C is plotted below the X-axis, and inductive reactance, X_L is plotted above the X-axis, as shown in Figure 12.7.

The general method of using the rectangular plotting method is like this:

1. Plot the magnitude of resistance on the X-axis pointing to the right of the origin.
2. Calculate each reactance value.
3. Subtract the smaller reactance value from the larger reactance value. The type of reactance remaining is the net reactance. It is always the type that has the largest reactance value.

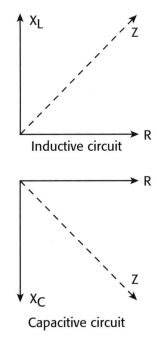

Inductive circuit

Capacitive circuit

FIGURE 12.7 Relationship among resistance, reactance, and impedance using the rectangular method of representation.

4. Plot the remaining reactance value, setting the direction according to the type of reactance.
5. Express the rectangular form of the impedance using the plotted values for resistance and reactance.

Example 12.9

Plot the following values and express the resulting impedance using the rectangular method: $X_L = 200$ ohms, $X_C = 100$ ohms, $R = 50$ ohms.

Solution

1. First plot the magnitude of resistance on the X-axis, in this case, 50 ohms.
2. Since the value of each reactance is already given, just subtract the smaller reactance from the larger reactance, in this case 200 ohms minus 100 ohms, leaving 100 ohms. The answer is X_L, since that is the larger reactance.

3. Plot the remaining reactance on the Y-axis, taking care to plot the correct type, in this case 100 ohms of X_L, which is plotted above the X-axis.
4. Plot the resistance magnitude on the X-axis.
5. Finally, express the impedance in terms of resistance and reactance, in this case, $R = 50$ and $X_L = 100$ (see Figure 12.8).

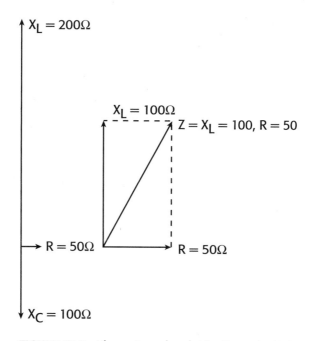

FIGURE 12.8   The rectangular plot for Example 12.9.

Example 12.10

Draw the rectangular plot for a circuit with the following values: $X_C = 1\ k\Omega$, $X_L = 470\ \Omega$, and $R = 80\ \Omega$.

1. Since the circuit is overall capacitive, we subtract the capacitive reactance value from the inductive reactance value, or $1{,}000\ \Omega - 470\ \Omega = 530\ \Omega$ of X_C since it is the largest reactance.
2. Plot this value on the graph.
3. Plot the value of resistance, 80 ohms.
4. Express the impedance in terms of resistance and reactance, $R = 80\ \Omega$ and $X_C = 530\ \Omega$ (see Figure 12.9).

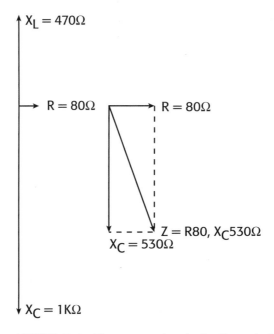

FIGURE 12.9 The rectangular plot for Example 12.10.

An Advanced Concept: the j Operator

This subject will certainly not come up on the A+ exam, but as a PC tech you *may* someday hear the concept discussed. For that reason, it is being briefly presented here.

To better describe the type of reactance, in a type of shorthand, the lowercase letter j is used. It is called the *j operator*, and is used to indicate the type of reactance. You probably already know the concept of a number line, with positive numbers shown to the right-hand side of the zero point on the line and negative numbers shown to the left-hand side of the zero point. The numbers are considered *real numbers*. To describe the reactance values, the j operator is often used. The j operator is said to represent *imaginary numbers* since they don't lie on the real number line. Remember that reactance is the action of temporarily storing energy that is given back to the circuit again, not actually dissipated or converted to heat. When the reactance is inductive, the j operator is considered positive, since X_L is above the X-axis line. The designation −j is used to describe capacitive reactance, drawn below or in the negative direction since X_C is drawn below the reference resistive X-axis. Figure 12.10 shows this idea.

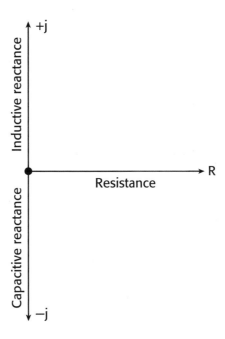

FIGURE 12.10   Use of the j operator.

Example 12.11

Express impedance for the circuit used in Example 12.9 by using the j-operator.

Solution

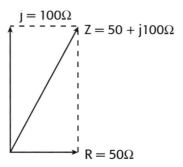

FIGURE 12.11   Impedance for Example 12.9 expressed using the j operator.

Example 12.12

Express the impedance for the circuit of Example 12.10 in terms of the j operator.

Solution

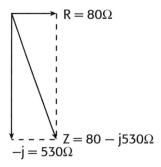

$R = 80\Omega$

$Z = 80 - j530\Omega$

$-j = 530\Omega$

FIGURE 12.12 The j-operator solution to the circuit in Example 12.10.

12.7 RL CIRCUITS

There are many uses for a simple circuit consisting of only a resistor and an inductor, called an RL circuit. The next few sections will present some of the more common uses for such circuits.

12.7.1 Output Taken Across the Inductor

There are two basic series circuits using a resistor and an inductor. Figure 12.13 shows one such circuit, with the input applied across both components and the output taken across the inductor. What happens to the output as the input frequency is varied from low to high is shown in Figure 12.13. Notice that the output rises since the inductive reactance rises with frequency. So the circuit acts as a frequency-dependent voltage divider, with more of the total voltage developed (dropped) across the inductor, as the frequency increases.

12.7.2 Output Taken Across the Resistor

When the output from the circuit is taken across the resistor, the greatest voltage is developed across the resistor when the input frequency is low. This is because the inductor offers the least reactance to the input at low frequencies, leaving the most voltage left to drop across the output resistor. Refer to Figure 12.14.

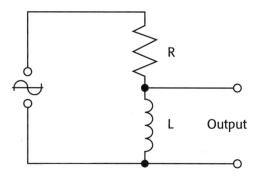

FIGURE 12.13 An RL circuit with the output across the inductor.

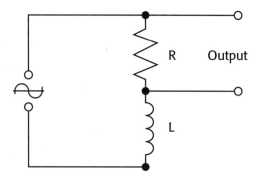

FIGURE 12.14 An RL circuit with the output taken across the resistor.

12.8   RC CIRCUITS

There are also many practical uses for simple circuits consisting of only a resistor and a capacitor, known as RC circuits. The next few sections present some of the more common examples of RC circuits.

12.8.1   Output Taken Across the Capacitor

A resistor-capacitor circuit with the output taken across the capacitor is shown in Figure 12.15. The output is largest at low frequencies since the capacitive reactance is greatest at low frequencies—the capacitive reactance is ideally infinite. The circuit can be viewed as a voltage divider with the resistor value fixed and the capacitive-reactance value frequency dependent. At very high frequencies the cap looks like a short circuit. Since no voltage can be developed across a short, the output is zero. At DC (the lowest frequency), the capacitor looks like an open circuit, and so the maximum voltage is developed across it. Voltage leads current in this type of circuit. ELI circuits are explained in Section 12.8.3.

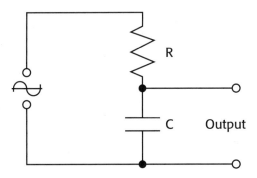

FIGURE 12.15   An RC circuit with the output across the capacitor.

12.8.2   Output Taken Across the Resistor

When the output is taken across the resistor, the greatest voltage is developed at high-input frequencies. This is because the capacitor offers the least reactance at high frequencies. Think of the circuit as a frequency-dependent voltage divider with the resistor value constant and the capacitive-reactance value frequency dependent. The voltage leads the current in an ELI circuit such as this. Refer to Figure 12.16.

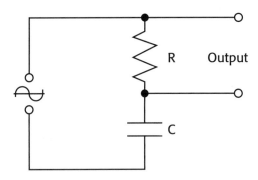

FIGURE 12.16   An RC circuit with output across the resistor.

12.8.3   Eli the Ice man

There is a simple pneumonic device (trick to help remember) for keeping track of whether a circuit is a leading or lagging one in terms of current. It is *Eli the Ice man*. E stands for voltage, L stands for inductor. *ELI* shows that E, voltage is before, or leads the letter I, current in an L (inductive) circuit. *ICE* signifies that I, current is before or leads voltage, E, in a C (capacitive) circuit. So you can easily remember the relationships between voltage and current in circuits by remembering old Eli.

12.9   RLC CIRCUITS

As you have probably guessed, RLC circuits consist of a resistor, an inductor, and a capacitor. There are many practical uses for these types of circuits, which are presented in the following sections. Figure 12.17 shows a few examples of RLC circuits.

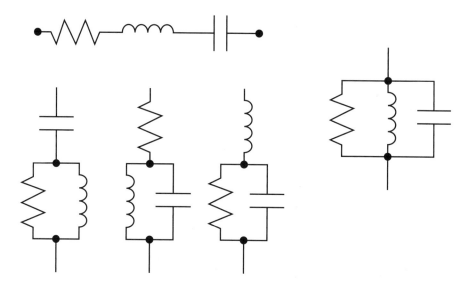

FIGURE 12.17   Some RLC circuits.

12.10   RESONANT CIRCUITS

There is a special case in which the value of inductive reactance equals the value of capacitive reactance in a circuit or component. This special condition is known as *resonance*, and most often happens only at one particular frequency, depending on the component values used. There are two basic circuit arrangements, series resonant circuits and *parallel resonant circuits*. Equation 12.16 defines the resonant frequency of a circuit.

$$f_R = 1 \div (2\pi\sqrt{LC})$$ (12.16)

12.10.1 Series Resonant Circuits

A capacitor and inductor in series or in parallel form the basic *series resonant circuits*. At resonance, the values of capacitive and inductive reactance have equal value but opposite phase, and therefore cancel each other out. The result is pure resistance. This resistance can be in the form of a resistor and/or the resistance of the circuit wiring, plus the equivalent series resistance of the capacitor, and the winding resistance of the coil. The result is the lowest amount of impedance (the X terms cancel and drop out of the equation, leaving $Z = \sqrt{R^2}$, which is of course, just R). So a resonant circuit operated at its resonant frequency acts purely resistive.

As a result, the current value through the circuit is at its maximum. Above and below resonant frequency, the circuit current is less. Refer to Figure 12.18.

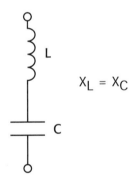

$X_L = X_C$

FIGURE 12.18   Series-resonant circuit.

12.10.2 Parallel Resonant Circuits

An inductor and capacitor in parallel form the *parallel resonant circuit*, also known as a *tank circuit*. This is because the circuit action is commonly described as being similar to putting water into a tank. The current first flows into one component, then out of it and into the other, and so on, back and forth. This back and forth current is called *circulating current* or *tank current*, as opposed to the current that flows in the rest of the circuit. The current external to the tank is known as *line current*. Refer to Figure 12.19.

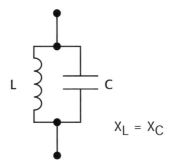

FIGURE 12.19 A parallel resonant "tank" circuit.

At resonance, the impedance of a tank circuit can be many times the impedance as when it is off resonance. And the voltage developed across the tank components can be magnified many times. This magnification factor is known as "Q," for *quality factor*. It is also defined as the ratio of impedance to resistance in a circuit. Since inductors are the least "perfect" of the two reactive components in the tank circuit, the inductor's Q is used.

Equation 12.7 defines the Q of an inductor.

$$Q = X_L \div R_W \tag{12.17}$$

where R_W represents the coil's winding resistance.

At resonance, the tank circuit has the least line current drawn, since the circuit impedance is at maximum value. The circulating current will be Q times the line current, as shown in Figure 12.20.

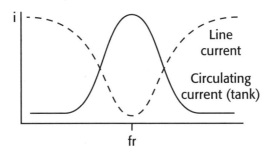

FIGURE 12.20 A graph of the currents flowing in a parallel "tank" circuit.

Resonant circuits are used almost universally in radio and other types of high-frequency communications equipment. They form the basis of the so-called *tuner* in most radios. They have the capability of either passing or rejecting a range of frequencies and so are very useful circuits. More on filters is presented in the following sections.

12.11 PASSIVE FILTERS

Passive filters are so called because they do not employ any amplification, as vacuum tubes or transistors do. Still, they are used to perform very useful functions as filters. Filters are frequency-selective circuits that pass or reject certain frequencies or ranges of frequencies. The general types of passive filters include; low-pass, high-pass, band-pass, and band-reject filters.

12.11.1 RL High-pass Filters

Remember the series resistor-inductor circuit presented in section 12.6.1? The output is highest at high frequencies, so this circuit is commonly called a high-pass filter, due to its circuit action. This type of filter circuit is commonly used in high-frequency amplifiers to prevent the desired frequency from passing back through the positive power supply connection to ground. TVs, radios, and all sorts of radio frequency equipment use this type of circuit. Refer to Figure 12.21.

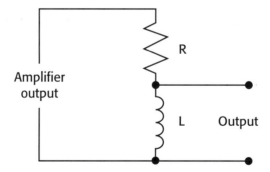

FIGURE 12.21   An RL high-pass filter used to block frequencies below the desired cutoff frequency in an amplifier.

The filter's cutoff frequency is found using:

$$f_C = R \div (2\pi L) \qquad\qquad (12.18)$$

12.11.2 RL Low-pass Filters

When the output is taken across the resistor, the circuit action is different. This time the greatest output will be at low frequencies, since the inductor will drop the most voltage at high frequencies, leaving less to be dropped across the resistor.

Such circuits are commonly used in certain DC power supplies in order to attenuate (greatly reduce in amplitude) the high-frequency components called ripple. A circuit like this is also used for the speaker crossover feeding the woofer (low-frequency speaker) in a stereo speaker system. The high frequencies, which the woofer is not capable of reproducing, are wasted and tend only to heat the woofer's voice coil if allowed to pass through to it. Woofers are large speakers with a lot of mass making up the cone, the surface that moves in and out and pushes on the surrounding air to make sound. The large cone mass prevents the cone from vibrating at high frequencies, so the high-frequency energy is wasted. So the RL low-pass filter is used to keep the high frequencies out of the woofer. The inductor is the "coil" that can be seen in speaker crossover networks. Refer to Figure 12.22.

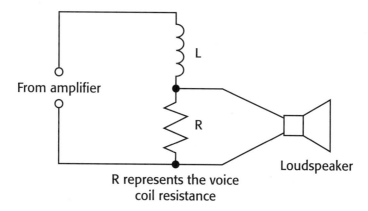

FIGURE 12.22 A loudspeaker's voice coil driven by an RL low-pass filter that prevents high frequencies above the filter's cutoff frequency from reaching the speaker.

12.11.3 RC High-pass Filters

When the output of this circuit is taken across the resistor, the circuit performs as a high-pass filter, as shown in Figure 12.23. This is the type circuit used in a conventional high-power level (operating on the output from the amplifier) speaker crossover for a tweeter (the high-frequency speaker.) It is also commonly used between different parts of a circuit or between two circuit stages to remove DC, since DC cannot pass through a capacitor. The RC filter's cutoff frequency is found using:

$$f_C = 1 \div (2\pi RC) \tag{12.19}$$

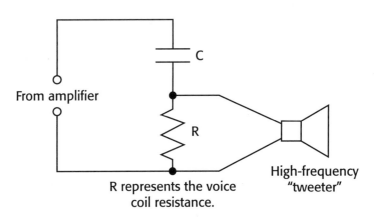

From amplifier

C

R

R represents the voice coil resistance.

High-frequency "tweeter"

FIGURE 12.23 An RC high-pass filter used as part of a speaker crossover system.

12.11.4 RC Low-pass Filters

When the output is taken across the resistor, the circuit performs as a low-pass filter. Such a circuit is commonly used in power supplies to remove ripple components, and is cheaper to implement than the equivalent RL low-pass filter. This type circuit is also used to "bypass" to ground most high-frequency noise present on lines on a PC motherboard or any other type of circuit board. This accounts for the presence of so many small-value bypass capacitors connected to chips on circuit boards. They serve to clean up the power and signal lines.

12.11.5 Band-pass Filters

There is a special filter that will pass a band or range of frequencies centered on a certain design frequency (the frequency at which they are designed to operate), and attenuates all other frequencies. The equation used to calculate the so-called center frequency is:

$$f_C = 1 \div (2\pi \sqrt{LC}) \tag{12.20}$$

Another name for band-pass filter is a *resonant filter circuit*.

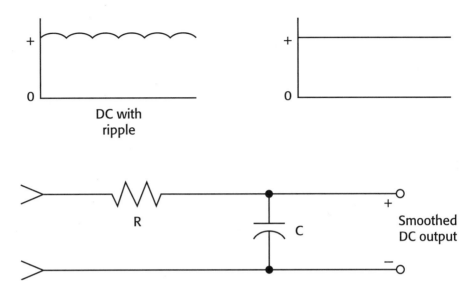

FIGURE 12.24 An RC low-pass filter in a power supply, used to reduce ripple in the output.

12.11.6 Band-reject Filters

Another type of filter is used to reject a band of frequencies centered on a design frequency. This is the band-reject filter. Figure 12.25 shows two forms of this type filter. One application for this type of filter is to remove interference at a particular frequency. Designs similar to these were used to allow "premium channels" to be seen on cable TV networks. The interfering signal was added by the cable company to disrupt the TV set's synchronization circuits. By adding the special filters in line with the cable TV signal, the interfering signal was removed, allowing normal viewing of the channel.

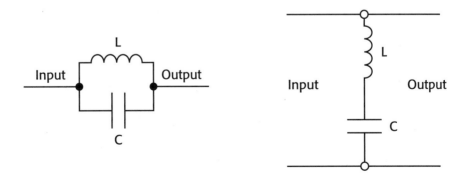

FIGURE 12.25 Two forms of band-reject filters.

12.12   POWER FACTOR

Not all the power delivered to a load is used, depending on the type of load it is. The only power actually used by the load is by the resistive components and is called *true power*. It is computed using Ohm's law,

$$P = I^2R \qquad\qquad (3.6)$$

If there is any reactance present in the load, the power delivered by the source will be greater than that used by the load. This is because power is temporarily stored and then returned to the circuit by reactances. So although the load does not use it, the source must still provide this value of power. The ratio of power used, called true power, measured in watts to power apparently used, called *apparent power*, measured in volt-amperes (VA) is known as the power factor of the circuit. The ratio, expressed as a number between 0 and 1 and represents W/VA. A non-reactive load has a power factor of 1 (all power supplied is used), and a purely reactive load, either capacitive or inductive, has a power factor of 0. Most real AC circuits have a power factor of between 0 and 1, meaning they consist of both resistance and reactance.

Since the commercial power generating and distributing establishments (the electric power companies) have a great monetary interest in not having to build their delivery infrastructure to a greater degree than necessary, they take special steps to ensure a power factor of as close to 1 as possible. They do this by periodically performing a neighborhood power survey and installing the appropriate value of capacitors across the power lines. These are used to compensate for the predominantly inductive reactance of electrical equipment, mostly those containing motors that have coils of wire. The capacitive reactance of the capacitors is used to cancel out the inductive reactance, thereby presenting the power companies lines with a real power factor. These power-factor-compensating capacitors can be seen by careful inspection of power poles. Figure 12.26 shows a bank of such capacitors mounted on a commercial power pole. Note that these are not transformers; capacitors used on power poles are usually rectangular in shape whereas most transformers are cylindrical.

In effect, the compensating capacitors are used to resonate the power lines at a frequency of 60 Hz.

12.13   SUMMARY

- Alternating current (AC) flows one direction in a circuit and then reverses. This process occurs over and over again.
- The magnitude of a waveform is known as its amplitude. Amplitude of an AC waveform can be expressed as peak, peak-to-peak, average, or RMS. RMS is the

FIGURE 12.26 A bank of six power-factor-compensating capacitors mounted on an electric utility power pole cross beam.

world standard form of expressing AC values. Unless specifically designated otherwise, always assume an AC value is in RMS units.

- The rate at which AC repeats per second is known as its frequency, and is measured in hertz (Hz).
- The length of time AC takes to repeat one cycle is known as the period.
- The physical length of an AC waveform is known as its wavelength.
- Two equations are used to relate period and frequency, $t = 1 \div f$ (12.1) and $f = 1 \div t$ (12.2).
- The most common AC waveform is the sine wave. There are other types of AC waveforms as well.
- In most AC circuits the effects of capacitance, inductance, and resistance are all present.
- The opposition to AC offered by an inductor is known as inductive reactance and can be found using the equation: $X_L = 2\pi f L$, (12.4).
- The opposition a capacitor offers to an AC current is called capacitive reactance and can be computed using: $X_C = 1 \div 2\pi f C$, (12.5).
- The net opposition to AC in a circuit is known as impedance, Z, and can be computed using: $Z = 1 \div \sqrt{(X_L - X_C)^2 + R^2}$ (12.6).

■ Ohm's Law for AC circuits includes these handy equations:

$$V = IX_L \qquad (12.7)$$

$$V = IX_C \qquad (12.8)$$

$$V = IZ \qquad (12.9)$$

$$I = V \div X_L \qquad (12.10)$$

$$I = V \div X_C \qquad (12.11)$$

$$I = V \div Z \qquad (12.12)$$

$$P = I^2 X_L \qquad (12.13)$$

$$P = I^2 X_C \qquad (12.14)$$

$$P = I^2 Z \qquad (12.15)$$

■ The rectangular method can be used to diagram phasors representing resistance, R, and reactance, X, magnitudes and directions. From these, the impedance, Z, can be found.

■ The j operator is used to describe and diagram reactance values. +j values signify inductive reactance, and –j values signify capacitive reactance values.

■ Using the rectangular form, a circuit can be described in terms of its resistance and reactance values like this: $R +/–J$.

■ A phasor is a line with a length representing a value's magnitude and a direction, representing resistance or reactance. Phasors are used to make rectangular plots of resistance and reactance, using the rectangular form.

■ ELI the ICE man is a pneumonic device for remembering the relationship between voltage and current in inductive and capacitive circuits. ELI means current lags voltage in an inductive circuit. ICE means current leads voltage in a capacitive circuit. Inductive circuits are called *lagging* circuits, and capacitive circuits are called *leading* circuits.

■ Passive filters use resistors, capacitors, and inductors to form frequency-dependent circuits. Such circuits are called passive filters.

■ Passive filters come in the form of low-pass, high-pass, band-pass, and band-reject filters.

■ Low-pass filters pass low frequencies below the cutoff frequency and attenuate frequencies above the cutoff frequency. For RC filters, the cutoff frequency is found using $f_C = 1 \div (2\pi RC)$.

■ High-pass filters pass frequencies higher than their cutoff frequency and attenuate frequencies lower than their cutoff frequency. For RC filters, use the equation $f_C = 1 \div (2\pi RC)$.

- Band-pass filters pass a band or range of frequencies centered on their resonant frequency, found using: $f_R = 1 \div (2\pi \sqrt{LC})$ (12.7).
- Band-reject filters reject or attenuate frequencies centered on their resonant frequency. For LC resonant band-reject filters, use Eq 12.16.
- The Q or magnification factor of a reactive circuit is given by: $Q = X_L \div R_W$ (12.17) where the reactance of the inductor and the inductor's winding resistance value is used, for circuits having either only inductive reactance or both inductive and capacitive reactance.
- The power factor of a circuit relates how reactive or resistive the power is. Power factor is expressed as a number between 0 and 1, which represents the ratio of real power over reactive power. A power factor of 0 means a purely reactive circuit that dissipates no real (true) power. A power factor of 1 represents a circuit that is purely resistive and dissipates all power in the form of heat.
- Power is only temporarily stored in the reactive devices, then returned to the circuit, with no losses. Most real circuits show a power factor of between 0 and 1, and have some real power dissipated, measured in watts (W), and some reactive power, measured in volt-amperes, reactive (VARs), which is stored and returned to the circuit. The resultant power seen by the power line, called apparent power, is volt-amps, VA.
- True power, measured in watts, is what we pay for. Actually, we pay for power used over time, measured in kilowatt-hours.
- Commercial power companies use banks of capacitors to compensate for the inductive reactance caused by many electric motor-powered appliances used by consumers. The idea is to try to get to a power factor value of as close to 1 as possible.
- In reality, unless a circuit consists purely of resistive devices, it will show a power factor of between 0 and 1.

12.14   KEY TERMS

AC	Power factor
Amplitude	Reactance
Band-pass filter	Reactive power
Band-reject filter	Resonance
High-pass filter	Series-resonant circuit
Low-pass filter	True power
Parallel resonant circuit	VA
Passive filter	VAR
Period	Wavelength

12.15   EXERCISES

1. The length of time one cycle of an AC waveform takes to repeat is called its . . .
   a. period
   b. amplitude
   c. second
   d. Wavelength

2. The "height" or magnitude of a waveform is called the waveform's . . .
   a. period
   b. amplitude
   c. time frame
   d. wavelength

3. The distance a waveform takes to repeat one cycle is known as its . . .
   a. period
   b. wavelength
   c. amplitude
   d. distance

4. Which is the "universal" means of referring to AC values?
   a. Peak value
   b. Peak-to-peak value
   c. Average value
   d. Rms or effective value

5. Find the period of a 200 Hz waveform.

6. Find the period of a 20 kHz waveform.

7. Find the frequency of a waveform that has a period of 50 ms.

8. Find the frequency of a waveform that has a frequency of 10µs.

9. A waveform with a long period has what kind of frequency?
   a. Low
   b. High

10. A high-frequency signal has what kind of wavelength?
    a. Long
    b. Short

11. An AC waveform can have a DC offset to it.
    a. True
    b. False

12. Calculate the value of inductive reactance that a 1 mH inductor offers at 50 kHz.

13. Calculate the value of inductive reactance that a 35μH inductor offers at 2 MHz.

14. Calculate the value of capacitive reactance that a 100μf capacitor offers at 120 Hz.

15. Calculate the value of capacitive reactance that a 10μf capacitor offers at 50 MHz.

16. Calculate the resonant frequency of a circuit consisting of a 20μf capacitor in series with a 2mH inductor.

17. Calculate the resonant frequency of a circuit consisting of a 100μf capacitor in parallel with a 1 H inductor.

18. Capacitors on power poles are used to do what?
    a. Increase the voltage
    b. Decrease the voltage
    c. Improve the power factor
    d. Compensate for appliances using motors
    e. Both c and d

19. Give a common example of the use of a low-pass filter.

20. Give a common example of the use of a high-pass filter.

21. Give a common example of the use of a band-pass filter.

22. Give a common example of the use of a band-reject filter.

23. Give an example of the use of a resonant circuit.

13 | Transformers

13.1 INTRODUCTION

This chapter introduces electrical components known as transformers. They are very versatile. In many cases, transformers can perform several different tasks simultaneously—something no other single electronic component can do. Although transformers, because of their weight and high cost, have been widely replaced with other, cheaper components, they are still sometimes used today. Technicians should be aware of the different types and uses of transformers.

All transformers operate as if they have two coils of wire placed physically, and more importantly, magnetically, close to each other. A change in current in one coil causes a change in voltage in the other coil. If the other coil is also connected to a complete circuit, then the voltage change will produce a changing current in that circuit as well. Normally, the coil to which the input signal or voltage is applied is called the *primary winding*, or simply the *primary*. The output signal or voltage is taken from the other coil, called the *secondary winding* or *secondary*. Refer to Figure 13.1 for a diagram of a typical transformer schematic symbol.

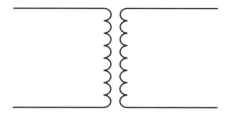

FIGURE 13.1   Transformer schematic symbol.

13.2   TURNS-RATIO

A transformer's turn-ratio is the number of turns of the secondary winding divided by the number of turns of the primary winding, or Equation 13.1:

$$TR = N_S \div N_p \qquad (13.1)$$

where N_S is the number of secondary winding turns and N_P is the number of primary winding turns.

When the turns-ratio is greater than 1:1, the transformer functions as a voltage *step up* transformer. For example, a transformer with a 2:1 turns-ratio would deliver approximately twice the voltage on the secondary winding as is applied to the primary winding. One with a turns-ratio of 1:3 would produce approximately one-third of the primary voltage on the secondary winding, and would be classed as a voltage *step-down* transformer.

The turns-ratio is indicated on a transformer's schematic symbol as shown in Figure 13.2.

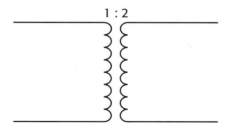

FIGURE 13.2   Transformer schematic showing turns ratio indicated above the core.

13.3 TYPES OF TRANSFORMERS

Transformers are classified in many ways; the next few sections present the more common transformer types. Common classifications include the intended frequency range, and overall circuit application.

13.3.1 Frequency Range

Transformers are built differently, depending on the frequency or more accurately, the range of frequencies they are intended to operate over. The following sections will present several of the most popular types of transformers, based on their intended operating frequency range.

Low-frequency (Power) Transformers

Low-frequency transformers, often called *power transformers*, are specifically engineered and manufactured for applications ranging from commercial power generation down to appliances that plug into the wall outlets that are supplied by commercial power generation. Typical power line frequencies used around the world are 50 and 60 Hz. At these frequencies, in order to achieve unity coupling of the primary and secondary windings, a relatively large mass of highly permeable magnetic material is used to form the windings and provide sufficient magnetic coupling. Unity coupling means that all of the magnetic flux lines produced by the primary winding will cut through the secondary winding.

Power transformers are designed to handle large amounts of electrical power and either change voltage to a higher or lower level, or to provide electrical isolation from the power source without a change in voltage. One example is the transformers used in linear power supplies in radios, TVs, stereo receivers, battery chargers, and so on (see Figure 13.3). Another example is the large power transformers used by commercial power distribution systems to raise and lower line voltage. Commonly seen examples of these power transformers are mounted on commercial power poles, and are called *pole pigs* in the industry. They are used to drop the roughly 21 kV or higher voltage of residential power lines down to 240/120 VAC supplied to our homes and businesses. The size depends on power and voltage rating, but many pole pigs are roughly 12 inches in diameter by 24 or so inches high. They include windings cooled by oil and usually incorporate a cooling system made up of either fins or tubes that radiate the heat lost from the core and windings.

There are two main types of core styles used for power transformers, laminated iron and toroidal. The older laminated-iron type of core consists of many individual sheets of an iron alloy, each lacquered and dried, and then stacked to form a sandwich of layers. The windings are wound inside these pieces of laminated iron. The sheets have shapes that look like the letters "I" and "E. The laminations

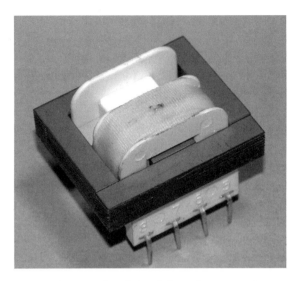

FIGURE 13.3 Printed circuit board mount style transformer made with "E" and "I" core lamination pieces.

are glued together, and a metal band is placed around them to keep them aligned. Finally, the band ends are formed into mounting tabs. Sometimes the laminations can become loose and vibrate slightly, at the frequency of the current. This vibration can often be heard in older equipment containing shell-type transformers. This would cause an audible *hum*. Another type of hum is the electronic version, which can be caused by a transformer's current frequency being radiated into another nearby circuit. A transformer can also pickup stray signals from nearby circuits, thereby introducing hum into the signal passing through the transformer. An old TV servicemen's trick to quiet mechanically noisy transformers involved driving small wooden wedges into the core. Sometimes transformers can also cause nearby magnetically affected parts to vibrate as well. Refer to Figure 13.4 for an illustration of this type of power transformer.

The second type of power transformer core uses a doughnut-shaped coil form called a *torus*. The windings are wound on this toroidal core and the magnetic field established by current changes in the windings tends to be more thoroughly contained within the core as opposed to the older laminated-iron cores. Toroid cores are mechanically solid so they don't vibrate mechanically as the older laminated-core types sometimes did. Toroidal cores have traditionally been costlier than the older core styles, but that is changing, and the toroid cores are the preferred type for

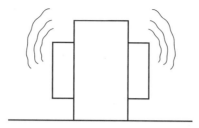

FIGURE 13.4 Power transformers sometimes mechanically vibrate at the AC frequency.

inductors and transformers. The toroid core material consists of a granulated iron alloy with an insulator and glue. The whole mixture is pressed into a form and then heated. When cured, it forms a rigid heavy coil form. Figure 13.5 shows a toroidal power transformer.

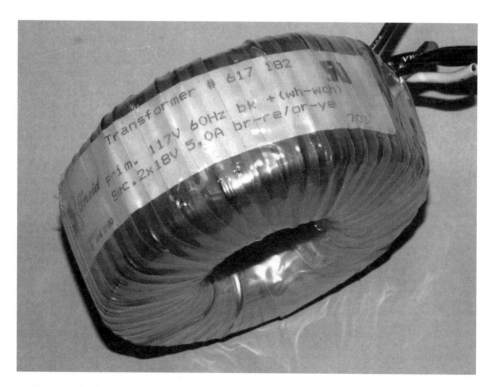

FIGURE 13.5 A toroidal power transformer.

Intermediate-frequency Transformers

As frequency increases beyond perhaps 20 kHz or so, the losses of a laminated core become excessive. The core style used at these frequencies is usually a toroid. The exact alloy mixture used in these toroids depends on both the desired frequency of operation and the power level to be handled. Intermediate-frequency (IF) transformers are commonly used in radios and communications equipment of all sorts. They are used in televisions, citizen's band (CB) and military radios, amateur (ham) radios, auto radios, and so on.

Some IF transformers are used as part of tuned circuits, those which can select or tune a particular frequency while rejecting others both higher and lower in frequency than the desired one. This type of circuit is known as a band-pass circuit. The small holes in the top and bottom IF "can" transformers used in a radio receiver circuits provide access for a tuning tool, which can adjust the transformer by sliding a magnetic slug into or out of the coil form. Performing this operation was part of a radio's alignment procedure when radios still used this type of transformer.

Radio-frequency Transformers

As frequency increases beyond roughly 1 MHz, the losses in even some toroid transformers become excessive and the core material must be of yet other types of materials. There are several methods used to overcome the core losses in radio frequency (RF) transformers. One method is to use relatively nonmagnetic materials such as aluminum and brass. Since these materials have little if any magnetic properties, they appear as essentially invisible to the magnetic flux lines and don't affect the transformer's performance. The other method is to make the wire used to form the windings large enough, and therefore stiff enough for them to support themselves. This winding style is known as an "air core." The use of solid copper wire or tubing of about 14 gauge or larger will provide sufficient mechanical rigidity to keep the turns in alignment. RF transformers are used in all sorts of communications gear, and especially in high-frequency, high-power transmitters. The largest windings are used in the high-power output stages of transmitters used for commercial radio and TV stations. Figure 13.6 shows the windings of a high-power, 2 kW-class radio antenna matching circuit.

13.3.2 Circuit Application

Transformers are also categorized by the intended circuit application for which they are designed. The next few sections detail some of the more common applications.

FIGURE 13.6 Large RF transformer windings.

Power Transformers

Power transformers, as explained earlier in the chapter, are used in the distribution systems of commercial electric power companies. They are also used in linear power supplies. More on power supplies is presented in Chapter 14, "Power Supplies." Some of the finest "high-end" stereo system power amplifiers still employ power transformers in their output sections. Refer to Figure 13.7 for a picture of power and *output* transformers used in a stereo amplifier. Refer to the next section, "Impedance Matching," for more on output transformers.

Impedance Matching

Transformers are often used to match the impedance of one circuit to another. In order to get the maximum power transfer between two circuits, the output impedance of the first circuit must closely match the input impedance of the second circuit. This is often not the case, so some method must be used to accomplish a match. A transformer may be constructed so that its primary impedance "looks

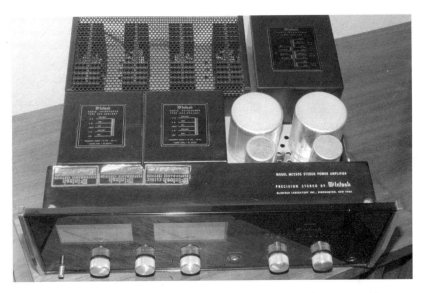

FIGURE 13.7   The large, rectangular black items in the right rear and front left of this stereo amplifier are the power and output transformers.

like" the value of impedance the output of the circuit connected to the primary winding wants to "see," and at the same time the transformer's secondary winding will present the input of the second circuit with the proper value of impedance that it wants to "see." This is accomplished by winding the transformer to present the desired value of impedance to each circuit. The formula used to determine this is Equation 13.2:

$$Z_R = (N_S \div N_p)^2 R_{Load} \tag{13.2}$$

where Z_R = the impedance reflected back through the transformer and what the circuit connected to the transformer primary winding would "see," $N_S \div N_p$ = the transformer turns-ratio, and R_{Load} = the input resistance of the circuit connected to the secondary winding of the transformer—the load.

The impedance of the circuit connected to the primary winding is called the *reflected impedance*, which is a function of both the transformer's turns-ratio and input impedance of the circuit the secondary winding is connected to.

Typical impedance-matching applications for transformers include input-or output-matching and inter-stage (between stages) transformers. A common use for impedance-matching transformers is the need to use 75-ohm coaxial cable on a 300-ohm impedance TV antenna. Figure 13.8 shows a 75–300 ohm TV matching transformer.

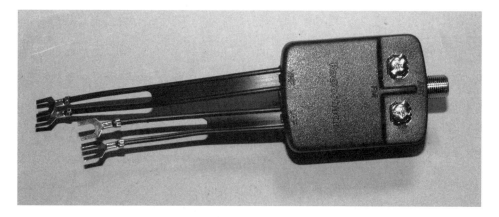

FIGURE 13.8 A 75 to 300 ohm TV antenna line impedance-matching transformer.

Another once-popular use for impedance-matching transformers is the matching of a tube or transistor output amplifier stage to the much lower impedance of loudspeakers. Typical tube output impedances range around several thousand ohms. Transistor amplifiers tend to have much lower output impedances, typically several hundred to perhaps a thousand ohms. In order to efficiently drive the speakers with low distortion (unintended discoloration of the amplified signals), some form of impedance matching is required. Up through the 1980s, some high-end (expensive) stereo amplifier manufacturers still preferred output transformers to accomplish this impedance-matching job. Two of the large transformers in Figure 13.7 are impedance-matching output transformers in a high-power stereo audio amplifier.

13.4   TYPES OF WINDINGS

Transformers are made using several distinct types of windings. The next sections present the major winding types used.

13.4.1   Voltage Step-up

When someone refers to a *step-up* transformer, they mean that the voltage is raised or stepped up from input to output. When a voltage higher than the input voltage to a transformer is desired, the transformer is wound with more turns in the secondary than the primary. A transformer with twice the number of turns in the secondary compared to the primary will provide roughly twice the value of primary voltage out. This type of transformer is used at the generating stations of commercial power companies. They are used to raise the generators' output voltage from several thousand volts to a much higher voltage, sometimes even to a million volts. As the voltage is raised to this level, the amperage required to deliver the same amount of power is reduced by the same ratio. Lower current requires smaller wires, and results in less line loss. An extreme version of a voltage step-up transformer is called a *Tesla coil*, named for its inventor, Nikola Tesla, 1856–1943, a famous inventor, engineer, and scientist. With the financial backing of the rich banker J.P. Morgan and the industrialist George Westinghouse, Tesla was instrumental in getting alternating current developed and ultimately adopted for commercial use. That is why we enjoy AC electricity delivered to our homes and businesses. Before Tesla prevailed and his AC devices became the norm, the prevailing form of electric power was DC. The main proponent of the DC system was Thomas Edison, the famous American inventor. Other important inventions Tesla made include a telephone repeater, the rotating magnetic field principle, the polyphase alternating-current system, the induction motor, alternating-current power transmission, wireless communication, radio, and fluorescent lights. He held more than 700 additional patents. A typical Tesla coil operates on household voltage and can step that voltage up to hundreds of thousands or even a million volts. Typically used for dramatic displays of sparks today in engineering and physics labs, Tesla originally developed them as a means to send large amounts of electrical power over distance. One infamous result of his experimenting with such ideas from his laboratory in Colorado Springs, Colorado, was the rather unpleasant shocks suffered by housewives hanging wet laundry to dry outside on wash lines. Tesla's high-powered transformer power system was the culprit.

13.4.2 Voltage Step-down

When a transformer's turns-ratio is less than 1:1, it functions as a voltage *step-down* transformer. This is the type of transformer that was traditionally used in most communications and home entertainment systems' power supplies until around the 1990s. 120 or 240 VAC was converted down (sometimes converted higher as well) to the voltages required by vacuum tubes and later transistor circuits. Vacuum tube circuits such as those in televisions required AC voltages of 50, 12, 6.3, and 3 to power the tube filaments, or "heaters," and 200–600 VDC for the anodes or *plates* in some sets. Later transistor-based circuits required voltages in the range of roughly 6–20 V. Some transformers used *multitaps* on a single secondary winding to provide the various AC voltages for the vacuum tube-based equipment, while others used separate windings for each desired voltage. Figure 13.9 shows the schematic diagram of a multioutput voltage power transformer.

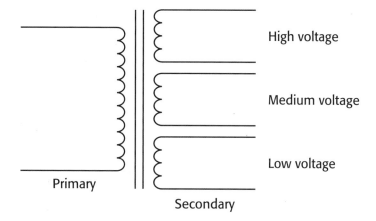

FIGURE 13.9 Multiple output voltage power transformer schematic.

Another use of a step-down transformer is in an AC arc welder. They need to supply a relatively low voltage of around 15–20 VAC, yet provide a large amount of current to make the heat for welding metal. A large step-down transformer using a heavy core and large gauge wires in the secondary is used. Welders operating on 240 VAC house current can provide 50 A or more, enough to weld up to ¼ inch steel plate. Figure 13.10 shows an example of such a welder.

FIGURE 13.10 Electric arc welder.

Another common use of a step-down transformer is a gun-style soldering iron. The copper tip is part of a single turn secondary winding. It is relatively large and carries a great deal of current. The primary of the transformer is designed to operate on normal household voltage. Refer to Figure 13.11 for a picture of a 100-watt soldering gun.

FIGURE 13.11 A 100-watt soldering gun.

13.4.3 Isolation Transformers

Isolation transformers are used to electrically isolate the output circuit from the input circuit (usually the commercial power line) while still providing the same output voltage as the line voltage. In this case, if a person should come into contact with both terminals of the secondary of the transformer, at least there would be a current limit provided by the transformer. If the person should come into contact with one side of the transformer's secondary winding, and the person was physically grounded, there would be no direct path back through ground to the commercial power line on the input side. Isolation transformers are essentially power transformers with safety as their main attribute. It should be noted however, that all true transformers, that is those with separate primary and secondary windings, offer some degree of electrical isolation from the input side. Isolation transformers are rated for operating voltage, frequency, and load current. In addition, most offer some degree of noise attenuation, since the high-frequency noise components present on the commercial 50- or 60-Hz waveform will be decreased from primary to secondary simply because the transformer has been engineered to transfer energy around the intended (much lower) power line frequency. Isolation transformers use an approximate 1:1 turns-ratio.

13.4.4 Autotransformers

A special type of transformer, called an *autotransformer*, has only a single winding and at least one connection, called a *tap*, partway along the winding. The portion of the winding from the cold end (ground) to the tap forms the primary portion, and the entire winding forms the secondary winding; the secondary transformer winding functions "automatically" due to the tapped connection. This type of transformer offers no electrical isolation and is used only in special applications. One application is the traditional "coil" in an automobile ignition system. It functions as a very high-voltage step-up transformer and is cheaper to manufacture than a conventional transformer as it uses less material and requires fewer manufacturing steps. Figure 13.12 shows the schematic symbol of an automotive coil autotransformer. Figure 13.13 shows an autotransformer, commonly called a *Variac®*, which can be used to test PCs at input voltages higher or lower than the normal 120 VAC. This unit functions as a continuously variable turns-ratio transformer.

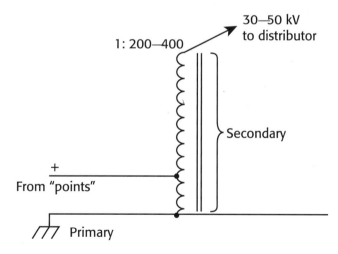

FIGURE 13.12 An automobile "coil" autotransformer schematic diagram.

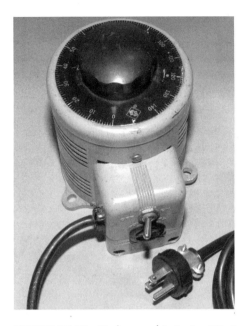

FIGURE 13.13   Variac used to test a PC at voltages that are higher or lower than the normal AC line voltage.

13.5   TRANSFORMER RATINGS

Transformers, depending on specific type, can be rated for many parameters, including:

- Intended frequency of operation
- Power handling capability
- Size and shape
- Turns-ratio
- Impedance ratio
- Primary voltage
- Primary current
- Secondary voltage
- Secondary current
- Primary or secondary with or without center taps
- DC resistance of primary and secondary
- Self-resonance of primary and secondary
- Inclusion of a Faraday shield

■ Tunable windings by slug or sliding contact
■ Overall efficiency

13.6   TRANSFORMER EFFICIENCY

Efficiency of a transformer refers to how much power applied to the transformer is passed on to the circuit connected to the secondary winding. Small transformers have efficiencies of around 85–90%, and the very large transformers used by commercial power generation and distribution companies are closer to 98% efficient. This is because at the level of power used by the commercial power companies, the extra cost for designing in better transformer efficiency is repaid quickly by the savings from lower power loss (power lost is mostly as heat). The largest transformers employ more efficient core materials, better winding styles, and better core construction methods, resulting in higher overall transformer efficiency.

13.7   SAFETY CONSIDERATIONS

Since transformers are inductive devices, the same types of concerns must be considered when working with them. Don't operate a transformer with excess voltage applied to it, nor allow excess current to be drawn from it. Never touch the exposed leads when connecting or disconnecting a transformer, as the inductive counter voltage (CEMF) of even a "low-voltage" transformer can be lethal.

Unless specially designed for it, transformers should not be operated in moist or wet conditions. Transformers should be operated only over their intended frequency range or they can be quickly damaged. Transformers designed for power supplies are specified to work in a specific type of load circuit. Using them with an incorrect type of load will damage them.

13.8   TYPICAL PC TRANSFORMER APPLICATIONS

PC transformer uses include the small transformer in most dial-up modems, which is used to provide DC isolation from the telco (telephone company), yet pass audio frequency signals (AC) both ways. The various coils one can see by examining any modern PC motherboard are actually inductors, part of switching power supplies. Other PC-related uses for transformers involve using an autotransformer as part of diagnostic testing of the power supplies. These special transformer units are usually referred to by one company's trade name for the device, Variac. Names for the same type of device by other companies over the years include Powerstat® and Adjust-A-Volt®. They are available at most large electronic parts supply stores, and the two most commonly seen units in the USA are those designed for either 120 VAC or 240 VAC input. They have a control knob that rotates a sliding carbon wiper along an uninsulated edge of the autotransformer winding, and makes continuous contact with the winding. The output voltage is taken across the portion of the total winding between the ground or common end and the wiper contact. This way the voltage can be adjusted with the knob position from 0–140 VAC in the case of the 120 VAC input types, and roughly 0–265 VAC or so with the units designed for 240 VAC input. The ability to vary the output voltage allows for device testing at input voltages other than the usual ones. For instance, a PC can be tested to see what voltage range will allow normal operation. So it is possible to test a PC's power supply over a range of perhaps 85–135 VAC, a rough range that the PC's switching power supply should normally accommodate with no problems. So a PC that is suspected of improper operation at slightly higher- or lower-than-normal input power line voltage can be quickly and easily tested. For instance, the author has a notebook computer whose small switching power supply is rated for 100–240 VAC input at a maximum current draw of 1.5 A. The supply provides 19 VDC at a maximum current of 2.64 A. If it were desired to test this supply to see if it could indeed provide the rated output voltage and current over the specified input voltage range, it would be necessary only to use a Variac, which could supply the required range of AC voltage at the specified current value, and see if the notebook computer would operate properly. A DC digital voltmeter or DVM is a good way to monitor the DC voltage delivered to the notebook PC while performing this test. Some of the more capable and expensive Variac units include an AC ammeter and AC voltmeter to monitor what the Variac provides to the load (in this case, the notebook's power supply). Figure 13.14 shows an older model Variac with an ammeter and voltmeter included.

FIGURE 13.14 A Variac with ammeter and voltmeter to monitor load conditions.

13.9   SUMMARY OF TRANSFORMER TYPES

Transformers are made to do the following jobs, depending on how they are designed and constructed:

- Step-up voltage.
- Step-down voltage.
- Provide DC electrical isolation from secondary to primary windings.
- Provide impedance matching between a circuit connected to the primary winding and one connected to the secondary winding.
- Provide two equal-amplitude but 180-degree out-of-phase signals to a circuit requiring these signals.
- Combine two equal-amplitude, 180-degree out-of-phase signals into a single signal.
- Provide a voltage step-up or step-down with a single winding (an autotransformer.)
- Be part of a tuned circuit, such as in communications gear, including radio transmitters and receivers.
- Block DC but pass AC.
- Provide a variable (user selectable) AC output voltage for testing power supplies and other uses.
- Provide any combination of the functions described in this list depending on how the transformer is designed and used.

13.10   SUMMARY

- Transformers consist of at least one tapped winding, and more commonly consist of two distinct and electrically isolated windings.
- The input winding is called the primary winding and the output winding is called the secondary winding.
- If a transformer has more turns on its secondary winding than its primary winding, it is known as a voltage step-up transformer.
- A transformer with fewer turns on its secondary than its primary is called a voltage step-down transformer.
- A transformer with an approximate 1:1 turns-ratio is called an isolation transformer, and is used for safety and to limit line noise and other forms of interference riding on the power line.
- Transformers are built to have unity coupling from the primary to the secondary. This means each magnetic flux line created by the primary current will cut through the secondary winding and induce a voltage in it.

- Transformers can perform many jobs, including changing voltage, providing equal-amplitude out-of-phase signals, combining two equal-amplitude signals into one signal, and block DC while passing AC. They can also be part of a tuned circuit and provide electrical isolation. No other single electrical component can do all of these things.
- Transformers can be made with laminated-iron cores, powdered iron or ferrite cores, or no core at all (air core), depending on the frequency range and power level they must work at.
- Small transformers are about 85–90% efficient, and large transformers used by commercial power companies are close to 99% efficient.
- Linear (older design) power supplies are typically transformer based.
- Nikola Tesla, with the financial backing of George Westinghouse and J. P. Morgan, was largely responsible for the worldwide adoption of AC power systems.

13.11 KEY TERMS

Primary winding	Autotransformer	Coupling
Secondary winding	Variac	Hum
Turns-ratio	Tap	
Transformer	Core	

13.12 EXERCISES

1. Transformers are designed to do which jobs?
   a. Increase voltage
   b. Decrease voltage
   c. Transfer power
   d. Provide electrical isolation
   e. All of the above

2. A transformer that provides higher voltage from the secondary winding than is supplied to the primary winding is called which type of transformer?
   a. Step-up
   b. Step-down
   c. Step-through
   d. Blocking

3. A transformer that is designed to couple a signal from one circuit to another and match the impedances of each circuit as well is called…
   a. a step-down transformer
   b. a step-up transformer
   c. an impedance-matching transformer
   d. a power transformer

4. A transformer with a connection partway along its winding(s) has what type of windings?
   a. Complete
   b. Incomplete
   c. Tapped
   d. Connected
   e. Shorted

5. A transformer can be designed to produce two equal-amplitude but 180-degree out-of-phase signals from a single input signal.
   a. true
   b. false

6. A transformer can be designed to combine two equal-amplitude but 180-degrees out-of-phase signals into a single larger signal.
   a. true
   b. false

7. A transformer designed to operate over power line frequencies is called…
   a. a power transformer
   b. a radio transformer
   c. an inductor
   d. a radio-frequency transformer

8. Transformers can get warm or even hot during normal operation because of . . .
   a. losses in the windings
   b. losses in the core
   c. nuclear radiation
   d. solar energy
   e. Answers a and b make the most sense

9. Switching power supplies usually do not incorporate transformers.
   a. True
   b. False

10. Transformers can help to select one frequency over another, and the frequency of operation can be varied, when they are a part of what type circuit?
    a. An electric arc welder
    b. A small linear power supply
    c. A radio tuner circuit
    d. A battery charger
    e. None of the above

11. Transformers can be designed to perform which of the following functions, all at the same time?
    a. Provide a secondary voltage increase or decrease over the primary voltage
    b. Provide electrical isolation from the input circuit to the output of the transformer
    c. Be part of a tuned circuit
    d. Combine two signals into one
    e. All of the above

12. A transformer with a variable-tap connection and a single winding is commonly called a…
    a. true transformer
    b. fixed autotransformer
    c. Variac
    d. diode
    e. resistor

13. A type of transformer without any core would most likely be found where?
    a. In an AC arc welder
    b. In a high-frequency transmitter
    c. In a small battery-operated receiver
    d. On a commercial electric power pole
    e. In an automobile

14. Small transformers are typically how efficient?
    a. 40–50%
    b. 50–60%
    c. 60–75%
    d. 85–90%
    e. 95–98%

15. What is the turns-ratio of a transformer that is designed to operate on 120 VAC and provide 20 VAC from the secondary winding?
    a. 10:1
    b. 20:1
    c. 1:20
    d. 6:1
    e. 1:10

16. Transformers are rated for the frequency range over which they are intended to operate.
    a. True
    b. False

17. Linear power supplies used for TVs and home entertainment equipment once used transformers exclusively.
    a. True
    b. False

18. A Tesla coil is an example of what?
    a. Very low-voltage output transformer
    b. Very high-voltage output transformer
    c. A transformer designed for small appliances
    d. A transformer used by the power companies on the power poles

19. Impedance matching transformers are commonly used in which applications?
    a. A notebook PC's power supply
    b. A desktop PC's power supply
    c. Connecting a TV antenna's 300-ohm line connection point into a 75-ohm coaxial cable
    d. On a dial-up modem card to interface with the telco line
    e. None of the above

20. Transformers can be the cause of both mechanical vibration and electrical hum.
    a. True
    b. False

21. A certain transformer is designed to operate on 120 VAC. It is designed to provide 12 VAC at 1 A to a load. Assuming a 100% efficient transformer, how much current is drawn in the primary circuit when this transformer is operating into a load that draws the maximum specified current?

22. Assume the same transformer specifications as in Question 21, but assume this time that the transformer is only 85% efficient. How much primary current is drawn?

23. What is the turns-ratio of a transformer that is designed to operate on 48 VAC and provide 10 VAC from the secondary? Assume a perfect transformer.

24. Describe a testing procedure using a PC and a Variac.

25. What is the danger of touching exposed wires connecting a low-voltage transformer?

26. What impedance is "seen" looking into the primary of a transformer with a primary-to-secondary turns ratio of 10:1 and a secondary load of 5 ohms?

14 Power Supplies

14.1 INTRODUCTION

Power supplies are common to nearly every piece of electronic equipment, from TVs and stereos to microwave ovens and PCs. They all perform the function of converting one type of current to another or one voltage to another or both. This chapter details the essential things a PC technician needs to know about the most common power supplies.

14.2   GENERAL TYPES OF POWER SUPPLIES

Power supplies are designed to perform specific jobs, and the general categories that power supplies fall into are the following:

- AC to AC, usually a transformer.
- DC to AC, known as an *inverter* circuit.
- DC to DC, known as a DC-to-DC converter.
- AC to DC, the type most often found in a PC and other computer-related equipment.

14.3   THE HALF-WAVE RECTIFIER CIRCUIT

Since the majority of PC-related equipment uses the fourth type of power supply, AC to DC, this chapter will concentrate on this type. You have already learned about transformers and the various jobs they are used for. The essential job they do in an AC-to-DC power supply is to change the incoming AC voltage from the power company into a desired higher or lower voltage. Since PCs and most PC-related equipment use transistors and other semiconductors, which are designed to operate on relatively low-voltage DC, the next job after converting the incoming high-voltage AC (in the 100–240 VAC range) into a lower voltage (15–24 VAC) is to change the AC into DC.

The AC-to-DC conversion is accomplished by one or more rectifier diodes, relatively simple solid-state devices that allow only a one-way path for current. The first rectifier diodes were a type of vacuum tube; modern rectifier diodes are semiconductor devices. The simplest of these rectifier circuits uses a single diode connected to the output of the step-down transformer. The output of the diode is a pulsating half-wave remnant of the complete AC sine wave. For this reason, the circuit is known as a *half-wave rectifier.* It is only used for the most undemanding circuits, as the output has such a large degree of ripple voltage in it. Ripple is the AC that remains after rectifying the complete AC sine wave. A large-value electrolytic capacitor is usually connected across the circuit output in order to filter or reduce a good portion of the AC ripple. Since there is still so much ripple, however, this circuit is used only for very low current demands by the load circuit; otherwise the ripple rises to a high level. Figure 14.1 shows the IV curve for a typical silicon rectifier diode. Figure 14.2 shows the schematic of a simple half-wave rectifier circuit with a filter capacitor connected and the typical output waveforms with and without the filter capacitor.

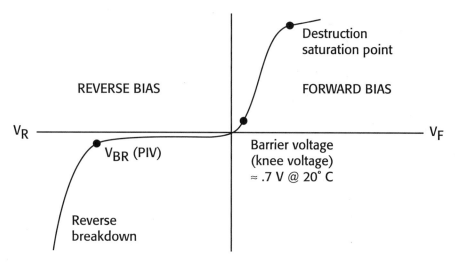

FIGURE 14.1   Silicon rectifier diode IV characteristic curve.

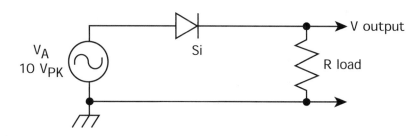

FIGURE 14.2   Half-wave rectifier circuit with representative output waveforms.

14.4   THE FULL-WAVE, CENTER-TAPPED CIRCUIT

A better circuit than the simple half-wave rectifier circuit is called a full-wave rectifier circuit, since the output consists of portions of the full input AC waveform, and is more efficient. It requires two rectifier diodes, however, and a special center-tapped transformer. The transformer provides the desired voltage change as well as providing two equal-amplitude waveforms 180 degrees out of phase with each other. The two diodes conduct on alternate portions of the AC waveform, since when the top of the transformer's secondary voltage goes positive, the bottom end of the secondary winding goes negative, relative to the center tap. So the diodes conduct first one, then the other, then the first, over and over. They feed the output 180-degree pulses of current to the load in the same direction. Since the transformer is tapped halfway along the secondary, the output waveform is only a little less than one half of the total secondary voltage. The reason the output voltage is a bit less than one-half of the secondary voltage is because of the forward voltage drop of the diodes. The majority of rectifier diodes are made from the semiconducting element silicon. Silicon diodes show a forward voltage drop of around 0.7 volts. So the output voltage peaks of the full-wave rectified output waveform are down some 0.7 volts from the peak value of the AC waveform applied to them. Since both halves of the input sine wave are rectified and appear as current pulses on the output, this circuit is about twice as efficient as the simple half-wave diode rectifier circuit. However, the same transformer secondary output voltage, the full-wave, center-tapped circuit provides only a bit less than one half of the peak output voltage compared to the half-wave circuit. Plus, a center-tapped transformer costs more than one without the center tap. So a more cost-effective circuit was developed, as discussed next. Refer to Figure 14.3 for schematic and output waveforms for a full-wave, center-tapped rectifier circuit.

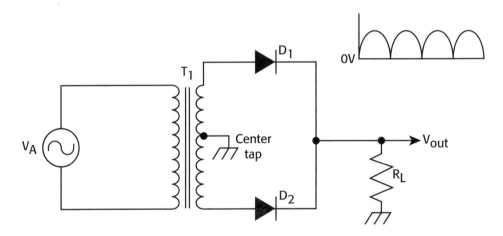

FIGURE 14.3 Schematic and waveforms for the full-wave, center-tapped rectifier circuit.

14.5 THE BRIDGE RECTIFIER CIRCUIT

By the 1960s or so, the price and availability of reliable semiconductor diodes made it cost effective to develop a newer rectifier circuit that would replace the larger, less efficient, and more costly vacuum tube rectifier circuits. This new circuit offers the best of both the half-wave circuit and full-wave center-tapped circuit. It provides an output waveform that is easier to filter into DC than the half-wave circuit, and provides nearly as much DC output voltage. It does not require a more expensive center-tapped power transformer, but it does use four rectifier diodes instead of one or two. The circuit is called a *bridge*, because the arrangement of the four diodes looks like a bridge.

The circuit functions when two diodes, effectively in series, conduct together on alternate half cycles of the input waveform from the other two diodes. This arrangement provides a full-wave output. Since there are two diodes in series each half cycle, their forward voltage drops add. So roughly 1.4 volts are dropped as they conduct. Still, this additional voltage drop is small, and is simply figured into the requirements when specifying an appropriate power transformer for the circuit.

Virtually all linear DC supplies use a full-wave bridge circuit due to the cost-effectiveness and reliability it provides. Refer to Figure 14.4 for a diagram of a bridge rectifier circuit.

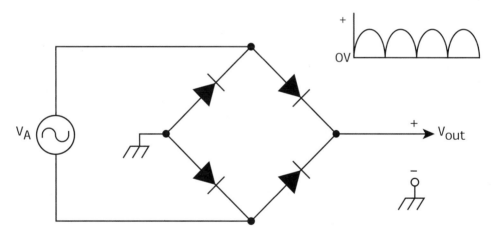

FIGURE 14.4 Bridge rectifier circuit.

14.6 LINEAR REGULATED DC SUPPLIES

Since the load connecting to a power supply can change its current demands, the voltage from the supply can change as well. This is due mainly to the power transformer's changing voltage drop as the current delivered by the secondary winding changes, due in turn to load current demand changes. Remember, a transformer is made with wire windings, and wire has resistance. So as current through the windings increases or decreases, the voltage dropped due to this winding resistance changes as well. Ohm's law applies here (see Equation 14.1):

$$V_{Drop} = I_{Load} R_{Wire} \tag{14.1}$$

So if a constant voltage to the load is required, another approach is needed. This entails a process known as *voltage regulation*. The simplest version involves a special type of diode known as a voltage-regulating *Zener diode*, named after the discoverer of the process by which the device works. The effect utilized in this type of circuit makes use of the characteristic IV curve of a Zener diode in the reverse breakdown region. Once biased into the Zener region, any increase in voltage across the Zener causes it to conduct more, and this serves to stabilize the output voltage. The result is not perfect regulation, but it is fairly good for such a simple circuit. The Zener diode connected in parallel with the output in this fashion is known as a *shunt circuit*. Zener diodes are available in ratings of up to 100 watts of dissipation. Those rated at more than a few watts are commonly mounted on heat sinks. Zener diodes are rated for their regulating voltage, V_Z, as well as their power rating and mounting style. Zener diodes can be made with breakdown (regulating) voltages from about four volts to several hundred volts. Figure 14.5 shows some Zener diodes. Figure 14.6 shows a typical Zener diode characteristic IV curve.

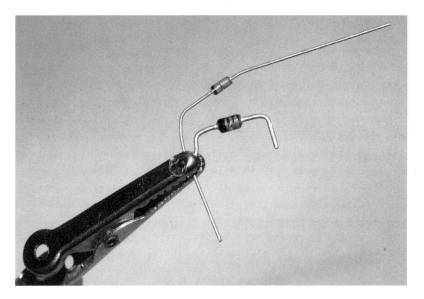

FIGURE 14.5 Zener diodes.

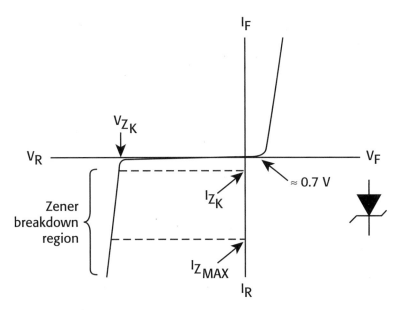

FIGURE 14.6 Zener diode characteristic curve.

14.6.1 Simple IC Regulators

Beyond simple shunt Zener diode voltage regulator circuits, circuits employing transistors as current pass devices were developed. These offered more accurate voltage regulation as well as increased current handling capability.

The next evolution was to integrate the pass transistor with several other components, creating a single-chip regulator. These became popular around the 1970s and are still widely available today. One of the more popular designs is called a "three-pin regulator," as it has three connecting pins, one for unregulated voltage into the chip, one for regulated voltage out of the chip, and one for ground. These are available in positive or negative varieties. More elaborate designs are available that offer both positive and negative output regulated voltages. Other designs offer the capability to provide a variable output voltage with the use of only a few simple additional components such as resistors, capacitors, and a potentiometer to adjust the voltage. Figure 14.7 shows a 3-terminal positive regulator circuit.

14.6.2 High-current Regulated Supplies

There are two general schemes for providing more regulated output current. One uses chip regulators designed for higher output, in the 10 A range, although these

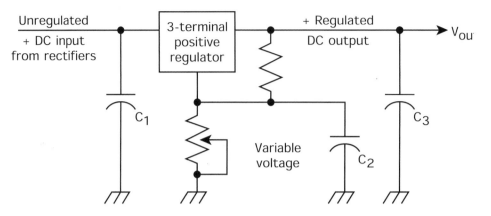

FIGURE 14.7 Three-terminal positive variable voltage regulator circuit.

units usually offer less attractive regulation and ripple specs than most of the lower current units. The other uses a small-current chip regulator to drive one or more high-current series pass transistors. Figure 14.8 shows a schematic of a high-current regulator circuit using pass transistors to boost the output current capability. *Series-pass transistors* are units designed to handle currents in excess of 1 A. They are commonly mounted on large heat sinks capable of dissipating large amounts of heat, thus keeping the pass transistor in its safe thermal operating range.

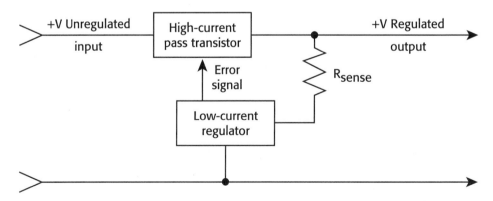

FIGURE 14.8 High-current regulator circuit using pass transistor to boost the output current capability.

14.7 SWITCHING SUPPLIES

Linear regulated supplies offer good regulation specs and can provide low ripple—but at a cost. The costs are the need for a transformer to lower the incoming AC line voltage, and relatively large amounts of capacitance required to filter the low-frequency ripple voltage. Transformers are relatively large, expensive devices. They can radiate magnetic fields and can pick up stray magnetic fields from other parts of the circuit in which they are used. Large electrolytic capacitors are expensive, subject to a large amount of leakage current, and take up a large amount of space on a chassis.

To overcome these shortcomings, a new type of regulated power supply was engineered: the switching supply, or *switcher*. A switcher does not require a transformer. It rectifies the incoming AC line voltage directly, which is then applied to a regulator stage. The regulator stage is turned on and off, and the rate of conduction, (pulse rate) or the amount of on-to-off time (*pulse width*), is varied in order to maintain a constant output voltage. The switching action, on and off over and over, is different from a linear supply. In a linear supply, the main conducting device, a pass transistor, is always on and conducting, and the degree of conduction is changed in the regulating process. A linear supply always dissipates the difference between the input voltage and the load voltage, times the load current (see Equation 14.2):

$$P_{Diss} = (V_{In} - V_{Out}) \div I_{Load} \qquad (14.2)$$

In a switcher, the pass device is either fully conducting or fully off, so it dissipates very low power. When it is off it appears as a very high resistance, so no current flows through it. When it is conducting, it appears as a near-perfect closed switch, with very low resistance, so again, it dissipates very little power.

14.7.1 Switching Noise

Switching a device on and off is a nonlinear action, and as explained earlier, nonlinearity generates new frequencies, called *harmonics*. This is essentially undesired noise and must be filtered out of the DC output before applying it to the load (the PC circuits). The addition of a hash filter, a circuit designed to greatly reduce the switching noise and harmonics, is required, something not needed with a linear supply. Still, the advantages of using a switching supply over a linear type for most applications outweigh the added cost. Switchers are smaller, lighter, and more efficient than a linear supply of equivalent output capability. This is the type of supply used in PCs and most PC-related equipment.

14.7.2 Wide Input Voltage Range

Switching supplies can operate over a very wide input voltage range, generally of 2:1 or more. It is common for a switcher used for a PC supply to be able to work over a range of perhaps 100–240 VAC, something a transformer-based linear supply can not do.

14.7.3 EMI Filters

Most power supplies also use an electromagnetic interference (EMI) filter at their input to greatly reduce the amplitude of frequencies significantly higher than the line frequency. If allowed to enter the power supply, they could cause damage to the supply or data loss in the computer.

14.7.4 Small "Block" Supplies

There are many types of small switchers used to power notebook PCs and peripheral equipment such as printers, scanners, et cetera. For example, the author uses a notebook PC powered by a switcher that measures approximately $4 \times 1\frac{1}{2} \times 1$ inches and is rated for an input voltage range of 100–240 VAC and an output voltage of 2.6 A @ 19 VDC. It fits in the palm of your hand and even after several hours of operating, is only warm to the touch. Refer to Figure 14.9 for a picture of a small switching block or *brick* supply.

FIGURE 14.9 Small block notebook PC switching power supply.

14.7.5 Typical PC Power Supplies

The typical PC power supply is enclosed in a metal chassis and mounts internally. Various form factors (exact size, shape, and layout) have been used over the years, as the PC has continued to evolve. Some of the major PC power supply form factors include: PC/XT, AT, ATX, ATX+, and NLX. Most new tower PCs use the ATX+ form factor.

ATX+

An ATX+ also known as ATX12V, refers to an enhanced power supply with an extra 4-pin connector called P4 that plugs into a mating socket on the motherboard. It is used to supply 2 lines each of +12 V and ground to an on-the-mothreboard switching voltage regulator circuit which converts the +12 V DC down to +3.3 V DC, to power the new processors which would otherwise max out the existing + 3.3 V DC lines. Refer to Figure 14.10 for a picture of a P4 power supply plug.

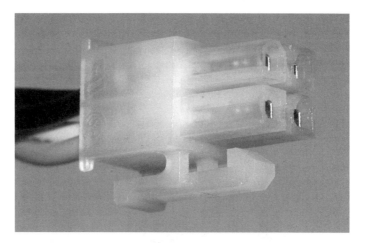

FIGURE 14.10 A PS power supply plug provides extra power for Intel P-4 and AMD processors.

EPS

There is also an even more heavy-duty power supply style designed for high-end servers, running multiple processors and RAID arrays, called EPS or enhanced power supply systems. They include a 6 or 8 pin extra plug to provide even more power for these more demanding systems.

Refer to Figure 14.11 for a picture of a group of secondary voltage regulators below the P1 socket on a motherboard.

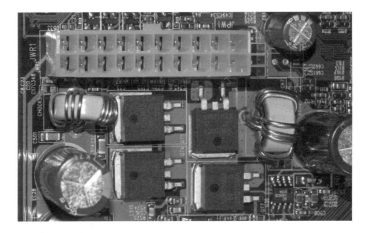

FIGURE 14.11 Voltage regulators below the P1 socket on a motherboard.

All of these PC supplies use at least one fan; some use two. At least one is mounted inside the supply and the second, if used, can be mounted either internally or externally. Refer to Figure 14.12 for a picture of an ATX power supply with two fans.

FIGURE 14.12 ATX PC power supply with two cooling fans (one is inside the power supply case).

The electrical specs of the ATX supply are shown in Table 14.1.

TABLE 14.1 ATX Power Supply Specifications

AC Input	115V/ 230V	7A/4A	60/50Hz				
DC Output	+3.3 V	+5 V	+12 V	−5 V	−12 V	+5 VDC SB (standby voltage)	Total
	28 A	30 A	15 A	0.3 A	0.8 A	2.0 A	300 W

Sizing PC Power Supplies

In order to provide safe, reliable PC operation over extended times and varying temperature conditions that are commonly encountered in a home or business environment, a PC power supply must be properly sized. There are a couple ways to do this. First, one can painstakingly add up all the existing and planned expansion cards and figure the current requirements per voltage required for each, plus add the electrical power demands of the motherboard populated with the main processor, chipset, and RAM, plus add in all the fans on both the motherboard and case, and then figure a conservative amount of derating before finally selecting a suitable supply.

The second method involves studying manufacturer's product specs and seeing what wattage supplies they use for a given set of installed components. So you can get a good "feel" for which wattage supply is typically used in a "low-end," "mid-sized," or "high-end" PC system. Pay attention to the features servers use, such as number and size of case fans, as well as power supply wattage ratings. Servers are expected to operate "24/7/365" without downtime, so everything inside is very conservatively rated. The benefits of more-than-adequate cooling can be achieved by the home PC builder/user as well. Larger cases offer not only more expansion room, but generally better cooling and spaces for more case fans. Don't overlook using fans to cool hard drives. Any hard drive operating at 7,200 rpm or higher gets hot enough after an hour of use that it is uncomfortably hot to the touch. They need good airflow to avoid either bearing failure or electronic chip destruction. It is unadvisable to mount two hard drives right next to each other as the heat from each will combine and fry parts. Leave an empty drive bay between two drives to allow for cooling airflow. There are dedicated fan/heat sinks available to cool drives. Some of the better-designed cases offer cooling fans placed directly in line with the hard drive bays and are much preferred over the cases that do not offer this feature. Don't rely on the one or two power supply fans to cover the cooling needs of the entire PC. There is also no need to overdo the cooling. A very basic computer does well with only the required CPU fan and the power supply fan. A slightly better machine will do fine with one case fan, and a souped-up machine usually does fine with two case fans, an intake and an exhaust fan. Make sure to follow the CPU manufacturer's requirements. For example, for its CPUs of over 3 GHz, Intel® specifies a separate vent at the side of the case with a plastic tube leading to the CPU fan.

14.7.6 Other Types of Power Supplies

There are other types of power supplies besides the AC-to-DC type presented. The next section describes some of these other types of power supplies.

AC-to-AC Supplies

An AC-to-AC supply consists of a transformer, and as presented earlier, it can be used to provide voltage step-up, voltage step-down, or isolation.

DC-to-AC Supplies

A DC-to-AC power supply, called an *inverter*, consists of a DC source, usually a cell or battery, followed by an electronic circuit called an inverter, which essentially "chops up" or turns the current on and off at a desired rate, say 60 times a second (60 Hz). This square wave is then applied to a voltage step-up transformer, which provides the approximate desired output voltage to operate the AC load. Such inverters are commonly used to power small household AC appliances from the 12.6 VDC batteries used in recreational vehicles (RVs) and automobiles.

DC-to-DC Converters

A DC-to-DC power supply, called a DC-to-DC *converter* is like a DC-to-AC inverter with a rectifier and filter section added to return the higher-voltage AC output from the transformer back into smooth DC. A common example of this type of supply is found inside most photoflash units used with 35 mm and modern digital cameras. The battery consists of two or four "AA" cells. They power the inverter, and the high voltage of around 300 VDC is used to fire a photoflash tube when the camera's shutter button is depressed. The small transformer used in such circuits often vibrates slightly at the inverter's chopping frequency, and as the "energy storage" capacitor charges up, and the electrical load lessens, the frequency can be heard to increase, and the camera operator knows the flash is getting ready to fire again. Other common DC-to-DC converters include adapters that power DC-powered equipment such as laptops from a car, boat, or airplane.

14.8   BACKUP POWER SUPPLIES AND RATINGS

PCs need to be protected from a variety of power-line related problems. First, the commercial power can stop entirely, which is called a *blackout*. Such an abrupt, unplanned power outage will usually cause lost data on the affected PC. The power line voltage may sag to a low value, called a *brownout*, which again can cause disastrous data loss. The power line may transmit very short-duration high-voltage "spikes," which can damage internal PC components, and cause data loss as well. Finally, the line voltage may have other, undesired frequency components "riding" on it, such as a high radio frequency due to the close proximity of the PC to a radio transmitter, or other electronic equipment that is radiating interference. All of these potential problems must be addressed in what is generally termed *power line conditioning*.

There are two main types of backup power supplies used to run PCs. The first is called an *uninterruptible power supply* (*UPS*). In this type, the AC line runs a battery charger that constantly powers a sealed lead-acid "gel cell" or other type of sealed battery, usually rated at 12.6 volts. This battery powers an inverter, which chops the DC into a pulsating DC square wave, which powers a voltage step-up transformer. This AC sine wave approximation, called a *simulated sine wave*, powers the PC. When the commercial power fails, there is no change, as the PC is always powered by the internal battery and inverter circuit. This will run until the battery reaches some low-voltage point where the inverter can not maintain the required output voltage. How long this lasts is called the backup supply's *run time*, and is a function of the battery size and condition as well as the load placed on the supply by the PC.

The second type of backup supply is known as a *standby power supply* (*SPS*). In this type, there is an additional switch section, which normally connects the input AC line voltage directly to the output receptacles to power the PC and peripherals. Like the UPS, there is a battery charger that keeps a float (maintenance) charge on the battery to maintain its readiness. When the commercial power fails, the switch section very quickly connects the output to the inverter, powered by the battery, and it will power the loads for its rated run time. In order for this type of supply to properly function, the switch must change over within one-half cycle of the input waveform. Considering the normal line current frequency in the USA of 60 Hz, this gives a period of:

$$t = 1 \div f$$
$$= 1 \div 60 \ Hz$$
$$= 16.67 \ ms$$

So one-half cycle at 60 Hz is approximately 8.33 ms. As long as the SPS can achieve the changeover in less than this time, the load (the PC) will not even know it. Most good SPSs can accomplish the changeover in about 5 ms. The author has used this style of backup system for many years without ever experiencing a failure.

14.8.1   Smart Backup Supplies

Most of the large PC backup power supply manufacturers also make a type of supply called "smart." In general, these types communicate with the PC they protect via a wired link, either via the COM port or USB port. When the power goes out and the PC is running off the backup supply, the supply can automatically send a signal to the PC to shut down cleanly, even to log off a LAN connection. These settings are done through a software program installed on the PC that is provided with the backup supply unit. The percent of charge left in the backup system can be chosen to trigger an automatic PC shutdown.

14.9   TROUBLESHOOTING

Troubleshooting is an art as well as a science. The next section presents some good techniques for troubleshooting power supplies.

14.9.1   Use All of Your Senses

Troubleshooting PC supplies involves several things. Pay attention to sights, sounds, smells, and other odd behavior of power supplies. Some will make a buzzing or clicking sound if they fail or are connected to a short. In such cases, unplug the supply, locate the short, and try again. If the supply works, then no permanent harm has been done to it. Inexpensive power supplies may fail with a short and not come back again.

14.9.2   Connect a Load

In all cases, make sure to have at least the minimum required load connected to the PC supply before applying power or it may be permanently damaged. The minimum load is usually considered to be a fully populated motherboard, meaning one containing a working processor, chipset, and RAM, plus at least one hard drive. CD, DVD, and floppy drives don't count, since they are not always spinning, and therefore don't draw current when idle.

Most good-quality PC supplies include a built-in load resistor to supply a minimum load. Inexpensive supplies may not, and if they are powered on with no load connected, they can be instantly and permanently damaged.

PC supplies should be checked with a DVM for proper output voltage levels only when operating into a proper load. Do not disconnect power supply cables in order to check voltages; instead, use the "back probing" technique to do this. This involves inserting the DVM's probes into the back of the power supply connectors to connect to the terminals.

14.9.3 Using a Variac

PC power supplies are specified as having the capability to provide the rated output voltages at the rated output current over a range of input voltage. A good way to test a PC supply's capability to do this is by varying the input voltage applied to it using an autotransformer specifically designed for this purpose. Variac is the name most commonly used for this type device. The PC is plugged into the Variac, and the output of the Variac can be varied from 0 to about 140 VAC. Most PC power supplies are specified to work properly over an input voltage range of about 85–135 VAC. So by monitoring the supply's output voltages with a DC voltmeter while slowly varying the input AC voltage to the supply, it can be tested under real-world operating conditions. Be sure the PC supply is actually powering the PC it is designed for at the time of testing; otherwise the test is meaningless.

Motherboard Monitoring Programs

There are two styles of motherboard monitoring programs: ones that get invoked under the CMOS setup routine and those that get invoked by clicking on an icon on the desktop. These programs are used to provide valueable feedback to the user on operating conditions such as cooling fan speed in RPM, case and processor temperature, and all power supply and motherboard voltages. Refer to Figure 14.13 for a picture of a hardware monitoring program which operates from the desktop.

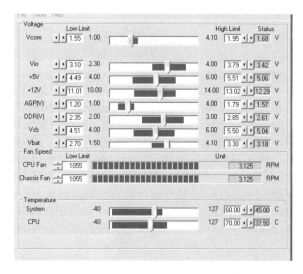

FIGURE 14.13 A hardware monitoring program that is invoked from the desktop.

14.10 SUMMARY

- Power supplies come in many types: AC to AC, DC to AC, DC to DC, and AC to DC.
- The type of power supply used by PCs is AC to DC.
- There are linear supplies and switching supplies. Most modern power supplies are the switching type.
- Switching supplies can operate over a wide range of input voltages and still provide well-regulated output voltage.
- The remnants of the AC input present on the output of an AC-to-DC power supply is called ripple and must be reduced.
- Capacitors and inductors are used as filter elements in power supplies to reduce AC ripple present in the output.
- AT power supplies provide +/– 12 VDC and +/– 5 VDC.
- ATX power supplies provide +/–12 VDC, +/–5 VDC, + 3.3 VDC, and + 5 VDC SB.
- Standby voltage is provided to the motherboard by ATX power supplies in order to allow wake-on-modem (WOM) and wake-on-LAN (WOL) functions to work.
- An ATX PC that is plugged into an AC wall outlet is still providing power to the motherboard even though the PC may be turned off.
- Never wear an ESD wrist strap when working on a powered unit, including an ATX-style PC that is plugged into the AC power socket.
- The half-wave DC supply is the simplest and least expensive, but is limited to applications requiring very low current; otherwise the amount of ripple in the output is far too high.
- The full-wave center-tapped rectifier arrangement uses a center-tapped transformer and two diodes to provide a smoother output but at a cost of a more expensive power transformer and one more diode compared to the half-wave design.
- The full-wave bridge rectifier arrangement is the best one since it does not require a center-tapped power transformer and its output is easily filtered. All modern AC-to-DC supplies use a bridge rectifier.
- PC power supplies are not user-serviceable and should normally not be worked on except for replacing the cooling fan.
- Dangerously high voltages are present in most switching power supplies.
- Switching power supplies should be operated while connected to a suitable electrical load, usually a fully populated motherboard and at least one hard drive.
- Use all your senses when troubleshooting.

- Backup power supplies are designed to give some run time after the commercial power goes out. They are rated in terms of VA, watts, or both, as well as run time for typical PC setups.
- Power line conditioning means removing or greatly attenuating potentially damaging AC line components.
- Spike protectors are the minimum protective circuit devices that should be used. They tend to wear out over time and lose their effectiveness.
- EMI filters reduce radio frequency energy that may be riding on the incoming AC line.
- Good backup power supplies provide for voltage spike protection and EMI filtering, and also provide runtime when the commercial power goes out.
- The best backup power supplies also provide for under- and over-voltage compensation.
- A Variac is a variable output voltage transformer that can be used to test PC power supplies for their capability of operating on slightly over- or under-AC line voltages.
- Some high-end backup power supplies also offer power-factor correction.

14.11   KEY TERMS

Diode

Half-wave rectifier

Full-wave, center-tapped rectifier

Full-wave bridge rectifier

Linear power supply

Switching power supply

Series-pass transistor

Voltage regulation

Ripple voltage

EMI

Voltage spike

UPS

SPS

14.12   EXERCISES

1. Which type of power supply is used in a photoflash unit?
   a. AC to AC
   b. DC to AC
   c. AC to DC
   d. DC to DC
   e. none of the above

2. Which type of power supply is the main PC power supply?
   a. AC to AC
   b. DC to AC
   c. AC to DC
   d. DC to DC
   e. None of the above

3. What percent load voltage regulation should a good-quality PC power supply provide?
   a. 20%
   b. 10%
   c. 5%
   d. 2%
   e. 1%

4. Which line voltage(s) are the most commonly used to power the worlds' PCs?
   a. 50 VAC
   b. 120 VAC
   c. 200 VAC
   d. 240 VAC
   e. Both b and d

5. Which line voltages are the most commonly used in the home in the USA and Canada to power PCs?
   a. 50 VAC
   b. 120 VAC
   c. 200 VAC
   d. 240 VAC
   e. 400 VAC

6. Which PC power supply voltages are used by the fans and drive motors?
   a. 12 VDC
   b. 5 VDC
   c. 3.3 VDC
   d. 2.0 VDC
   e. 1.0 VDC

7. Which PC power supply voltages are used by the logic boards on drives?
   a. 12 VDC
   b. 5 VDC
   c. 3.3 VDC
   d. 2.0 VDC
   e. 1.0 VDC

8. Considering a modern PC loaded with drives and memory, is it considered good engineering practice to rely solely on the power supply fan to provide total PC cooling?
   a. Yes
   b. No

9. What voltage(s) does the four-pin connector on an ATX+ PC power supply provide?
   a. +12 VDC
   b. +5 VDC
   c. 3.3 VDC
   d. Both a and b
   e. Both b and c

10. Modern processors run on less voltage than the older processors.
    a. True
    b. False

11. A processor that requires 3.3 VDC gets that voltage...
    a. directly from the main power supply
    b. directly from the power line
    c. from a secondary supply on the motherboard
    d. could be a or c, depending on the processor and motherboard

12. It's OK to work on an energized PC power supply since there are no dangerous voltages present.
    a. True
    b. False

13. The remains of the AC waveform still present on the output of a DC power supply is called . . .
    a. AC
    b. DC
    c. pulsating DC
    d. ripple
    e. hash

14. An EMI filter is used to . . .
    a. convert AC to DC
    b. convert DC to AC
    c. greatly reduce the amplitude of high-frequency noise present in the line power
    d. provide backup power in case of an outage
    e. none of the above

15. A spike protector is used to do what?
    a. Keep the PC running during a blackout
    b. Reduce high-frequency noise from the power line
    c. Greatly attenuate very short duration, high voltages
    d. Make up for low-input voltage conditions
    e. None of the above

16. Backup power supplies are rated for what parameters?
    a. Input voltage range
    b. Output voltage
    c. Volt-amperes
    d. Watts
    e. Can be all of the above

17. Some better backup power supplies communicate with the PC they power how?
    a. Through the AC line cord
    b. Through a serial (COM) port
    c. Through a USB cable
    d. Via a radio link
    e. Via the video cable
    f. Either b or c

18. A UPS differs from an SPS how?
    a. A UPS uses a switch section
    b. An SPS uses a switch section
    c. Only a UPS uses an internal battery
    d. Only an SPS uses an internal battery
    e. None of the above

15 | Tubes, Transistors, and Integrated Circuits

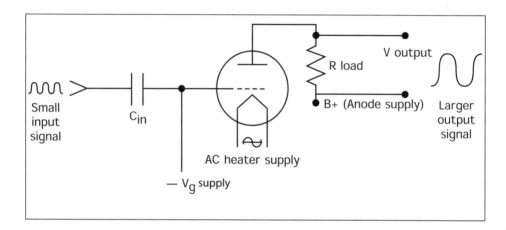

15.1 INTRODUCTION

You no doubt already know that nearly all modern electronic equipment uses devices generally termed integrated circuits (ICs) or chips. This chapter looks at the major types of ICs used in PCs and associated equipment, as well as a bit of background on the devices they evolved from.

15.2 VACUUM TUBES AND TRANSISTOR CIRCUITS

The following section presents vacuum tubes and transistors, electronic devices that led to the development of integrated circuits and ultimately made the modern computer possible.

15.2.1   Vacuum Tubes

Vacuum tubes operate by a principle called *thermionic emission*. The root word *therm* represents thermal (heat) and *ionic* refers to electrically charged atoms. So thermionic means using heat to produce electrical charges. A high-resistance wire, called the *heater* or *filament* is powered by electricity to produce a dull orange glow. This is similar to an incandescent light bulb, except that in order to achieve much longer filament life, tube filaments are run at a voltage lower than one that would produce a strong light.

The Vacuum Tube Diode

The most basic vacuum tube has just two electrodes. In addition to the filament, there is a second conducting metal structure called the *plate*. The plate and filament are mounted inside the evacuated tube, and connections from each electrode are made to the outside via conductors, usually in the form of metal pins that are inserted into a mating socket that stabilizes the mounted tube. Current powers the filament to heat up and produce ions and electrons. If the plate is connected to a high enough positive electric potential, it will attract the negatively charged electrons. This will happen only if the plate voltage is sufficiently high for a given tube type, as different tubes use different electrode spacing and plate shapes. The plate bias power supply is connected via its negative terminal back to the filament supply. Current will flow only when the plate is positive with respect to the filament. The more accurate name for a tube filament is *cathode*, meaning the negative electrode. Electrons flow across the distance inside the partial vacuum inside the tube to the plate, which is more accurately termed the *anode*, meaning the positive electrode. Current will flow only in this one direction through the tube circuit because if the anode is made negative, the electrons, which are negatively charged, will be repelled by the negative charge on the plate and no current will flow in the tube. The first job to which vacuum tubes were put was changing AC to DC, a process called *rectifying*, that you learned in Chapter 14, "Power Supplies." The first tubes, which contained a filament and a plate, are known as *diodes*, meaning "two-electrode tube," and were used for rectifying AC into pulsating DC. Refer to Figure 15.1 for a schematic diagram of a vacuum tube diode.

The Triode

If a third electrode, called a *control grid*, is added inside the tube, a three-electrode tube is formed, called a *triode*. The essential concept of this type of vacuum tube, first named the *Audion* by its inventor, the American scientist Dr. Lee DeForest, in

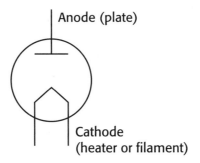

FIGURE 15.1 Vacuum tube diode schematic symbol.

1906, is this: a small grid voltage can control a large plate voltage. Triodes are still used in some specialty applications; such as high-end audio amplifiers used by guitar players and home audio fans called audiophiles (sound lovers). They feel strongly that the sound produced by a triode tube has a "warmer, sweeter" quality compared to sound reproduced by solid-state amplifiers.

Just as in the vacuum tube diode, in a triode tube the hot filament produces enough energy to cause free electrons and ions to be "boiled off" from it. The plate is charged to a relatively high positive voltage from the plate power supply, known as B+ on tube schematic diagrams. The positive plate voltage draws the negatively charged electrons to it, and they flow out of the plate, back through the plate supply and around to the common connection with the filament circuit. The control grid resembles a screen, made up mostly of holes, and it is physically located very close to, but not touching the filament. Remember that like electrical charges repel each other, so by placing a relatively small negative voltage on the grid, the electric field produces a repulsive effect and repels the stream of electrons which would otherwise flow from filament to plate. If a small negative control grid voltage is used, it causes a slightly decreased plate current. If sufficient control grid voltage is used, the effect will be to completely stop or cut off the plate current. This value is known as V_{gco}, for the grid cutoff voltage. This value is specific to each type of tube and for the value of plate voltage used. Control grid voltage can be set anywhere from zero to cutoff to start with. This is known as the *grid bias voltage*. A small signal voltage can be superimposed on the DC grid bias voltage by using a coupling capacitor, which will "pass" the AC signal while blocking the DC bias voltage.

Refer to Figure 15.2 for a schematic symbol of a vacuum tube triode tube.

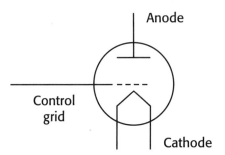

FIGURE 15.2   Vacuum tube triode schematic symbol.

Amplification

If the plate circuit contains a large value resistor, then the plate current will develop a relatively large voltage due to the plate current flowing through it. This is the output voltage, which can be many times larger than the small input signal that is applied to the tube's control grid. For this reason, vacuum tubes are said to be *voltage-controlled* devices. Refer to Figure 15.3 for a schematic of a triode vacuum tube amplifier circuit.

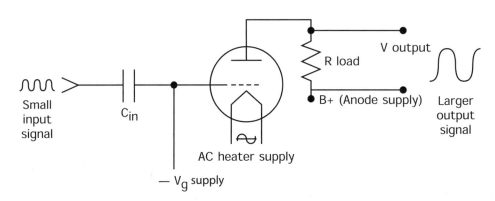

FIGURE 15.3   Triode amplifier circuit.

When a small electronic signal is applied to a device or circuit that produces a close approximation of the input signal, only larger in terms of voltage, current, or both (power), then the signal is described as having been *amplified*. The triode vacuum tube was the first device capable of electronic amplification. Dr. DeForest called it an Audion since the first use for the new triodes was amplifying weak received radio signals containing audio (sound) information.

Additional Tube Elements

In an attempt to get triodes to work at higher frequencies and provide more gain, an additional grid, called a *screen grid*, was placed between the control grid and the plate. This was done to electrically "screen" the plate from the cathode in order to reduce the cathode-to-plate capacitance, and allow the tube to be used at higher frequencies. The screen grid requires a positive voltage, somewhat less than the plate voltage, and normally draws very little current. Because of the extra bias supply needed, it requires a more complex circuit than the triode. This type of tube has four electrodes and is called a *tetrode*. Tetrodes provide increased gain over a triode.

As higher and higher plate voltages were used in tetrodes, in an attempt to get higher gain, a problem cropped up. As the electrons were accelerated faster by using higher plate voltages, they would strike the plate and instead of flowing out of the tube as plate current, they would knock other electrons that had already arrived at the plate off of it. This is known as *secondary emission*, the first emission being the electrons leaving the cathode. Secondary emission causes plate current to drop off beyond a certain plate voltage for a given tetrode.

The solution was to place another grid, this one between the screen grid and the plate, and to place a negative voltage on it. The system works like this: the secondary emission electrons bumped off the plate were now repelled back to it by the extra electric field force of the new grid, called a *suppressor grid*. This newer type of tube works well at higher plate voltages and high frequencies. It has five internal electrodes and is known as a *pentode*. Think of the Pentagon, the famous five-sided building in Virginia, USA as a memory aid to help remember the number of electrodes in a pentode.

The Decline of Tubes

As transistors grew in capability and popularity, and engineers become used to working with them, vacuum tubes were used less and less. General Electric Company (GE®) was the largest US manufacturer of vacuum tubes, and it closed its last tube factory in New Jersey around 1976. The majority of vacuum tubes made in the world today come from Germany (Telefunken®), Russia (Svetlana®), and China (Taylor®). There are still some specialty high-power RF tubes made in the USA by Eimac Industries® in San Carlos, California.

15.2.2 The First Transistor

The type of transistor that first gained widespread popularity is known as the *bipolar* type. Bipolar means the device uses two (bi) types of polarized material (polar) in its makeup. The two types of material are N-type and P-type semiconductor material. Silicon is one of the major semiconducting elements used to make many transistors, but there are actually many different semiconductor elements and compounds used to make a wide range of transistor types. The term bipolar actually refers to a family of transistors. Bipolar transistors were invented at Bell Laboratories in 1947. A team of three scientists, John Bardeen, Walter Brattain, and William Shockley produced the first working model of the bipolar transistor and the team jointly received the Nobel prize for its invention in 1948. Refer to Figure 15.4, the schematic symbols for an NPN and a PNP bipolar transistor.

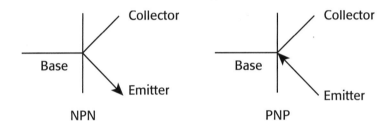

FIGURE 15.4 NPN and PNP bipolar transistor schematic symbols.

Other transistor families include *field effect transistors (FETs)*. One member of the FET family is known as *metal oxide semiconductor field effect transistor* (MOSFET). MOSFETs can be manufactured using either N-type or P-type semiconductor material for the conducting *channel*, depending on the type of circuit they are designed for. A special combination circuit using both types of devices, N-channel and P-channel, is called a *complimentary circuit configuration*, and the designation *CMOS* is given to them. You have probably heard the term CMOS used with PCs, specifically the BIOS configuration settings, which are stored in a type of CMOS-based IC. Actually, the CMOS configuration is used in many ICs in a typical PC including DRAM and parts of most microprocessors. The basic parts of a MOS transistor are labeled the *source, gate,* and *drain,* and are shown in Figure 15.5. These MOS transistor parts are roughly analogous in operation to the emitter, base, and collector of a bipolar transistor, but operate on different principles. MOS transistors are voltage-controlled devices, with the gate voltage, V_g, controlling the

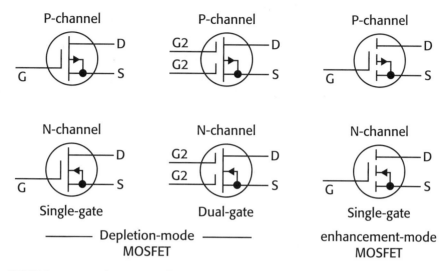

FIGURE 15.5 Various types of MOSFET transistor schematic symbols.

channel current, which if measured at the source terminal is termed I_S, and if measured at the drain terminal is termed I_D. This operation is actually more like the vacuum tube triode, another voltage-controlled device. Refer to Figure 15.5 for the schematic symbols of several types of MOSFET transistors.

15.2.3   Transistor Amplifiers

The first use the new transistor was put to was amplifying very small electrical signals, such as those received by a radio. Before the transistor was invented, the only way to make the infinitesimally small received radio signals great enough in magnitude to drive a loudspeaker was to use bulky and rather unreliable vacuum tubes, which produced much excess heat in the process. The new transistor could directly replace many of the smallest tubes used, and in time, replaced even many of the larger ones. Refer to Figure 15.6 for a picture of small bipolar transistors.

15.2.4   Transistor Switches

It was also quickly discovered that the transistor could function very well as a switch. The term *bias*, for electronic circuits, means to apply a voltage to specific terminals of a device in order to prepare the device to do a desired job. When biased into conduction, meaning a forward bias condition, a voltage applied to make the base-to-emitter junction of the transistor conduct, the bipolar transistor has a very low voltage drop across the two major current terminals, called the *emitter* and the *collector*. This voltage drop is known as V_{CE}, the voltage dropped from collector

FIGURE 15.6 Bipolar transistors.

to emitter. A small control current that flows from the base terminal to the emitter, designated I_B, directly controls the current that flows from the emitter to collector. This collector current, designated I_C, can typically be around 20–200 times the controlling base current, so it makes the transistor a very efficient electronic switch, as no filament, such as that used in most vacuum tubes, is required. Equation 15.1 shows the so-called base-to-collector current gain, designated by the Greek letter *beta* (β) of a bipolar transistor:

$$\beta = I_C \div I_B \tag{15.1}$$

For this reason, bipolar transistors are said to be *current-controlled* devices. Refer to Figure 15.7 for a representation of the current gain offered by a bipolar transistor.

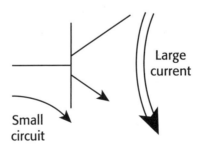

NPN bipolar transistor

FIGURE 15.7 Current gain in a bipolar transistor.

15.2.5 Oscillators

If some of the output signal of an amplifier is fed back into its input, there are two possible outcomes. If the output signal is applied to the amplifier's input in such as way that it is in phase with the input signal, the amplifier output will rise to a point at which it cannot increase anymore, limited by the power supply and the amplifying device's gain figure. *Gain* is the ratio of output signal magnitude to input signal magnitude. Gain is measured and expressed by the letter "A," and gain can be measured and expressed in terms of voltage (A_V), current (A_I), or power (A_p). If we are interested in the voltage gain for a given amplifier, the formula to use is shown in Equation 15.2:

$$A_v = V_{Out} \div V_{In} \qquad (15.2)$$

As long as the amplifier will provide a voltage gain of ≥ 1, then the circuit is capable of going into a condition known as *oscillation*, provided the signal from the output is fed back to the input in phase. Oscillation can be either a desired or an undesired condition. If it is undesired, which is the case in a normal amplifier, steps are taken when designing the circuit to ensure that any signal fed back will be out-of-phase with the input. This is normally done to ensure a stable amplifier, but does cost some gain. As they say, there is no free lunch.

If however, the intent is to produce a frequency at the output to be used for some purpose, then the circuit is designed to produce an in-phase feedback signal. The circuit will then go into oscillation at the design frequency; provided the amplifier has sufficient gain at the desired frequency. There are many ways employed to accomplish the feedback, depending on the type of amplifying device and the frequency desired. In general, a circuit called a *frequency-determining circuit (FDC)* is used in the feedback loop from output to input. The FDC will set the oscillator frequency by providing the proper amount of phase shift between the output and input to ensure the signal feedback will be in-phase with the input. Refer to Figure 15.8 for a schematic diagram of a simple oscillator.

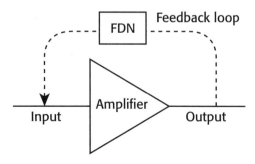

FIGURE 15.8 Simple oscillator circuit.

15.3   THE FIRST INTEGRATED CIRCUITS

Transistors had been around for roughly ten years. Small AM transistor radios were commonplace when Jack Kilby of Texas Instruments™ Incorporated (TI) invented the first integrated circuit (IC) in 1958. TI filed for a patent for their "solid circuit" on February 6 of that year. Robert Noyce independently invented a very similar design in January 1959 while working at The Fairchild Semiconductor® Corporation, a new start-up company in Northern California. Noyce's devices were called "unitary circuits." The Fairchild team also filed for a patent on the new design. The Fairchild team was aware of Jack Kilby's previous work and hoping to avoid a patent infringement issue, they used great detail when describing their invention. Their plan seems to have worked well, for on April 25, 1961, the patent office awarded the first integrated circuit patent to Robert Noyce. Jack Kilby's application was still in the process of being analyzed. Today, the credit goes to both men for having independently conceived of the idea of an integrated circuit.

Most ICs are made by a process called photolithography to "print" circuit components on the substrate (building block) material using light. This process allows for very high precision, and for very small circuit components to be fabricated in large quantities while keeping the quality high.

15.4   USES OF INTEGRATED CIRCUITS

ICs are classed as being analog, digital, or hybrid in design. Hybrid ICs contain both analog and digital circuits in the same package. The following sections present each of these IC types.

15.5   ANALOG ICs

Analog ICs include those designed as voltage and/or current regulators, modula-
· tors, demodulators, and audio amplifiers. Refer to Figure 15.9 for a schematic of a voltage regulator IC.

15.6   DIGITAL ICs

Entirely digital ICs work on digital signals as input and do not convert them to analog signals. Typical digital ICs include registers, which are used for holding and sometimes manipulating data, and also as counters, timers, and other devices. A

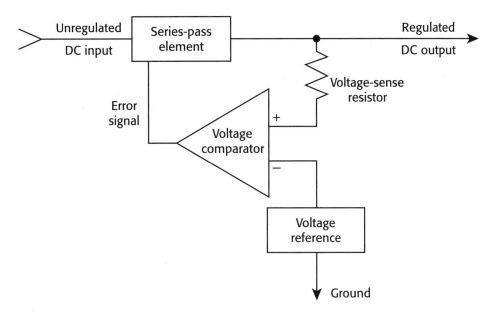

FIGURE 15.9 Basic elements of a linear voltage regulator IC.

register can be used to simply temporarily store a value, which could in a PC be a memory address, a byte of data, or some other value. A microprocessor is an example of a very highly integrated digital IC with millions of transistors making up registers, counters, memory cells, and so on.

15.7 HYBRID ICs

Hybrid ICs contain both analog and digital circuits and usually perform a bridge function between the two types of circuits. Typical examples of hybrid ICs include digital-to-analog converters (DACs), analog-to-digital converters (ADCs), and digital signal processors (DSPs). An example of a DAC is the RAMDAC present in many video cards. The video information is stored in the RAM portion and the DAC formats the output signal into an analog type suitable for display on an analog monitor. Refer to Figure 15.10 for a picture of a RAMDAC chip on a video card.

DSP chips as used on sound cards and as a part of the sound chipset on motherboards, take analog sound signals, digitize them, perform all sorts of manipulation on the signals digitally, and then reconvert them for an analog output. DSP chips are also commonly used in nearly all modern communications gear including cell

FIGURE 15.10   RAMDAC chip on a video card.

phones and military and amateur radio transceivers (combinations of transmitter/receiver, also known as walkie-talkies). Refer to Figure 15.11 for a picture of a DSP sound chip on a PC sound card.

For example, the DSP chip used on one of the Sound Blaster® Audigy® boards, part number E-MU10K2, is a programmable DSP chip that supports hardware sound acceleration (meaning the card has its own processor), which is a feature needed by the latest version of Microsoft DirectX®/DirectSound®/Direct3D®. This feature allows multiple sounds to be produced at the same time, while synchronizing with the on-screen action.

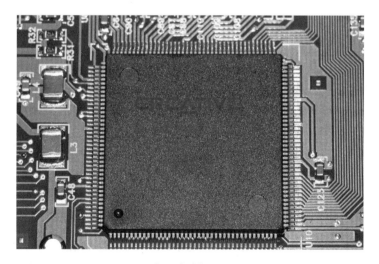

FIGURE 15.11   DSP sound card chip.

Audio converter chips include audio DACs, ADCs, sample rate converters, and CODECs (coder-decoders). These types of chips are used for home theater, professional, automotive, computer, and consumer audio applications. They are widely used in audio/video receivers, DVD players, digital mixing consoles, car audio systems, and PCs.

15.8 IC EVOLUTION

Integrated circuits have evolved from simple logic gates to microprocessors over about a 40-year time span. The following sections present a survey of this evolution.

15.8.1 Basic Logic Gates

Some of the simpler ICs used to perform digital logic functions are illustrated here. They are known as *logic gates*. Logic gates are always associated with their *truth table*, which is a listing of all allowable output states based on all allowable input states. The following shows the most common basic logic gates and their truth tables. In some cases, an analog hard-wired system using switches is also shown, so you can get a "feel" for the logic function being performed by the gates.

The Inverter or NOT Gate

The first type of logic gate presented, also know as an inverter, is the NOT gate. Refer to Figure 15.12.

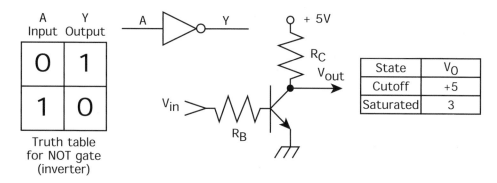

FIGURE 15.12 NOT gate, its truth table, and a discrete switch version.

The AND Gate

The AND gate is shown in Figure 15.13.

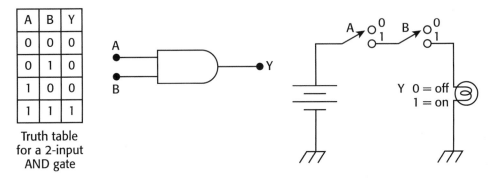

A	B	Y
0	0	0
0	1	0
1	0	0
1	1	1

Truth table
for a 2-input
AND gate

FIGURE 15.13   AND gate, its truth table and a discrete switch version.

The OR Gate

The OR gate is shown in Figure 15.14.

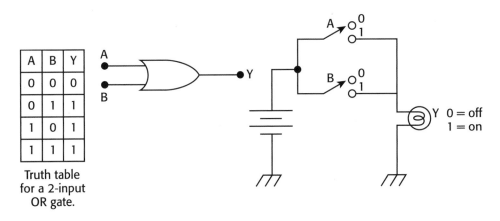

A	B	Y
0	0	0
0	1	1
1	0	1
1	1	1

Truth table
for a 2-input
OR gate.

FIGURE 15.14   OR gate, its truth table, and a discrete switch version.

The NAND Gate

The NOT-AND or NAND gate is shown in Figure 15.15.

The NOR Gate

The NOT-OR or NOR gate is shown in Figure 15.16.

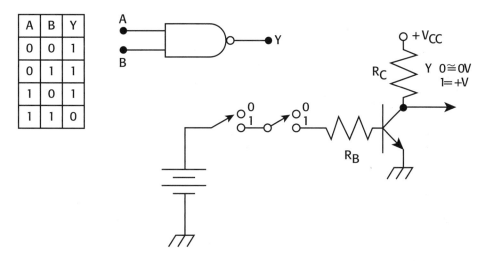

A	B	Y
0	0	1
0	1	1
1	0	1
1	1	0

FIGURE 15.15   NAND gate and truth table.

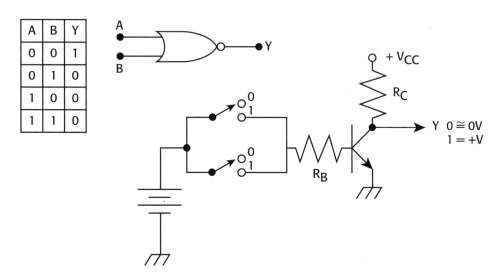

A	B	Y
0	0	1
0	1	0
1	0	0
1	1	0

FIGURE 15.16   NOR gate with truth table.

The XOR Gate

The Exclusive-OR or XOR gate is shown in Figure 15.17.

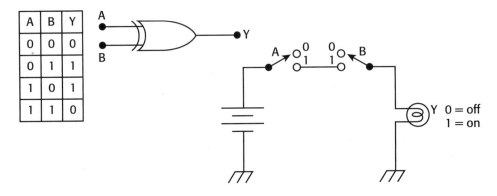

A	B	Y
0	0	0
0	1	1
1	0	1
1	1	0

FIGURE 15.17 XOR gate and truth table.

These logic gates will perform decision-type operations, based on the state of all the inputs as well as a trigger signal, also called a *strobe* or *clock pulse*.

15.8.2   The First Microprocessor

In 1969, Marcian "Ted" Hoff of Intel® Corporation of Sunnyvale, CA, took a new approach to circuit design. He combined the functions of about a dozen separate devices into a single unit built on a common *substrate* or building block of material. A Japanese calculator manufacturer named Busicomp had contacted Intel Corp. to design for them a set of ICs to be the heart of a line of programmable calculators. Mr. Hoff and his employee Stan Mazor decided to try something new and designed the first arithmetic chip, later called the 4004, which eventually led to ever more capable chips we now know as microprocessors.

Mr. Gordon Moore is reported to have stated that the microprocessor is "one of the most revolutionary products in the history of mankind." It was ranked as one of 12 milestones of American technology in a survey in *U.S. News & World Report* in 1982. The chip is the actual computer itself. The chip did not garner much notice until 1973. The chip had to be programmed, and of course required input and output hardware as well a power supply.

15.9   ICs IN THE TYPICAL PC

Early PCs used motherboards and expansion cards literally covered with small ICs that were not very integrated. This means that they performed only a few functions each, with a relatively small number of transistors—or the equivalent number of components similar in complexity to transistors—on each chip. Modern PCs have much fewer, but much more highly integrated chips on them. The equivalent "transistor count" of the latest processors from Intel and AMD® is up to over 55 million.

Refer to Figure 15.18 for a picture of an early IBM® original PC motherboard showing the may individual ICs populating the board. Refer to Figure 15.19 for a picture of a modern motherboard, circa 2003, and notice how few, but much larger (and more highly integrated) chips are present on the board.

FIGURE 15.18   Original PC motherboard showing the many discrete chips used.

The Chipset

The typical PC motherboard contains several large highly integrated chips called the *chipset*. These are usually dedicated microprocessors that perform several tasks, thus reducing the number of chips required as well as freeing the main PC microprocessor from these jobs. This increases overall computing efficiency, since each chip can work independently. A typical example of this is *DMA*, or *direct memory*

FIGURE 15.19 Modern motherboard showing few but more highly integrated ICs.

access. Using DMA, two devices within the PC can transfer data between themselves without going through the main processor. This frees up the main processor to perform other computations.

One of the popular chipset schemes is the use of two chips, one that interfaces directly to the main microprocessor, called the *north bridge* (short for north bridge controller) that interfaces or forms the "bridge" between the main microprocessor and the fastest motherboard components such as RAM, and another chip connected to the north bridge chip called the *south bridge*. The south bridge interfaces or bridges the fast front-side bus (FSB) with the slower devices and circuits. The north bridge runs at the system FSB speed, which can be as high as 800 MHz or more today. Some motherboard manufacturers even supply their boards with heat sinks and fans installed on the north bridge chip, since they tend to run rather hot.

Another popular naming scheme uses two chips called *integrated controller hubs* by their maker, Intel. For example, the 865G chipset chip (north bridge) interfaces directly with an Intel® Pentium 4® processor over a system bus that can run as fast as 800 MHz. It controls dual channel DDR memory, the AGP8X slot, and a Communications Streaming Architecture® (CSA) plus Gigabit Ethernet (GbE). It also interfaces to the slower chipset chip, called the ICH5/ICH5R (south bridge). That chip interfaces with the 6-channel audio system, the PCI slots, high-speed USB 2.0, Intel RAID technology chip (for the "R" version of the chip), BIOS, legacy ATA-100 devices, 10/100 connect interface, and dual serial ATA (SATA) controllers. Refer to Figure 15.20 for the 865G architectural layout.

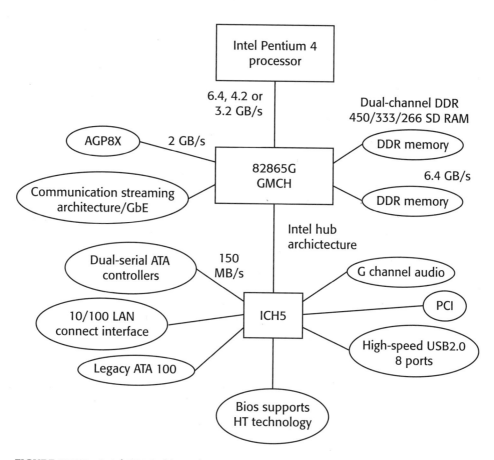

FIGURE 15.20   Intel 865G chipset layout.

A chipset scheme used by motherboards that supports Advanced Micro Devices® (AMD®) processors such as the ABIT® model NF7-S motherboard uses an nVidia® nForce2® chipset. nVidia is well known as one of the leading video card chipset manufacturers. They were also the first to offer integrated video as part of the main motherboard chipset. The nForce2 uses the north bridge/south bridge scheme as well.

As another chipset manufacturer example, Via Technologies, Inc.® offers chipsets for motherboards that accept AMD processors as well as those that accept Intel processors. A motherboard that uses the AMD Athlon™ XP processor might use the Via® KT800 chipset, which consists of the KT880 north bridge and the VT8237 south bridge. The set supports DDR 400 memory, AGP 8X, a 400 MHz front-side bus, built-in SATA/ATA RAID, legacy support for up to four parallel ATA devices, plus built-in audio, and when used with companion controllers, Gigabit Ethernet.

Audio Chips

Audio chips may be found on sound cards, which are preferred for their superior overall feature set and sound quality, or soldered to the motherboards of often less-expensive PCs. Some of the manufacturers of audio chips include Creative Labs®, maker of the de facto standard Sound Blaster products®, and Voyetra Turtle Beach™, Inc.

Audio Quality

Several criteria are used to specify overall audio quality. The frequency response is a rating of the upper and lower frequency limits that a unit can operate over. It usually includes a reference to how uniform or "flat" the response is, most often specified in decibels (dB). Human hearing runs roughly from 20 Hz to about 20 kHz. The upper end of our frequency response falls off after the age of about 18, making it harder to hear the high tones, and requiring our turning up the treble control. When referencing power levels, 3 dB means a difference of a factor of two; +3 dB is a sound twice as powerful as the reference level, and −3dB is a sound half as powerful as the reference level. Any decent sound card, amplifier, et cetera, should not only state the advertised frequency response, but also how much the power level can vary over the specified frequency range. For instance, a sound card with a response rated at 50–15 kHz, +/−3dB means the power level produced by the card over that frequency range can vary by a factor of two above and below the reference power level. That, in fact, is not such a good response if you are looking for high-fidelity sound reproduction. In fact, many sound cards will not even specify the amount the frequency response will vary. Check out ads for loudspeakers, amplifiers, and so on, and look for the decibel difference the specified frequency response can vary. Add to this the fact that some audio product manufacturers are less than honest, and you can start to see why it is so difficult to shop for sound products. Look for reviews by well-respected publications such as *PC World*,

PC Magazine, Maximum PC, and *Consumer Reports.* Actual listening tests will also give you a good idea of a piece of sound gear's real value.

Total Harmonic Distortion (THD)

When two or more frequencies pass through a piece of equipment, they can influence each other in a process called *mixing.* The result is a changed signal, and that degree of change is known as *signal distortion. Total harmonic distortion* (THD) is a rating of how much, expressed as a percentage, distortion (undesirable changes in the sound) are produced by a piece of audio gear. The lower this distortion is, the closer to true sound reproduction without coloration you'll hear. Look for components with low THD figures. In a high-quality home stereo amplifier, distortion figures of considerably less than 1% are common. In inexpensive computer audio equipment, 10% distortion is a much more likely number.

Signal-to-Noise Ratio

Signal-to-noise ratio, (S/N or SNR), is the ratio of the level of signal to the level of noise produced by the equipment that processes it. S/N is expressed in dB. An inexpensive sound card may have a S/N of only 90 dB, while a high-end card may offer 106 dB. The higher this ratio is, the cleaner the sound will be.

Output Power

A specification that is widely advertised for home stereo systems and particularly their amplifiers is *output power.* An output power rating of 25, 50, or 100 watts per stereo channel is not at all unheard of today. However, the small sound boards in a typical PC supply only a watt or two per channel, so they must be used with amplified speakers in order to fully appreciate their sound reproduction capabilities. The proper way to specify output power from audio amplifiers is in RMS watts per channel, with both channels (or in the case of multichannel systems such as Dolby® Digital EX, all channels) driven, and specified as to the impedance of the speakers the output power was tested at. In addition, the output power is supposed to be specified over what frequency range, and to what degree it will vary, expressed in dB, over that range. Furthermore, the percentage of output power that is THD should be included. Unfortunately, most sound card makers do not specify the power capabilities of their products this accurately. Most embedded sound chips on motherboards do not even mention the output power. In fact, many manufacturers outrageously overate their products' specs. One should do careful research and listening tests before purchasing sound reproduction products. A famous example of this, and one that helped spur the FTC ruling on amplifier power ratings back in the 1970s, was a certain electronics retailer's preference for stating amplifier power as "instantaneous peak power, one channel driven." A way they sometimes measured this was to turn off one amplifier channel, thus easing power supply demands, crank the remaining channel's gain to max, connect an oscilloscope to the output, and drop a record player's needle onto the record from several inches above the platter. It didn't matter how bad the output looked on the 'scope, how it sounded, or over what frequency range the output was tested, all that was measured was raw "whomp" power.

Home Stereos

One of the best ways to ensure adequate amplification power and low distortion is to use your existing home stereo system or home theater system. The amplifier and speakers will be far superior to all except the most expensive "PC speaker" systems.

15.10 SUMMARY

- Integrated circuits can consist of several to many transistors, resistors, and small-value capacitors on a single substrate material.
- Most ICs use photolithography to "print" circuit components on the substrate.
- ICs are categorized by their complexity in terms of the equivalent number of transistors they contain.
- ICs are classified as digital, analog, or hybrid, based on the types of circuits they contain.
- Digital ICs are the type most used in computers.
- Analog ICs are typically those dealing with audio, radio frequencies, and power regulation, such as voltage and current regulators.
- Hybrid ICs have parts that are digital and parts that are analog. One example is the RAMDAC chip used in some video cards. It contains digital memory (RAM) as well as a digital-to-analog converter (DAC.) Another is the digital signal processor (DSP) chip. A DAC is used on sound cards and is the main sound-related chip in the chipset included on motherboards with integrated sound.
- The most complex ICs built so far are microprocessors, with current chips having a transistor count of over 55 million.
- CMOS technology is used to build many of the ICs used in computers. They are extremely sensitive to static electricity and special care must be used in their handling to avoid damage due to ESD.
- Logic gates are circuits whose output state depends on the type of gate, the condition of the input signals, and in some cases, the previous condition of the gate's output.
- The most common logic gates are the inverter (NOT gate), AND, OR, NAND, NOR, and XOR types.

■ Registers are electronic circuits that can store and manipulate digital numbers. They are essential parts of microprocessors. Registers are the main type of circuit comprising RAM chips.

■ Motherboards use chipsets consisting of one or two dedicated processors called bridge controllers. One scheme uses a north bridge to connect directly with the main PC microprocessor via the front side bus. It interfaces with the fastest components on the motherboard, such as RAM. It also talks to the south bridge, another dedicated controller that interfaces to the slower components such as PCI slots, ATA devices, Ethernet, and legacy I/O ports, plus Firewire and USB 2.0 ports.

■ Many motherboards include built-in audio, video, and RAID chips.

15.11 KEY TERMS

Digital	Logic gates	Transistor
Analog	Microprocessor	Output power
Diode	North bridge	S/N ratio
Triode	South bridge	Grid
Tetrode	Integrated controller hub	Anode
Amplification	Frequency response	Cathode
RAMDAC	THD	

15.12 EXERCISES

1. What was the name of the first type of vacuum tube made?
   a. Diode
   b. Triode
   c. Tetrode
   d. Pentode
   e. Heptode

2. Diodes are used mainly to do what?
   a. Amplify
   b. Rectify
   c. Nullify
   d. Perform logic functions
   e. None of the above

3. The first triode was named by its inventor as the what?
   a. Auditron
   b. Audion
   c. Audition
   d. Athlon
   e. Voice-A-Thon

4. The first integrated circuit came about by a company working to do what?
   a. Create the first TV
   b. Make the first practical radio
   c. Make a single unit to replace many individual circuit elements
   d. Satisfy an order by a foreign company
   e. Both c and d

5. The group of integrated circuits that supports the main microprocessor in a typical PC is called what?
   a. The parts
   b. The chipset
   c. The whole nine yards
   d. The resistor set
   e. The firmware

6. What is the main function of a north bridge chip?
   a. Communicate with the slowest parts of the PC
   b. Communicate with the fastest parts of the PC, including RAM
   c. Communicate directly with the outside devices, such as the keyboard and mouse
   d. Communicate wirelessly over the Internet
   e. Allow direct connections with Microsoft

7. What functions does a south bridge chip serve in a PC?
   a. Communicate with the slowest parts of the PC
   b. Communicate with the fastest parts of the PC, including RAM
   c. Communicate directly with the outside devices, such as the keyboard and mouse
   d. Communicate wirelessly over the Internet
   e. Allow direct connections with Microsoft

8. What does DMA mean?
   a. Done Measuring Amplitude, a signal by the PC that it is finished with a measurement
   b. Direct Memory Access, a way for devices to transfer data without going through the main processor

    c. Digital Mouse Adjustment, the applet that calibrates your mouse for you

    d. Disney Music Authoring, a certified way to send audio files

    e. Nothing

9. Sound in a PC can be provided how?

    a. By a sound chip embedded on the motherboard

    b. By a sound card installed in a motherboard slot

    c. By an external sound box, connected to the USB port

    d. By a wireless box in another part of the building

    e. Both a and b

10. THD is a measurement of what?

    a. How good sound is

    b. How much distortion is generated by a piece of audio gear

    c. How much output power a sound card will generate

    d. The range of frequencies over which a sound card will provide useful output

    e. None of the above

11. Frequency response is a measure of what?

    a. The range of frequencies over which a device or circuit can operate

    b. The range of frequencies over which a sound card or chips can produce useful output

    c. The number of times per month the PC is "down"

    d. The number of cycles per second at which the processor runs

    e. None of the above

12. The audio output power produced from a sound card or chip should be specified how?

    a. Watts per one channel

    b. Watts per channel, with all channels operating

    c. Watts per channel with all channels operating, and over what frequency range

    d. Watts per channel, with all channels operating, and over what specified frequency range, and the impedance value of the speakers the other values were tested with

    e. All of the above, plus how much distortion is in the output

13. What is the recommended method to provide the best sound from a PC?
    a. Use a monitor with included speakers
    b. Use the small PC speakers available at flea markets
    c. Use the small PC speakers supplied by the PC maker
    d. Use inexpensive earphones
    e. Use your existing home stereo or home theater system

14. Some motherboard manufacturers supply heat sinks and even fans for the north bridge chip.
    a. True
    b. False

15. S/N stands for what?
    a. Saturday Night, a time to get together with friends for some friendly PC gaming
    b. Serial Number
    c. Signal-to-Noise ratio
    d. Standby, No 'power
    e. All of the above

16 Communications

16.1 INTRODUCTION

Communications involves many types of equipment. Much of that equipment is in some way tied to computers. This chapter presents some of the major communication methods and equipment used.

16.2 THE COMMUNICATIONS PATH

The communications path consists of a sender, the medium, and a receiver. The traditional radio broadcast path is from one or more transmitting sites through the air to one or more (usually a great number of) receivers. Starting around the turn of the last century, this led first to "spark" transmission, used by the military and ships at sea, and later to what we now call commercial AM broadcast radio. AM stands for *amplitude modulation*, a method used to impress or add an information signal to another signal, called a carrier wave. The *carrier wave* is a high-frequency radio signal capable of radiating from an antenna and traveling long distances. Broadcasting refers to sending the signal to everyone, rather than to a specific intended receiver.

Guglielmo Marconi, 1874–1937, is credited with the invention of modern radio. Around the 1930s a man named Edwin Armstrong made significant improvements in both radio receivers and the method used to impress the information on the transmitted high-frequency radio signal. The new circuit is called the *superhetrodyne receiver* circuit. Armstrong's improved method of impressing the information is called *frequency modulation* (FM). Figure 16.1 shows a typical broadcast radio communications path.

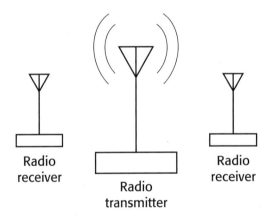

FIGURE 16.1   Radio communications path.

16.3   WIRED LINKS

Wired connections to and from PCs take many forms. The next sections look at the most common of these forms.

16.3.1   PC Port Connectors

PCs can be connected to other equipment via wired or radio links. The wired link consists of two or more PCs connected with some form of wires. Early PCs used a serial (COM port) or parallel port cable to connect two PCs together to allow file transport. Serial (COM ports) and parallel connections (LPT for *line printer*) on PCs were originally used for connecting printers and later mice (COM ports). The latest PCs can also use USB 1.1 (11 Mb/s) or USB 2.0 (480 Mb/s) to connect two PCs together. Ethernet, established in the 1970s, first used coaxial cables to connect two or more PCs in a network. The general trend with Ethernet has been moving away from using coaxial cables, which are thick, heavy, and expensive, and which required more expensive hardware, to smaller, lighter, and cheaper cables that closely resemble telephone cables.

Sound Card Connectors

Most PCs today come with some form of sound hardware, either a chipset on the motherboard or on an add-in card. In either case, the most common sound connectors include line-in, line-out speakers, and MIC-in. Line-in is used to connect signal sources to be amplified or processed by the sound circuitry. Line-out/speakers is used to provide stereo output for an external amplifier (as in a home stereo system) or amplified speakers, or very small, nonamplified PC speakers. Better-quality sound circuits also include outputs for rear channel speakers, a front center channel speaker, and even a subwoofer channel. The group of sound connectors also includes a two-row, D-shaped 15-pin connector that doubles as a *MIDI* (musical instrument digital interface) and joystick connector. MIDI port connectors are usually used to connect a digital keyboard and digital recording hardware such as mixers. The newer versions of the MIDI port connector look like round PS/2-style connectors, and use a separate connector for MIDI-in and MIDI-out.

16.3.2 Ethernet

Ethernet has become the de facto standard for high-speed wired connections between PCs. Originally invented by Robert Metcalf of Xerox® Corporation on May 22, 1973, it allows two or more PCs to be connected together and share data and resources. The minimum requirements for setting up two PCs connected by Ethernet are as follows:

Two PCs running *network operating systems* (*NOSs*): A NOS is an operating system that provides the software and protocols necessary for the computer to communicate with another computer.

Each PC must have a *network interface card* (*NIC*), or equivalent network hardware built into the motherboard: A NIC designed for use in Ethernet networks is called, appropriately enough, an *Ethernet adapter*.

A special Ethernet cable called a *crossover cable*:   It is so called since some of the wires used for sending and receiving data are "crossed over" on one end of the cable. This allows for data sent and data received to be connected to the proper pins on the other PC. The normal type of Ethernet cable, used in networks with a central connecting device, is known as a *straight-through* cable. In this type of cable, none of the wires are changed in position on the plug from one end to the other.

Ethernet Cables

The cables used to link PCs today are known as *Cat-5* cables, short for *Ethernet category five*. They consist of eight solid copper wires, each individually insulated, with four sets of two wires each twisted together in pairs. The other name given to this type of cable is *unshielded twisted pair* (*UTP*). The connectors on the cable ends resemble the RJ-11 and RJ-14 types used with telephones, only they are larger and are known as RJ-45 plugs. Figure 16.2 shows some Cat-5 cable with the wire leads exposed.

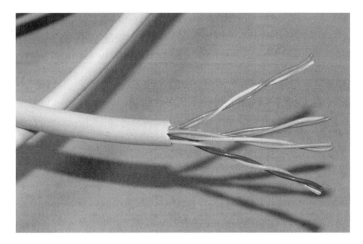

FIGURE 16.2   Cat-5 cable wire ends.

Nearly all newer PCs include an Ethernet port connector, known as an RJ-45 jack. It is included either as an integrated unit on the system motherboard, or as an expansion card plugged into a PCI motherboard slot. Figure 16.3 shows straight-through and crossover Ethernet cable ends. Figure 16.4 shows examples of each type of Ethernet port connector. On notebook PCs, they are also offered as either built-in or added via a PC Card (also called PCMCIA or CardBus) adapter.

FIGURE 16.3   Cat-5 Ethernet cables showing crossover and straight through cable ends.

FIGURE 16.4   PC Card Ethernet NICs.

A Three or More PC LAN

To use three or more PCs in an Ethernet LAN requires using either an Ethernet hub or a switch, a central box to which all the networked PCs attach. For multitasking use, a switch will provide better overall data throughput, and prices have fallen so fast it doesn't make much sense to settle for a hub anymore. The most common forms of hubs and switches are the 10/100 varieties that will work with NICs of either speed. Also required for this setup is a length of Cat-5 straight-through cable to connect each PC to the hub or switch. Some of the newest switches autosense the type of cable being connected, so they will work with either a crossover or a straight-through cable. Some other switches have mechanical switches to set the type of cable used.

10-Mbps Ethernet

Ethernet systems use *shielded twisted pair* (*STP*), unshielded twisted pair (UTP), or coaxial cable. STP uses a metal shield around the center-insulated conductors, which reduces interference to and from the signals carried by the twisted wires inside. One of the coaxial cable types is known as *thicknet* or RG-8, and has an outside diameter (OD) of approximately ½ inch. It has a central solid conductor surrounded by an insulating dielectric material. Later Ethernet used a slightly smaller diameter cable, about ¼ inch OD, called *thinnet,* or *RG-58.* Both thicknet and thinnet cable installations require the use of resistive terminators at each end of the system, or data loss and unreliable operation due to signal reflection would occur. This type of Ethernet is also known as *10BaseT.*

The maximum specified data rate for these early Ethernet systems was 10 Mbps (10 megabits per second). Twisted pair cables use RJ-45 connectors, which look like the type used on standard telephones, only larger.

100-Mbps Ethernet

This is the most popular version in use today, and is also known as *100BaseT.* It uses either UTP or STP cables. Both 10BaseT and 100BaseT can exist together on the same network. Common NICs are rated as 10/100 to reflect the fact that they can operate at either speed, depending on the speed of the other NICs, hubs, switches, and so on existing on the network. 100BaseT Ethernet is also known as *Fast Ethernet.*

Fast Ethernet should be used with Cat-5e (Cat-5 enhanced) or better cables; Cat-5e uses more twists per unit length of cable than Cat-5, effectively increasing the degree of shielding, and therefore allowing less crosstalk. *Crosstalk* refers to the amount of one signal carried by one wire inducing a signal in an adjacent wire in a cable. It amounts to interference, and will cause data corruption and the need for resending data, which slows down overall throughput. So the better shielded a cable is the less crosstalk there will be at a given frequency. Cat-6 has less crosstalk than Cat-5e cable.

100BaseFX

100BaseFX uses fiber-optic cable, offering even higher bandwidth due to the high frequency of light.

1000-Mbps Ethernet or Gigabit Ethernet

A new version of Ethernet, Gigabit Ethernet uses both fiber-optic and twisted-pair cable. Still relatively expensive, it is slowly gaining popularity.

Here are the specifications on some of the most popular Cat cables used for new Ethernet connections:

Category 3: Unshielded twisted pair (UTP) cables and associated hardware for use up to 16 MHz.

Category 4: UTP cables and hardware for use up to 20 MHz.

Category 5: UTP cables and hardware for use up to 100 MHz.

Category 5e: UTP cables and hardware for use up to 100 MHz. Can be used on Gigabit Ethernet systems.

Category 6: UTP cables and hardware for use up to 250 MHz. Can be used on Gigabit Ethernet systems.

Categories 1 and 2 were not recognized for new cable installations. They could still exist in old telephone applications for voice only, but do not have sufficient bandwidth for high-speed computer use. The Telecommunications Industry Association (TIA) working groups has not yet standardized Category 7 UTP cables.

The proposed standard is Category 7 (Class F). Cat-7 is rated for up to 600 MHz. Not only will it work well with Gigabit Ethernet, but it also allows for longer cable runs without significantly more data loss than Category 6.

Ethernet systems using UTP cables do not require terminations to be used.

Windows Will Detect Your NIC

Newer versions of Microsoft® Windows® including 2000 and XP will automatically detect the LAN cards (another name for NICs) and start the Network Wizard that will walk the user through the typical setup scenarios. Hardware problems can be troubleshot using Windows Device Manager.

Half-duplex Mode

Half-duplex mode means a unit can only send or receive data in a single operation, but not both at once.

Full-duplex Mode

Full-duplex mode means a unit can both send and receive data simultaneously, similar to people talking to each other on the telephone. Your NICs should be capable of full-duplex mode. If your Ethernet network uses switches instead of hubs, and your NICs are full-duplex capable, if they are 100 Mbps NICs, then you can achieve 200 Mbps in full-duplex mode.

Build Your Own Cables

You can save a great deal of money, especially when a large number of Ethernet cables are required, by building your own. Materials required include a roll of the proper type of cable, cable end connectors, and a cable connector crimper tool. All of these items are available at any large computer outlet, plus many mail-order electronics parts and computer outlets.

16.3.3   Other Wired Systems

Other wired systems that were once in real competition but in time have lost out in favor of Ethernet include ArcNet and token-ring technologies. Those types are now considered obsolete.

Fiber distributed data interface (FDDI) is a network topology used for large, high-speed networks that use fiber-optic cables and a physical ring topology.

16.3.4   Wired Topologies

The three types of network topologies used today are the following:

Bus: Each computer on the network connects to the next computer in a straight-line, or linear style. An example is 10Base2 Ethernet.

Star: Each computer on the network connects to a central access point or device, such as a hub.

Ring: Each computer on the network connects to the other computers in a ring or loop. The "ring" exists only inside the central access point, called a *multistation access unit (MSAU)*.

16.4   WIRELESS SYSTEMS

Wired systems require running cables through walls or along baseboards, et cetera. To avoid the necessity for this, wireless systems were developed. They consist of NICs with built-in transceivers, and a central unit, usually a switch or combination router and switch that also has the capability to send and receive data over a radio link.

16.4.1 Transmitters

Transmitters are electrical circuits that use a high radio frequency to send a signal over a distance to a receiver. They combine some form of information (in the case of Ethernet, this is computer data) with a high-frequency radio signal called the carrier wave. The data is impressed on the transmitted carrier wave by a process known as *modulation*. So it is the modulated carrier wave that is amplified and sent to the transmit antenna. The antenna radiates the modulated signal to a distant point.

It should be noted that PCs generate many radio signals, since the frequency of the data stream inside a PC is in the radio frequency spectrum. As has been pointed out in other sections of this text, toroid chokes, which are doughnut-shaped ferrite beads, are commonly used to help prevent unwanted RF radiation from a PC. They are used on video and keyboard cables and many other types of cables connected to a PC. Refer to Figure 16.5 for a picture of a ferrite toroid choke used to cancel RF radiation from the wires connecting a PC motherboard to the front panel switches.

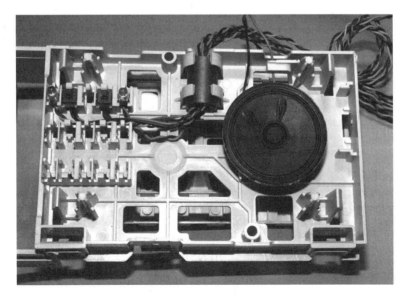

FIGURE 16.5 A ferrite toroid bead used to suppress RF radiation from the leads connecting the front panel switches to the PC motherboard— the tube-shaped device in the top-center of the photo.

16.4.2   Receivers

A receiver does the opposite of a transmitter. The weak transmitted signal strikes the receiver's antenna and generates a very weak current. This tiny electric current is amplified and the carrier wave is discarded and the information is retrieved in what is known as a *demodulation* or *detection* process. The recovered signal is then processed and converted into the appropriate levels (amplified) and presented to the PC as data.

16.4.3   Transceivers

A transceiver is a combination transmitter and receiver in one unit. Often, several electronic stages in a transmitter and receiver can be shared, saving money in the process. All wireless Ethernet units used today consist of transceivers combined with other circuitry such as switches and routers.

16.4.4   Antennas

Antennas are used to both send and receive radio signals. They are said to act reciprocally. They consist of conductors arranged to be electrically resonant at the desired frequency of operation. To a great extent, the size of an antenna relates to both its intended frequency of operation and its overall gain. Low-frequency antennas are large; very high-frequency antennas are physically smaller. The sizes of antennas used for wireless Ethernet are on the order of 4 to 5 inches. The frequency of operation for 802.11b and 802.11g is in the 2.4 GHz range and for 802.11a is in the 5 GHz range (802.11 is a set of standards for wireless computer networking). In general, the higher the frequency, the shorter the antenna used.

Antenna Polarity

Antennas have a specific polarity, which refers to how the radio frequency (RF) leaves the antenna. More specifically, it refers to the major radiation fields produced by a transmitting antenna, or how well a receiving antenna responds to those fields. Often, a single antenna is used for alternately transmitting and receiving. This works because an antenna has *reciprocity*, which means that how it acts with respect to frequency and radiation patterns holds true for both transmission and reception.

Antennas are tested and rated in terms of where the major amount of radiation occurs, either vertically or horizontally, or a combination. Special antennas used for communication with satellites often use circular polarization. The antennas used for wireless devices use mostly vertically polarized antennas, so when installing or adjusting antennas on this type of equipment, it is usually best to keep the antenna pointed vertically; this will help ensure the strongest signal.

16.4.5   Wireless Ethernet

The Wireless Ethernet Compatibility Alliance (WECA) is more commonly known as the Wi-Fi Alliance or just Wi-Fi. For more in-depth information, visit this site: *www.wi-fi.org.*

All types of 802.11 wireless networks have two basic components:

- Access points
- NICs equipped with radio transceivers

An *access point* is a small device that uses an RJ-45 port to connect to a 10Base-T or 10/100 Ethernet network. It contains a radio transceiver plus data-encryption software and management software.

The access point converts regular PC Ethernet signals into wireless Ethernet signals that it transmits, broadcast style, to wireless NICs on the network. It does the reverse to change received wireless signals back into PC-style Ethernet signals from wireless NICs to the PCs.

Wireless NICs

Many new PCs come with wireless NICs installed; otherwise they can be purchased for around $60. They are also available as PC Cards for notebook PCs.

16.4.6   Other Wireless Systems

Wireless systems include 802.11a, b, and g versions, Bluetooth and Ultrawideband. They all offer different combinations of frequency range, maximum usable distance, and maximum data transfer rates.

Bluetooth

Bluetooth™ is a very low-power radio link system designed to link computers and peripherals as well as other types of units. Version 1.2 is not only intended to be backwards compatible with these products, but it also works with a number of mainstream products such as mobile phones, headsets, PDAs, MP3 players, cameras, and laptops in a consumer's personal area network.

As an example of Bluetooth use in automobiles, the 2004 model Toyota Prius uses Bluetooth technology in a unique way. After registering your Bluetooth-enabled cellular phone using the multi-information display, you can use hands-free calling. A maximum of four cellular phones with built-in Bluetooth can be registered. By using the car's touch-screen panel and/or steering wheel controls you are able to make and receive calls.

Wi-Fi Hot Spots

So-called public access wi-fi hot spots are locations where a user with a wi-fi-enabled PC can log onto the Internet. Coffeehouses, so-called Internet cafes, bookstores, college campuses, and even some gas stations now offer the chance to log onto the Internet free for their customers. These public access *wireless access points* (*WAPs*) make it easy to get connected while doing one's daily business while out on the town.

16.4.7 A Comparison of Wireless Technologies

Refer to Table 16.1 for a comparison of these various wireless systems.

TABLE 16.1 Comparison Chart of Wireless Technologies

Designation	Transmit Power	Maximum Range	Maximum Data Transfer Rate	Frequency Range
802.11a	32 mW	175 ft	54 Mb/s	5.15–5.825 GHz
802.11b	32 mW	350 ft	11 Mb/s	2.4–2.483 GHz
802.11 g	32 mW	350 ft	54 Mb/s	2.4–2.483 GHz
Bluetooth	1–100mW	30–100 ft	1 Mb/s	2.4–2.483 GHz
Ultrawideband	155 mW	33 ft	480 Mb/s	3.1–10.6 GHz

It should be noted that the claimed ranges and maximum data rates are subject to the conditions at the actual installation. Structures inside of buildings, such as metal supports, electrical wiring, pipes, and metal cabinets can do three things to radio signals: absorb, reflect, or refract them (bend their path). If you have ever been listening to the radio in a moving vehicle when it went over a bridge with a metal superstructure or through a tunnel, and have noticed the loss of radio signal, you have experienced these effects. In the real world the distances achieved can be far less than the claimed maximums.

16.4.8 The Need for Encryption

When the public access hot spots first started, it was soon realized that other people equipped with wireless PCs could "eavesdrop" on other peoples' wi-fi traffic and steal personal information. So the need to encrypt data was realized and soon methods were developed to prevent others from viewing wi-fi data indiscriminately. The *wireless equivalent protocol* (*WEP*) was developed. It was good, but not totally secure because it could be hacked. The improved version, called *wi-fi protected access* (*WPA*), is much more secure. Used with most new wireless-g equipment, it is better because it changes its encryption keys (security codes) at regular intervals, which defeats hackers trying to break into your network.

Even more protection is offered by using a hardware firewall. These are included with most wireless network gear nowadays. They serve to block network traffic into or out of your network unless you enable it. Standard wireless units include a modem, firewall, and router in a single package, and the cost has steadily fallen to less than $75 for these units.

Here are some measures to take to protect your data on a wireless network:

- Turn off the broadcast feature.
- Change the default IDs that come with the unit.
- Enable encryption on the network.

Windows Firewall Protection

Windows XP offers a rudimentary software firewall that can be enabled. The default is for the firewall to be disabled. But enabling this firewall causes problems for home networks when users attempt to share files, drives, and programs. So the hardware firewall solution is preferred.

16.5 COMMERCIAL COMMUNICATIONS

Commercial communications encompasses radio, TV, facsimile (fax), and other forms, using wired, radio, and even satellite communication links. The people and companies that advertise on the TV or radio stations support broadcast services to our homes for TV, radio, and so on. There are even commercial "spy" satellite services that sell very detailed photos taken from their satellites orbiting the Earth to people doing research for plants and minerals, searching for long-lost ancient cities, et cetera.

Commercial television signals in the USA follow the National Television Standards Committee (NTSC) rules and regulations. All nonmilitary communications in the USA is governed by rules and regulations established and enforced by the Federal Communications Commission (FCC).

16.6 MILITARY, AMATEUR, AND CB COMMUNICATIONS

Communication is essential for the military and also enjoyed by average folks who enjoy "playing radio." The next section presents these uses.

16.6.1 Military Communications

In general, the military uses frequencies other than those allocated by the FCC to commercial, amateur, and other users. But in reality, the military has the capability and legal permission to operate on any frequency, mode and power level they deem necessary. They use wired systems, radio links, and both terrestrial repeaters (devices that receive and retransmit signals in order to extend their range) and space-based satellites. They even have the capability to communicate with submerged submarines using high power, long-wavelength radio signals.

16.6.2 Rather Exotic Radio Research

One interesting government-funded research project in progress is based in Alaska and is known as Project HAARP, which stands for the High-frequency Active Auroral Research Project. It consists of a very large array of antennas each powered by its own powerful amplifier, and pointed straight up to the sky. The antennas are driven in a phased array (all members working together), which produces an extremely powerful radio signal. The basic idea is to try to heat up and therefore modify the *ionosphere*, the layer of atmosphere that refracts (bends) radio signals back to Earth. The ionosphere consists of atmospheric layers of rarified gasses, stripped of their outer electrons, leaving charged atoms called *ions*. These ions can, depending on their density and height above ground, reflect or refract radio signals. The state of the ionosphere is highly dependent on the output of the Sun, which is highly variable. There is a roughly eleven-year solar cycle of intense sunspot activity, which tapers off and repeats again. During years of the most intense sunspot activity, the frequencies that will sustain long-distance radio communication via reflection and refraction by ionospheric layers of the atmosphere increases dramatically, to over 50 MHz. During the lowest activity years, the peak frequency is closer to 15 MHz. The ionosphere also "charges up" over the course of the day and dissipates at night. So to help overcome these vagaries, the government, which is so radio dependent, would like a means to control the ionosphere more or less at will. For more information on this project see *http://www.haarp.alaska.edu.*

16.6.3 Amateur Radio Communications

Amateur radio operators, also known as *ham radio operators*, or *hams*, communicate with each other via radio. The FCC licenses them after they first successfully pass examinations testing their electronic theory knowledge as well as their knowledge of FCC radio rules and regulations. There are currently three classes of amateur radio license.

The Technician Class license holders must pass a relatively easy exam that does not require any Morse code test. Technician Class license holders may enjoy limited access to the ham radio frequency bands. This is the entry-level license class and is intended to offer an easy entry into ham radio, and hopefully encourage the holders of this license to gain practical experience and knowledge that will enable them to progress to higher license levels in the future. Doing so will grant them further access to ham bands and operating modes.

The General Class license requires passing a more demanding theory test, plus tests on the rules and regulations defining operation of their class of license. It also requires passing a Morse code receiving test at a speed of five-words per minute (WPM).

The Amateur Extra Class license requires the most exhaustive theory test, plus the rules and regulations defining this class of operator privileges, plus passing a 5-WPM Morse code receiving test.

Hams can operate, depending on the class of FCC license held, using Morse code, voice, radio teletype (RTTY), slow-scan television (SSTV), and a host of other modes, including many new digital types. Some hams like to specialize in finding and talking to (called working) other hams in far away rare countries. A country is considered *rare* if there are few if any active hams who reside there, or if the country's government bans ham radio operation. China is a good example of a country considered rare for 40 plus years, since their government disallowed ham activity of any kind. Other rare countries include small uninhabited islands. They are frequently the hosts to ad-hoc groups of hams on a "dx-pedition" to temporarily put the island "on the air" for other hams to work. These so-called DX-ers (DX meaning long distance) often operate their stations at odd hours of the day or night in order to take advantage of the favorable radio conditions at those times, which facilitate the propagation (travel) of radio waves between the two stations. Many things influence radio propagation at the frequencies used by DX-ers (roughly 1.8–30 MHz). What the Sun is doing, how it influences the ionosphere (the layer of atmosphere from about 20–200 miles above the earth), and what the geomagnetic field is doing all affect radio propagation. So I think you can see that to develop DX-ing skills requires learning new scientific principles, in various areas. Many hams have reported that they went into engineering or other technical fields because of an early interest in ham radio.

Many DX-ers enjoy exchanging *QSL cards*, which are postcards verifying their contacts, and often containing information about the ham operator's station equipment and other information. Figure 16.6 shows examples of some typical QSL cards received by the author over the course of his 40+ years of ham radio.

FIGURE 16.6 QSL cards verifying contacts with distant ham operators around the world.

Some hams even bounce signals off the moon to communicate with other hams. This signal path is called *EME* for Earth-Moon-Earth. Many Space Shuttle missions have had licensed ham operators on board who used ham radio to communicate with school children in order to help motivate them toward science and in particular, space travel, astronomy, and ham radio.

Many hams have small radios installed in their vehicles and use ham radio to socialize with their ham buddies while commuting to and from work. Often they utilize ham radio repeaters, small transceivers mounted on high peaks, to relay their signal over mountains.

For more information, refer to the ham radio section in Appendix A, "Electronics Careers and Hobbies."

16.6.4 CB Radio

CB, short for *citizens band radio*, is now an unlicensed radio service for the general public that requires no test. CB radios are popular with long-distance truckers, who keep in touch with their fellow drivers. They also like to warn each other of police roadblocks, bad roads, traffic or weather conditions, et cetera. Some people like to have a CB radio in their vehicles in case they need to call for help in case of an emergency.

CB-ers are very limited as to output power, under five watts, and are not supposed to talk to other CB-ers by using signals "skipped" off the ionosphere, as hams regularly do. But the truth is that most CB-ers "work skip" anyway and may also operate illegal power amplifiers called "kickers" to boost their signals. The FCC is in a constant hunt for these illegal CB-ers, and when they locate the offenders, often fine and sometime even jail the operators.

In contrast to the illegal CB operators, there are many CB groups whose main goal is to aid stranded motorists, hikers, et cetera. They monitor the CB channels and listen for those in need, and then advise emergency workers of the need for assistance. CB channel 9 is supposed to be reserved for those requiring assistance in an emergency, such as vehicle accidents, lost travelers, and others, and not for general chitchat.

CB radios (the legal kind) can be purchased for less than $200 for a *base station*, and some hand-held units for about half that price. Mobile units intended to be installed in vehicles are also available.

Many former CB-ers have become disillusioned with the careless operating practices of some CB-ers and have moved up to amateur radio. They find it is generally a much more professional group of folks, often willing to share their technical expertise in electronics and radio operating practices.

16.7 DIGITAL VERSUS ANALOG TECHNOLOGY

Radio communications and communications in general has been undergoing a change from all analog to nearly all-digital circuits since around the 1960s when the first transistors became widely available.

In general, digital circuits are cheaper to produce, easier to guarantee uniformity for, and offer a greater degree of reliability than analog circuits.

16.8 RF INTERFERENCE PREVENTION

Interference to PCs can happen anytime there is a source of radio transmissions close to the PC. Some standard methods of preventing radio frequency (RF) interference are as follows:

- Keep cordless telephones away from the PC.
- Try moving the PC to a different location.
- Use toroid ferrite beads on all cables connecting to the PC (Figure 16.7).
- Use AC line filters on the PC power cord after carefully coiling up the PC power cord to minimize interference pickup from the cord.
- Make sure the metal I/O shield is in place and secure on the rear of the PC.
- Always operate the PC with its metal case cover installed and with all case screws securely tightened.

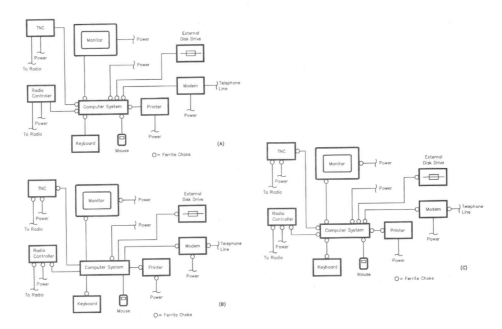

FIGURE 16.7 Typical placement of ferrite beads to prevent interference in a PC system.

- Avoid using PCs with cutouts in the case since these defeat the shielding.
- Do not power the PC from the same branch circuit used also by heavy machinery such as power tools, washing machines, and others.
- Use a good-quality backup power supply that incorporates an EMI filter as well as a voltage spike protector.
- Avoid using extension cords on keyboards or mice, since this increases the chance for RF interference.
- Avoid long extension cords on the PC power cord.
- Avoid the use of wireless keyboards and mice in strong RF environments.
- Try to correct interference at its source, if possible, as it is usually easier than trying to fit all affected equipment with filters.

16.9   SUMMARY

- PC communications can be established using either wired or wireless means.
- Wired methods include using the serial (COM) or parallel ports, which are cheap, relatively easy to set up, and extremely slow.

- Other PC communication connections are the mouse, keyboard, sound, and video port connectors. Mice have evolved from using the COM ports, to bus expansion cards, to integrated PS/2 mouse connectors and USB connectors on the motherboard.
- With newer Windows versions, it is easy to set up a two-PC LAN using an Ethernet NIC in each PC and a crossover cable.
- For three or more PCs in a LAN, use a hub or switch, Cat-5 Ethernet cable, and a NIC in each PC. Set up the protocols yourself or use the Network Wizard in the latest Windows versions to partially automate the process.
- Wireless PC technologies include 802.11a, b, and g, plus Bluetooth and Ultra-wideband.
- 802.11 technology is used for applications such as connecting multiple PCs wirelessly to a common access point (a centrally located wireless transceiver).
- Bluetooth is a very low-power RF method of connecting peripheral devices to a PC, such as wireless keyboards and mice. It has a very limited range due to the low power used.
- Ultrawideband uses a huge chunk of radio spectrum space to greatly increase the data transfer rate to a maximum of 480 MB/s.
- Wireless Ethernet is usually considered easier to set up than wired Ethernet since it does not require drilling holes through walls and floors or laying out cable.
- Wired Ethernet does not offer as high a speed of data transfer as wired Ethernet, since the vagaries of radio transmission through and around different building materials and other obstructions make it hard to calculate without a "trial and error" approach in each different installation.
- Wireless users need to be aware that their data is being broadcast to and from their PCs, and they should always set up WEP or even more secure WPA encryption before using a wireless system to swap sensitive data.
- Wireless systems are subject to interference from other radio sources.
- Take appropriate steps to prevent or alleviate radio frequency interference.
- Amateur radio operators are licensed by the FCC and enjoy worldwide communications as a hobby.
- The Citizens Radio Service (CB radio) was established by the FCC in an effort to make short-range hobby-type communications available to the general public without requiring a license.
- The military uses communications to deploy its forces, keep tabs on the enemy, monitor weather conditions, and so on.
- One government-funded radio research study is called Project HAARP; it shoots radio energy at the ionosphere in an attempt to modify it to facilitate long-distance communication at will, rather than relying on the whims of Mother Nature.

16.10   KEY TERMS

Wired	Transceiver	802.11
Wireless	Ethernet	Bluetooth
Transmitter	ArcNet	Ultrawideband
Receive	Cat-5	Antenna

16.11   EXERCISES

1. What is another name for a PC's serial ports?

2. What was the main use for a PC's parallel port?

3. List another name for a PC's parallel port.

4. What are the maximum data rates for USB 1.1 and USB 2.0 respectively?

5. List all the most common connectors used on a sound card.

6. What is the basic difference between an Ethernet hub and an Ethernet switch?

7. List the basic requirements for setting up a two-PC LAN using Ethernet.

8. List the basic requirements for setting up a three- to five-station PC LAN using Ethernet.

9. What is the most fundamental reason to use a hub rather than a switch for multitasking systems?

10. What is the highest claimed data transfer rate for Ultrawideband technology?

11. What is the highest claimed data transfer rate for 802.11b technology?

12. How do a building's wiring and other internal structures affect wireless range?

13. Why is using cut outs on PC cases not a good idea?

14. Describe the use of ferrite toroid beads to reduce RF interference.

15. List at least two uses of Bluetooth technology.

16. What governmental agency licenses amateur radio operators and polices the CB radio service?

17. Describe at least two uses for which the US government uses radio technology.

17 Building Power Wiring

17.1 INTRODUCTION

We power our marvelous toys, TV, stereo system, washing machine, dryer, and oh yes, our computers, from AC power provided to our homes and places of business by a commercial power company. This chapter presents the fundamental process of how AC power is generated, distributed, and delivered to us.

17.2 THE COMMERCIAL POWER SYSTEM PATHWAY

It is important to understand how the electricity that powers our homes and PCs is generated and reaches us. This electric power system pathway is presented next.

17.2.1 Power Generation

It all starts at the power company's generator plant. Large fan blades mounted on a rotating shaft, called a turbine assembly, turn the generators. The turbine is turned by water or steam, in most cases. In the hydroelectric form of power generation, falling water is forced through large diameter pipes that direct the water to flow through the turbine's blades, causing it to turn. In the case of coal, gas, or nuclear power generation, these various forms of fuel are used to heat water into steam, which turns the turbines. Once the electric power is generated, transformers step up the power to a much higher voltage, in some cases one million volts, and that electricity is transferred across long distances by wires held high off the ground by towers. Transformers were originally developed by Nikola Tesla, who pioneered the use of AC electricity. Figure 17.1 shows a large electric generator inside an electric power plant. Figure 17.2 shows wind-powered electric generators.

FIGURE 17.1 Large electric power generator in a power plant. Photo provided courtesy of Pacific Gas and Electric Company.

FIGURE 17.2   Wind-powered electric generators. Photo provided courtesy of Pacific Gas and Electric Company.

17.2.2   The Distribution System

The power distribution system consists of transmission towers, switching systems, and substations. The towers keep the dangerously high voltage power lines a safe distance away from people, animals, and buildings. Switching systems link various parts of the nationwide power grid and can be operated to share resources. In fact, their commercial power grids link the USA and Canada. In some instances, a failure of one part of the grid can bring down parts in distant locations, as has happened from time to time. Failures such as these are attributed to several factors: human operator error, equipment failure, and solar disturbances. Refer to Figure 17.3 for a picture of large high-voltage towers used to distribute electricity over long distances.

Human error is always a problem with complicated systems, and was the cause of failure in both the USA's Three Mile Island and the former USSR's Chernobyl nuclear power generating facilities.

FIGURE 17.3 High-voltage power distribution towers. Photo courtesy of Pacific Gas and Electric Company.

Equipment failure can also occur without notice, such as large power transformers burning out, or power lines breaking under icy or strong wind conditions. An example of this type of failure happened in August of 2003, due to equipment failure in a northern Ohio power grid. Before being diagnosed and repaired, it lead to the loss of power over much of the northeast USA, including Boston and New York City.

The Sun periodically spurts out great amount of solar material, consisting of highly charged atoms and electrons, at millions of miles an hour. These events are called *coronal mass ejections* (*CMEs*), and when the charges strike the Earth's surrounding magnetic shield, called the *magnetosphere*, the charges are funneled down to Earth along the magnetic field lines. When the moving charges strike the long commercial power lines, the lines react as huge antennas. The changes rapidly moving past the long conductors induce in them a huge over-voltage that can far exceed the levels the power system is designed to handle, causing breakers to trip. Sometimes the induced voltage is sufficient to cause widespread loss of power, such as what happened to the northeast USA and southeastern Canada in 1999 and 2003.

In 2003, solar eruptions called *flares* caused the failure of commercial power distribution system in Sweden. The stream of charges took about 19 hours to travel from the Sun to the Earth.

If all this makes you start thinking about buying a backup power supply for your PC, you are thinking correctly.

17.2.3   The Drop from the Mains

After the very high voltage used to transmit the power from the generating station is lowered by a sub-station, it is sent at roughly 20–40 kV along local streets to our homes and businesses. Step-down transformers, called *pole pigs*, are mounted on the power poles outside and provide 240 and 120 VAC to our homes. Figure 17.4 shows the overall scheme of the power distribution system. Figure 17.5 shows a power pole-mounted step-down transformer.

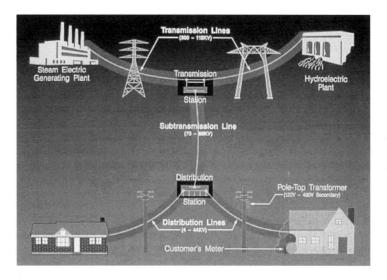

FIGURE 17.4   The power distribution system. Diagram courtesy of Pacific Gas and Electric Company.

The output of the power pole transformer feeds several homes or businesses. The wires dropping down from the power pole transformer are all called *drop lines*, which connect to a metal pole mounted on the roof or high up on an outside building wall. The lines then connect to the service entrance.

FIGURE 17.5   Pole-mounted step-down power transformer (pole pig).

17.2.4   **The Service Entrance**

The point at which the power lines connect to the home or business is called the *electric service entrance*, and consists of the power meter and distribution panel. The panel contains current control devices, fuses in older installations, and circuit breakers in new installations. Fuses haven't been used in distribution panels used on homes in the USA for at least 30 years. Circuit breakers offer the distinct advantage of being resettable after they trip, thus avoiding the need for replacement every time they blow, as do fuses.

The Power Meter

The lines then connect to a power meter, which measures the amount of power used by the customer. Employees of the commercial power company periodically come to read the meter in order to know how much to charge the customer. Figure 17.6 shows an electric power meter mounted on a power panel outside a home. Figure 17.7 shows a close-up picture of an electric power meter.

FIGURE 17.6   Power meter and distribution panel on a home.

FIGURE 17.7   Close-up picture of a commercial power meter.

The type of power meter used contains both current-sensing as well as voltage-sensing circuits, in the form of coils of wire. The current coil is wired in series and the voltage coil is wired in parallel with the power line. Refer to Figure 17.8 for the schematic diagram of a power meter.

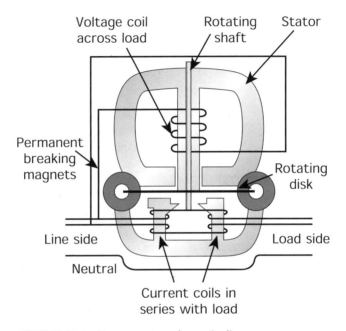

FIGURE 17.8 Power meter schematic diagram.

The Power Panel

The power panel distributes the electric current by using many, perhaps 20–30 or so individual parallel circuits for the average home. Each parallel circuit is known as a *branch* circuit, and is individually controlled by a fuse or circuit breaker.

17.2.5 Branch Circuits

There are two styles of branch circuits used in most homes, 240 VAC and 120 VAC. The 240 VAC branch circuits are used to power electric ranges or ovens, electric dryers, electric water heaters, and sometimes electric wall heaters and large air conditioners. The 120 VAC branch circuits power everything else, such as the lights and appliances. Most electrical codes do not allow mixing of lighting and baseboard outlet circuits. This might seem strange, but it is designed for safety. If a short circuit trips a breaker, it would be nice if the lights stay on. Otherwise, we might be stumbling around in the dark in case of a tripped breaker that controls an outlet—not a very good idea. Figure 17.9 shows a schematic diagram of a simple house branch circuit arrangement.

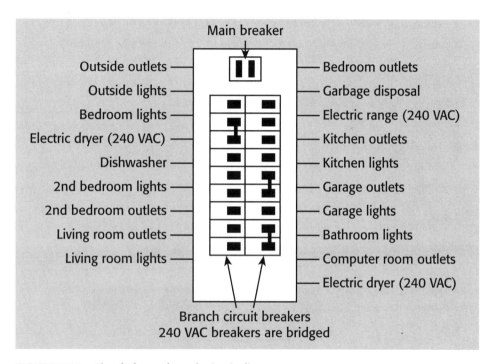

FIGURE 17.9   Simple home branch circuit diagram.

17.2.6 Fuses and Circuit Breakers

As presented earlier, fuses are one-time devices and most cannot be reset. There are special resettable fuses, but the fuses used in power distribution panels were generally not resettable. Circuit breakers trip open after a certain designed current value is exceeded for a specified amount of time. Both fuses and breakers are specified for a designed operating voltage and current. In 120 VAC applications, a single fuse or breaker is usually used on the line or "hot" side of the power line. For 240 VAC applications, two fuses or breakers are used. The breakers are mechanically linked, so they trip out in pairs. This is because with the 240 VAC line, there are two line or "hot" wires, with current 180 degrees out of phase, plus a neutral wire. Refer to Figure 17.10 for a schematic diagram of a 240 VAC branch circuit.

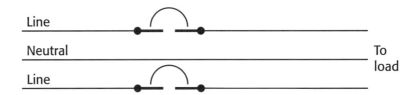

FIGURE 17.10 240 VAC line circuit breaker schematic.

17.3 WIRING PRACTICES AND CODES

Wiring codes and standard building practices concerning electrical wiring and equipment are dictated by both the National Electrical Code (NEC) and local building codes. Always employ a certified electrical contractor or electrician when having new electrical power lines installed or modified. It will help ensure that your house or business has reliable electrical service and doesn't suffer an electrically caused fire. Refer to Figure 17.11 for a picture of an NEC handbook.

FIGURE 17.11 NEC handbook.

17.4 SAFETY

The old adage of "if you don't know what you are doing, get a professional to help" is especially true when working with electricity and electric power wiring. There is no sense in getting injured or killed when professional help is available. Look in the phone book for local electricians or electrical contractors, rather than relying on your buddies or the salesperson at the local home supply store.

Some general safety tips when working around electrical circuits follows:

- Don't drink alcohol or use drugs prior to or during electrical work.
- Don't work in wet conditions.
- Don't rely on others to turn off power for you; check it yourself.
- Follow all manufacturers' recommendations when using or installing electrical equipment.
- Use the correct gauge of wire for the current required by the equipment.
- Use test meters or lights and not your hands to check for an energized circuit.
- Wear insulating boots when working on potentially energized circuits.
- Keep one hand in your pocket when touching exposed wire conductors.
- Learn CPR.
- Don't touch a person still attached to an energized circuit; you will be shocked or electrocuted as well. Use an insulated stick or board to try to dislodge the person affected from the live circuit.
- Call the commercial power company when you find a downed wire. Keep well clear of it.
- Know the phone numbers of the closest hospital in case you need it.
- In the case of blown fuses, never replace a blown fuse with a fuse of greater amperage capacity. This can cause fires!
- When a circuit breaker trips, determine the cause before blindly resetting the breaker.

17.4.1 Wall Outlets

Modern 120 VAC wall outlets have three connector holes, two for spade lugs and one round hole. They are wired using the line, neutral, and ground cable conductors. As presented in a previous chapter, inexpensive outlet testers are commonly used to test for correct wiring.

GFCIs

Anytime a wall outlet is within reach of a water pipe (which connects to earth ground), a ground fault circuit interrupter should be used. This device shuts off power to the outlet within a few milliseconds of sensing a ground fault, that is a current through the safety ground wire, which should normally never carry current. These GFCIs are used in bathrooms, kitchens, laundry rooms (close to water pipes) and for outside outlets. They are even available as plug-in additions to heavy-duty extension cords.

17.5   POWER LINE CONDITIONING

The power delivered to our homes and businesses is not pure; it commonly includes unwanted and potentially damaging components or other undesirable characteristics. Such unwanted components include frequencies other than the power line frequency of 60 Hz. They can be due to other electrical equipment operating nearby, such as power tools or large equipment with motors, nearby operating commercial broadcast, amateur radio, or CB transmitters, and under- or over-voltage.

Short-duration, very-high voltage spikes can occur due to nearby lightning strikes or heavy electrical machinery powered by motors switching on or off. Figure 17.12 shows examples of a 60 Hz power line waveform with undesired components present in the waveforms.

Several of the larger manufacturers of backup power supplies also make and sell dedicated units called *power line conditioners*. Most commercially made power line conditioners include sections that reduce the intensity (amplitude) of voltage spikes, frequencies outside the normal power line frequency (60 Hz in the US), and sections that can compensate for over- or under-voltage conditions, within limits. Simple "spike-protector" outlet strips use components known as metal oxide varistors (MOVs). These are about the size and shape of a US quarter, and have two leads that are connected across the power lines. They present a very high resistance at normal line voltage, but when the voltage rises to the MOV's *threshold voltage*, the MOVs resistance value drops to a much lower value. The result is the MOV dissipates the extra energy by converting it to heat. Some of the problems associated with MOVs include the fact that a normal 120 VAC home branch circuit consists of three, not two wires. It is entirely possible to get a circuit fault condition where there exists excessive voltage from any two of the three wires, line, neutral, and ground. Miswired outlets are not all that uncommon, and a nearby lightning strike could cause such an excessive voltage condition. So when purchasing spike-protector outlet strips, look for ones offering "three-way" spike protection. This designation means the strip uses three MOVs, connected across the line to neutral path, line to ground and neutral to ground. These cost a bit more but are worth it. The second MOV problem is they lose their effectiveness over time, and there is no easy way for the average person to test

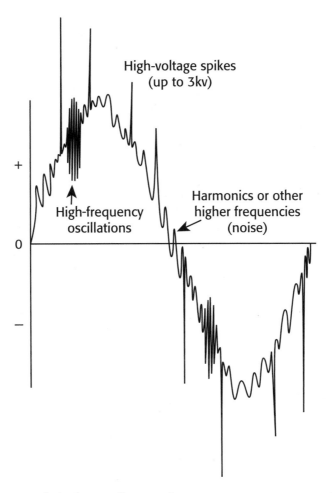

FIGURE 17.12   Undesired power line waveform components can pose problems for PCs.

them. When they finally fail, they do so as a short-circuit, which causes them to get very hot, and sometimes explode. Unfortunately, most inexpensive spike strips use a cheap plastic enclosure, which of course is not what you want to contain a potential firecracker of a device when it explodes. So spend a few extra dollars on a three-way spike strip in a metal housing. You will help ensure you do not burn down your home or place of business. You boss will thank you (maybe).

All of the better-quality UPSs and SPSs offer at least spike and radio frequency interference (RFI) filter sections. Figure 17.13 shows an SPS that powers and conditions the AC line for the author's office PC.

FIGURE 17.13   A backup power supply that protects and runs the author's office PC.

17.6   RF INTERFERENCE PREVENTION

The general rules to prevent one piece of electrical equipment from interfering with another (known as electromagnetic compatibility) include the following:

■   Ensuring proper circuit layout of both the equipment generating the radio frequencies as well as the equipment subject to interference.

■   Use of tightly-fitting metal covers with a minimum of holes, to prevent unwanted signals form exiting a transmitter or leaking into a receiver. Typically, military communications gear, built to very high standards, uses metal covers fastened securely with many screws to insure that signals don't leak in or out of the equipment.

■   Use of special filters on both transmitters and potential receivers.

■   Observing maximum permitted operating power and minimum spacing from transmitters to potential receivers.

■   Paying attention to equipment ground connections and using a "single-point ground" in RF-sensitive environments such as radio transmitter sites. A single-point ground system has all grounded equipment connected to a single common point, which helps prevent the chassis of different pieces of equipment from rising above ground to different levels.

■ Use of ferrite toroid beads to attenuate (reduce in amplitude) RF traveling on the outside of wires and cables.

■ Use "spike protectors" as a first line of defense for the power to your PC. They contain electrical devices known as metal oxide varistors (MOVs). MOVs effectively clamp down at a maximum voltage to reduce the level (amplitude) of line voltage applied to the equipment protected. A better approach is to use a UPS or SPS, which contains spike and EMI protection.

Examples of electrical shielding techniques used on PCs include the I/O shield around the ports on the rear of ATX-style PCs and the use of ferrite beads on cables such as video monitor cables. Other examples include the many ground screws used on ATX motherboards, and the redesign of the ATX standard over the older AT standard to better shield the boards from both giving off and receiving unwanted radio signals. Signals generated by all computers include clock and control signals as well as video signals.

Most well-designed transmitters include filters on their output called low-pass filters, which greatly attenuate harmonic signals that could potentially cause interference to other equipment. All well-designed receivers include high-pass filters that greatly attenuate frequencies below the intended operating frequency of the receiver.

When making connections using cables, always ensure that the joint is physically sound. If the cables are shielded, such as coaxial cables used in the early Ethernet systems, cable TV, or home entertainment systems, make sure the connections are tight and that the shields are continuous, with no breaks. Even one shield break in a shielded system will compromise its integrity, causing it to both radiate the signals it carries as well as making it susceptible to outside signals getting in. As an example of the extreme problems caused by poor shielding and the lengths taken to maintain good shielding, consider the cable TV industry in the USA. For years they struggled with interference to and from their systems. They originally used coaxial cable of roughly ¼ inch diameter, known as RG-59/U. It consists of a center conductor, an insulator called the *dielectric material*, a foil or braided shield, then finally an outer plastic cover. The industry went through years of trying to figure out the shield problem using this cable and finally modified it to what is used now, which is called *quad-shield coax*. It uses four distinct shielding conductor layers between the inside conductor and the outer insulating cover. Of course it costs a good deal more than the original single layer shielded cable, but in terms of overall system concerns, it is considered to be cost effective. Refer to Figure 17.14 for a diagram of quad-shield cable.

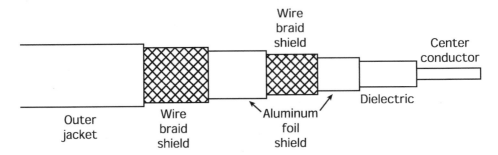

FIGURE 17.14 Quad-shielded coaxial cable.

17.7 HELPFUL HINTS

- Keep electrically noisy appliances, such as those using motors, on a different branch circuit from the PC.
- Use EMI filters on electrically noisy equipment, such as motor-powered appliances (washers, power tools, et cetera).
- Do not use long extension cords on PCs; this will cause insufficient line voltage at the PC, causing problems.
- Power your PC from a power line conditioner in case of very noisy AC power.
- Use an AC outlet tester to check for miswired outlets, and then correct wiring faults.
- Use an AC voltmeter to check the actual line voltage present. Do this at different times of the day and keep a log. Sometimes low voltage is a problem only at certain times of the day or night, depending on nearby equipment use.
- Use a DVM with a PC-interface cable and software to perform a long-term AC-outlet voltage test.
- Keep all PC covers in place and all cover screws in place and snug. This helps prevent electrical noise problems.
- Ensure that the mating surfaces of all metal covers are bare metal, to create a complete electrical enclosure with no gaps.
- Don't customize PC cases with nonmetallic windows; this will increase the likelihood of RF interference problems both to and from a PC.
- Use good-quality shielded cables where they are specified.
- Never operate a PC in a moist or wet environment.
- Turn off PCs when they are not going to be in use for long periods.
- Do not place objects on top of monitors or allow objects to block air inlets or outlets.

- Use ferrite beads on cables connecting to the PC to lessen the chance for RF pickup or radiation from those cables.
- Use good-quality backup power supplies to remove harmful AC line components.
- Remember that spike protectors wear out over time and lose their effectiveness.

17.8   WIRING CONCERNS FOR PC USERS

Improperly-wired AC outlets can cause all sorts of problems for PCs, including shock hazard to the operator, data loss, random reboots, freeze-ups, and so on. The easiest way to check for a properly wired AC outlet is to test it using an inexpensive outlet tester. These are available for about $10 at most hardware, electronic supply, and computer stores.

PCs should not be connected to the same branch circuit used to power heavy machinery, power tools, or any type of electrical appliance that produces a lot of electrical switching or other noise. If it is absolutely necessary to do so, the use of AC power line filters on all the offending electrical equipment power lines, as well as on the PC power line should be used.

17.9   SUMMARY

- Commercial AC power is generated by electromagnetism. A turbine turns a generator. Several means are used to turn the turbine, including moving water or steam. Steam is produced by heating water by one of several means, including the burning of coal or oil, or nuclear decay.
- Wind power is the use of the wind to turn blades connected to a generator.
- AC power is delivered over long distances by using very high voltage.
- AC voltage is stepped up or down by transformers.
- AC power is often not pure and can contain other energy riding on the waveform that must be removed or greatly reduced to avoid data loss, PC lockups, and other problems.
- Removing potentially harmful AC line noise components before they reach the PC is called power line conditioning.
- Backup power supplies come in two types; uninterruptible power supplies (UPSs) and standby power supplies (SPSs). UPSs always run off an internal battery-powered circuit; SPSs run off the line power and switch over to the internal battery-powered circuit once line power fails.

■ UPSs and SPSs are rated in terms of volt-amperes or watts plus the amount of runtime for typical listed PC systems.

■ Good quality UPSs and SPSs include spike protection and EMI filters.

■ Spike protectors will reduce the amplitude of very short duration high-voltage spikes that can damage a PC or cause lockups and data loss. They employ devices called metal oxide varistors (MOVs) that wear out and lose their effectiveness over time.

■ Better quality backup systems also compensate for slightly over- or under-voltage conditions.

■ EMI filters are used to reduce the amplitude of radio signals entering an appliance via its AC power cord. They should be used on any known noisy appliances as well as PCs.

■ An AC outlet tester is an inexpensive tool for finding out if an AC power outlet is correctly wired, and for pinpointing specific wiring problems.

■ Consider using a long-term PC-based test on the AC line voltage to determine exactly when power problems occur during the day or night. This test is useful for pinpointing power problems caused by other equipment that operates only at certain times.

17.10   KEY TERMS

Turbine	Circuit breaker	Power line conditioning
Service entrance	MOV	UPS
Power meter	UPS	SPS
Power panel	BPS	EMI
Branch circuit	Interference	
Fuse	EMI filter	

17.11   EXERCISES

1. Why is AC used for long-distance power transport and distribution?

2. What device is used to step-up or step-down AC voltage?

3. What are the commonly available AC voltages delivered to homes and small businesses in the USA and Canada?

4. What purpose does the service entrance serve?

5. What is the difference between fuses and circuit breakers?

6. What is a branch circuit?

7. Why should appliances using motors be on a different branch circuit from a PC?

8. What device can be used to lessen the effects of voltage spikes on the AC line?

9. What device can be used to reduce the amplitude of radio frequencies on the AC line?

10. Describe what an AC outlet tester does.

11. Why might you want to use a DVM connected to a PC for AC-line testing?

12. Why is it a bad idea to remove metal from the outside case panels on a PC?

13. What is the difference between a UPS and an SPS?

14. What is meant by *run time*?

15. What units are backup supplies rated in?

16. Who was Nikola Tesla?

17. What is a *pole pig*?

18. What function do capacitors mounted on power poles serve?

19. What is meant by "power factor"?

20. What is a GFCI?

18 Computer Electronics: Review and Final Insights

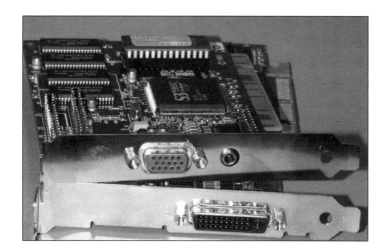

18.1   INTRODUCTION

Now that you have learned (hopefully) all the basic electrical circuits and how to measure and calculate resistance, voltage, current, and power, you have most likely gained a new perspective and appreciation of the complexity of PCs and associated equipment. This chapter presents a good review of the general electrical concerns one should have when working with PCs. It also provides some good troubleshooting hints.

18.2   THE MOTHERBOARD

As you know, the PC motherboard contains most of what makes a PC a computer: the main processor, chipset, RAM, and expansion cards; plus many integrated services such as keyboard and mouse port connectors, USB, Firewire (IEEE-1394) connectors, and so on. Working with motherboards requires great care and awareness of proper handling techniques. This section presents the most common concerns a PC tech should be aware of when working with motherboards.

18.2.1   ESD

A fully populated motherboard contains many different devices, most of which are extremely static sensitive, such as processors, RAM, and so on. Not only is the main microprocessor very sensitive to ESD but the chipset chips, being dedicated processors as well, are too. RAM is very sensitive to static, as are most expansion cards, such as video cards (they have dedicated video processors called *graphics processing units* or *GPUs*) and even sound cards (which contain digital signal processors). Use of at least a properly connected ESD wrist strap is essential when working on a non-powered PC. It is even better to use an ESD conductive mat connected to the PC chassis. Always place components such as expansion cards, RAM sticks, et cetera, on the ESD mat to keep them at the same electrical potential as the rest of the PC and the technician. Do not get up and walk around when carrying any boards or other small PC components without first placing them in antistatic bags.

18.2.2   Working on Powered PCs

It's risky to work on powered electrical equipment. However unlikely, it is possible to suffer electrical shock or, under severe and unusual situations, even death by electrocution. One possible scenario in which a tech might have to work on a powered PC is testing the output voltages from the power supply. In such cases, do not wear an ESD wrist strap, since doing so increases the chance of the tech's body becoming a part of a circuit from a powered device back to the grounded PC chassis. So in such cases, the tech should take extreme care when working not to touch powered circuits with bare hands or skin. Consider powering the PC from a GFCI to help prevent fatal electrical accidents. Also, techs should make sure not to touch any chips or circuit traces, since without wearing an ESD strap, doing so could cause ESD chip damage. Recall the tribolelectric series from Chapter 4, "The Electric Circuit and Voltage Generation."

Do not wear watches, rings, or other metal objects when working on any powered circuits, since metal is a good conductor, and doing so greatly increases the possibility of shock or electrocution.

Refer to the motherboard document for the location of all jumpers, switches, and connection pins to wires leading to LED indicators and pushbuttons on the PCs front panel. If the motherboard documentation is unavailable, most motherboards include silkscreened markings indicating these connections. Refer to Figure 18.1 for an example of silkscreened motherboard connector markings. Also, this information is often marked on a sticker on the inside bottom of the case.

FIGURE 18.1 · Front-panel connector markings silkscreened on a motherboard.

18.3   ADAPTER CARDS

Things to watch out for when working with adapter (expansion) cards include the need to remember that most of them include dedicated processors, so they are ESD sensitive. These include video, SCSI host adapters, NICs, and sound cards. Keep cards in their antistatic plastic bags until actually installing them in the PC. Place them back in a spare antistatic bag as soon as they are removed from a PC. Don't just casually chuck them onto a desktop or workbench, since doing so greatly increases the chance they will be damaged by static, get dirty, or have something heavy placed on them that will physically damage them.

When installing expansion cards, it's always a good idea to use a small flashlight to check out the motherboard expansion slots for any sign of foreign matter, dirt, et cetera. The author once had a PC that would not work right after installing a new SCSI host adapter. The adapter checked out fine in another PC. The problem turned out to be the presence of a very small rubber band, used to secure some of the power supply drive power connector lines, in the adapter card expansion bus slot. This prevented a single edge connector pad from making contact with the mating contact on the SCSI card. Once the rubber band was found and removed, the host adapter worked fine. So never assume something when it is easy to confirm with an eyeball check.

Keep cards clean and securely mounted. Use "TV tuner" spray and foam swabs to clean edge connectors. Avoid using cotton swabs for cleaning cards and sockets since they tend to shed cotton fibers, which cause problems. Never use an eraser on contacts, as this will quickly remove the extremely thin gold plating (only a few atoms thick) as well as cause a damaging triboelectric static charge. Always secure adapter cards in the PC chassis with the proper screw through the "L" bracket. This provides not only a mechanical connection, but also a secondary electrical ground connection for the card.

Some expansion cards, such as high-performance video cards, must be connected to the four-pin power connector called the ATX +12 V connector in order to properly power the video card's on-card GPU. These cards also carry dedicated fans for the GPU.

18.4   MEMORY

RAM is extremely static sensitive. Keep memory sticks in their antistatic bags until time to install them in the motherboard. Place removed RAM sticks in antistatic bags to protect them while in storage. Don't touch the edge connector metal contacts.

Make sure to properly install and remove RAM sticks. *SIMMs* (the single inline memory modules used on very old PCs) are installed by placing the stick into the socket at an angle, sliding the stick down into the socket, and then rotating the stick to a vertical position, making sure the end clips both snap closed.

Dual inline memory modules (*DIMMs*) have an even better system of mounting. They slide straight down vertically into their mating sockets, and are secured by automatically actuating end clasps. Make sure to get both ends of the DIMM sticks properly aligned into the slot and press down firmly and evenly along the length of the stick until it snaps into place.

Removing memory sticks is done in the reverse way of installing them: release the end clasps first and then remove the sticks.

Modern RAM runs hot, and many sticks come with installed heat spreaders made from aluminum or copper. These can also be added later in case the RAM sticks did not come with them.

18.5   CONNECTORS

There are many types of electrical connectors used on PCs. The A+ exam assumes technicians are very familiar with them, so here's a quick review of the main PC connectors.

18.5.1   COM Port Connectors

PCs used to have two serial *COM ports*, more accurately called COM port connectors on the rear panel. They were provided by an add-on I/O card and later via a super I/O card, which also contained the floppy controller logic as well as the hard drive controller logic. In modern PCs, all this is integrated into the motherboard. Most new PCs include only a single COM port connector. Refer to Figure 18.2 for a picture of COM port connectors.

FIGURE 18.2   COM port connectors.

18.5.2   Parallel Printer Port Connectors

The first PCs contained at least one and sometimes two connectors for connecting parallel printers. These were called p*arallel ports* and they were labeled as LPT1 and LPT2, which stands for "line printer 1" and "line printer 2." Actually, PCs can access up to three parallel printer ports, LPT1–3, provided they are equipped with three LPT physical port connectors. Most parallel printers have a proprietary connector on the printer itself, developed by *Centronix*, a printer company. Today, even though the company has long since gone out of business, the Centronix printer connector still survives. On the rear of a PC, the parallel printer port connectors are female 25-pin units, also called *D-sub 25* connectors. Refer to Figure 18.3 for a picture of a PC parallel printer port connector.

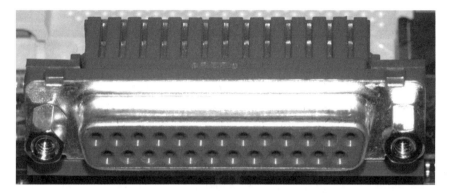

FIGURE 18.3 Parallel printer port connector (LPT).

18.5.3 Keyboard Connectors

The early PCs (designed to fit the XT and AT form factors motherboards) used a round keyboard connector called a DIN for *Deutsche Industrie-Normen*, consisting of five pins in a semicircle. PCs using the baby AT form factor motherboard all used this type of connector as well. The new ATs and ATXs use a smaller version of the DIN, called a 6-pin "mini DIN," and more commonly called a "PS/2" connector. The PS/2 connector was named by IBM with its 1987 introduction of the PS/2™ line of PCs. This type is used on most PCs that have an LPX, ATX, or NLX form factor motherboard. Refer to Figure 18.4 for pictures of the 5-pin DIN and PS/2 styles of keyboard connectors.

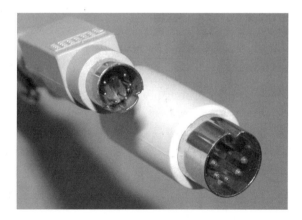

FIGURE 18.4 5-pin DIN and PS/2 style keyboard connectors.

The newest keyboards use USB connectors, and some USB keyboards also offer additional USB port connectors on the keyboard body. Refer to Figure 18.5 for a picture of a USB keyboard connector.

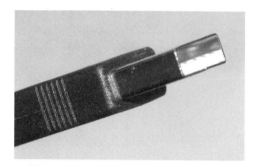

FIGURE 18.5 A USB keyboard connector.

18.5.4 Mouse Port Connectors

The first mouse connectors were serial, and connected to a serial COM port connector on the rear of the PC. This type of mouse is known as a *serial mouse*. Older versions of Windows recognized serial mice only when they were connected to the COM 1 or COM 2 port connectors. Later versions of mice that avoided using up all the COM port connectors were known as *bus mice*. They were sold with proprietary expansion cards that used a special PS/2-style connector, which was wired differently than the currently used PS/2 mouse connector. The currently used PS/2 mouse connector is integrated on the motherboard. By the way, inadvertently connecting a PS/2 mouse to the PS/2 keyboard connector or the keyboard to the PS/2 mouse connector will totally confuse the PC! A way to remember which goes where is the keyboard always connects using the connector closest to the motherboard. Modern computer mice also come in the USB interface as well as the wireless variety.

Wireless mice talk to a small pod that connects to a USB port on the PC. They are often sold as part of a set along with a wireless keyboard. Figure 18.6 shows an example of a wireless keyboard and mouse set.

18.5.5 Video Monitor Connectors

Video expansion cards output their signals to a monitor via connectors on the rear "L" bracket of the card. Depending on the video capabilities of the card, the connector can be a 9-pin D-shell, a high-density 15-pin D-shell, or a DVI connector for digital monitors. Some computers have an RCA connector for a "TV out" signal. Refer to Figure 18.7 for a picture of various video card connectors.

FIGURE 18.6   A wireless keyboard and mouse set.

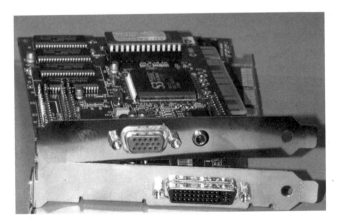

FIGURE 18.7   Various video card monitor connectors.

Inexpensive PCs integrate the video function on the motherboard and so you will find the video connector for those units mounted directly to the motherboard.

18.5.6   Sound Connectors

Originally, PCs attained the ability to reproduce complicated sounds, other than simple beeps via the built-in speaker, by the addition of a sound card. Creative Labs was the first company to achieve wide success with its line of Sound Blaster cards, first monaural only, and later stereo. Typical connectors on a sound card today include several ⅛-inch miniature phone jacks (female connectors) for line-in, line-out, mic-in, front speakers, and in some cases, Rear speakers. Additionally, some sound cards include a 15-pins-in-two-rows variety of D-shell connector for connecting joysticks or a musical instrument digital interface (MIDI) device such as a music keyboard or music processing equipment. Refer to Figure 18.8 for a picture of a sound card.

FIGURE 18.8 A typical sound card.

Standard phone plugs are ¼" thick. Miniature phone plugs are ⅛" thick. Subminiature phone plugs, found mostly as remote on/off jacks on portable cassette recorders, and on some wireless phones, are ³⁄₃₂" thick.

Just as with video, keyboard, and other component port connectors that have been integrated into the motherboard, sound chips are now commonly found on inexpensive motherboards, with their associated sound connectors also mounted directly to the motherboard. Some of the latest sound chips included on motherboards offer advanced features such as four- or five-channel sound. The A+ exam

expects you to know all the standard sound card or chip port connections. The
⅛-inch jacks used for the sound interface follow this color code: green—line
out/speakers, blue—line-in, red—microphone. On sound cards or embedded sound
chip designs offering more than two channels, there are additional ⅛-inch jacks
used, such as a black colored jack used for the rear sound channels. Some motherboards and sound cards allow for programming the use of the jacks for specific purposes. Refer to Figure 18.9 for a picture of integrated sound chip motherboard
connectors.

FIGURE 18.9   Integrated sound chip
motherboard connectors.

Joystick/MIDI Port Connector

Many new motherboards also contain a combination joystick/MIDI port connector,
used for playing games and interfacing musical instruments such as keyboards with
a MIDI interface cable. Figure 18.10 shows a joystick/MIDI port 15-pin connector.

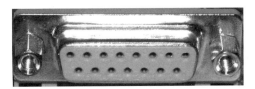

FIGURE 18.10   Integrated motherboard
joystick/MIDI port 15-pin connector.

18.5.7   AC Power Connectors

There is at least one and sometimes two AC power connectors on the rear panel
surface of PC power supplies. One is a recessed male connector that mates with the
female end of the power cord that plugs into the wall outlet. Some power supplies,

especially older AT-type supplies, also had a second power connector to power a monitor. Refer to Figure 18.11 for pictures of the two types of power supply power connectors.

FIGURE 18.11   Two styles of power supply power connectors.

18.5.8   Expansion Bus Slots

The most common interface between added expansion cards and a PC motherboard are called *bus slot connectors*, or *slots* for short. The next section summarizes these.

ISA Bus Slots

The original expansion bus slots found on the original PC were 8-bit ISA slots. They are made of black plastic. Later, with the introduction of the AT, 16-bit bus slots, also black in color appeared. The 8-bit ISA slot is now obsolete, but the 16-bit slot can still be found as a "legacy feature" on some new motherboards.

MCA Bus Slots

IBM released an improved bus slot design, called *microchannel architecture* or *MCA*, around 1987. It was the first plug-and-play system for the PC, and required no jumper settings. Running the required configuration setup disk was all that was required. IBM decided to charge a royalty for its new MCA system. Most motherboard makers, used to the free and open-architecture days (when IBM did not enforce its patents on the PC system) balked at paying for it, so they used other types of slots.

EISA Bus Slots

As a counter to the MCA system, a group of the largest "clone makers" formed a group that developed an improved bus system based on the old original IBM *ISA* (*industry standard architecture*) design. It allowed for backward compatibility with ISA cards but also allowed using the new, more-capable EISA cards. Depending on whose book you read, *EISA* stands for *Enhanced ISA* or *Extended ISA*. In either case, the general public was confused by the new bus styles and for the most part, stuck with good old ISA cards and motherboards.

VLB Slots

When Windows 3.1 came out around 1991, with its graphical user interface (GUI, pronounced "gooey"), it slowed down PC performance. Most users felt it was due to the increased video demands made by the OS. Several motherboard manufacturers developed proprietary "local bus" systems.

In an attempt to speed data transfer and establish a common, industry-wide "local bus" (one connected directly to the main CPU), Intel developed the *VESA local bus*. VESA stands for the Video Electronics Standards Association. Established around 1989, it provided for an additional slot, tan in color, to be added in-line with existing bus slots. Motherboards were made using VLB slots attached to the ends of not only ISA slots but also the new EISA slots developed by the "clone makers" once IBM came out with its new MCA bus slots around 1987 or so.

Unfortunately, there were major problems with VLB. Motherboard manufacturers were making boards with one, two, and three VLB slots in-line with ISA or other older bus slot connectors. The problem was that if more than one VLB card was used at a time, the data transfer rate dropped significantly. The problem was due to circuit loading. So VLB died quickly, after only about two years.

PCI Bus Slots

Intel realized the problem and fixed it with the introduction of the *peripheral component interconnect* or *PCI* bus slot. Made of white-colored plastic and offset back from the rear edge of the motherboard, PCI is like VLB, only without the circuit-loading problems. Most add-in cards for PCs still use the PCI bus style.

AGP Bus Slots

Intel also developed the *accelerated graphics port* (*AGP*). It is used exclusively by video cards. There are several versions of AGP, both in speed and data path width.

The connectors are tan and are set back even farther than PCI slots from the rear motherboard edge. They look to the author suspiciously like the old VLB slots!

PCI Express

This is a very new bus slot developed by Intel as well. It is just starting to appear in high-end motherboards in mid-2004. It is supposed to greatly increase data throughput and eventually eliminate PCI, AGP, and even COM and parallel ports on PCs. It is also written as PCI-X.

AMR/CMR Bus Slots

Audio/modem riser (*AMR*) and communications/modem riser (*CMR*) are names for the bus slots intended to support small, inexpensive cards, often of proprietary design, for low-end PCs. The cards support modems and sound functions. The slots are brown in color and are usually found on the rear motherboard corner the farthest away from the main CPU.

18.6 MOTORS

Motors are used to power the drives and fans. Drives include hard drives, floppy drives, CD, and DVD, as well as Zip drives, et cetera. Fans are used in the power supply and on the main processor. Fans are often found on the main chipset chip (north bridge or ICH1 chip) as well as on the video card processor chip on high-end cards. In addition to all these fans, most new PC cases include fans or places to mount additional (often up to three) case fans. In general, the motors run on 12 VDC.

18.6.1 Types of Fans

When buying additional case fans, always choose the "ball-bearing brushless" 12 VDC variety. Standard PC cooling fans for the case come in 80, 92, and 120 mm diameter models. In each size, there are many choices based on cubic feet per minute (CFM) of air moved as well as differences in voltage and current requirements. Different fans make different amounts of noise when running, and these ratings are usually listed in decibels (dB), a logarithmic method of comparing two measurements, usually power, voltage, current, or loudness. As far as practical uses, just remember a fan with a lower dB noise rating will be quieter than a fan with a higher dB noise rating.

Most fans used for cooling the main processor are rather thin and use blades, and are usually called *muffin* fans. Another type of fan is known as a *blower* or a *squirrel cage*, which is usually more efficient at moving the air compared to the muffin type. Figure 18.12 shows both types of fans.

FIGURE 18.12 A muffin-type fan and a blower used to cool microprocessors.

18.6.2 Hardware Monitoring

Cooling fans made for PCs come in two wiring styles: two-wire and three-wire. With the two-wire style, there is usually a red lead and a black lead. The red lead is the + DC wire and the black lead is the –DC wire. Fans with three wire leads use the third lead, usually a different color, such as white or yellow, to provide the fan's speed (in RPM, or revolutions per minute) information to a motherboard equipped with three-wire fan connectors. Then the PC operator can monitor each three-wire fan's RPM using either a hardware-monitoring program running on the operating system (OS) or a BIOS-based monitor, depending on the style of monitor that is supported by the system. Figure 18.13 shows an example of a BIOS-based hardware-monitor showing fan speed. Often the PC user can set a trip point for alarms for CPU temperature, case temperature, all voltages, and speed of all the fans in order to be promptly alerted to problems.

18.6.3 Fan Maintenance

There is not much to do for fans to keep them working well besides making sure they are properly connected to the power supply or motherboard and keeping them clean and turning freely.

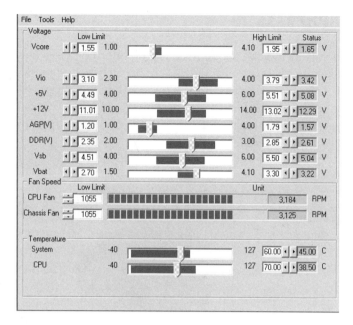

FIGURE 18.13 Fan speed shown by a hardware-monitoring program.

When cleaning the PC, make sure to wipe off all the fan blades and blow out the fans using canned air. This is a job best done outside, since it can be messy, and people with allergies to dust will not be pleased if it is done inside. Small cotton swabs or foam swabs dipped in window cleaner work well to clean the fan body and blades after blowing off with the canned air first. Also clean the air intakes on the metalwork next to the fan, using a soft cloth dampened with a good mild household cleaner. Another trick that can be done for very dirty fans is to shoot the canned air stream into the fan blades while the fan is running. Again, do this job outdoors, as it will surely be messy.

Before closing up the PC case, take a minute to make sure there are no loose cables that can come into contact with any fan blades, as this will either prevent the fan from turning, which leads to overheating, or it will cause a noise as the fan blades rub on the cables. If large bundles of wires are too close to the fan, they tend to restrict airflow. It can also lead to damaged cables if allowed to go on too long. Use nylon tie-wraps to secure cables and keep them away from fans.

Fan motors do seize up and when they do, the system can overheat, and the frozen motor will tend to draw excess current. Motor bearings that are beginning to fail will usually squeak and give themselves away. Do not ignore this timely repair forewarning.

18.7   PERIPHERALS

Peripherals include all the items that plug into the PC such as printers, scanners, mice, keyboards, external modems, hubs, switches, routers, digital cameras, et cetera. They all have cables that can become loose or damaged. Most use their own dedicated power supplies that can become unplugged or can fail outright. Some peripherals such as external USB Zip drives use the power from the USB cable. Figure 18.14 shows an external USB Zip drive.

FIGURE 18.14   External 250 MB USB Zip drive, which obtains its power from the USB connection.

18.7.1   Printers and Scanners

Most printers in the PC world are dot matrix. There are still dot-matrix printers being made today, although they are obsolete for home use. Many receipt and forms printers are of the dot-matrix type. More popular are ink-jet and laser printers. Most of these use a dedicated power supply, usually in the form of a small "brick" or "wall wart." Be sure to check these supplies if the printer appears dead before looking to the PC when troubleshooting. The most common printer connection types used with PCs include the older serial printers that connect to a COM port, parallel port printers, and USB printers. Older OSs required installing the proper printer driver first before the PC would recognize the printer; new OSs such as Windows XP usually include most printer drivers. The trick to this is if the printer was introduced after the version of OS you are running, you will need to find and install the correct driver program for the printer. Otherwise, the printer will either not be recognized at all, or not all printer features will be available.

Scanner interfaces include parallel, SCSI (pronounced "scuzzy"), and more recently, USB. Make sure you are using the latest driver for your scanner, usually available from the scanner manufacturer's Web site under the banner label of "support."

Check for loose or damaged scanner interface cables when troubleshooting scanner problems. Check to make sure the scanner's dedicated power supply is connected to the AC wall outlet and that power is actually present at the outlet. Use an AC outlet tester to perform this test. Make sure to power on any external SCSI devices before booting the PC they are attached to; otherwise, they will not be detected.

18.7.2 Mouse Mechanisms

The older style mice use a roller ball that rubs on two shafts, each connected to a potentiometer (variable resistor). Physical mouse movement is translated by the roller ball and "pots" into varying voltages, one for the X-axis, the other for the Y-axis. These signals are sent to the PC, which can determine the mouse position relative to the displayed area on the PC monitor.

Problems with this type mouse include dirty roller balls and potentiometer shafts. Potentiometers themselves get dirty, which can cause erratic cursor movement. Cables can become loose or damaged. Periodic maintenance for the roller-ball type of mice includes securing the mouse cable, and cleaning the roller ball and pot shafts with denatured alcohol or household window cleaner using foam swabs or cotton-tipped sticks.

Optical mice avoid the problems of using mechanical methods to translate physical movement into screen cursor movement. They use an LED and light sensor to sense movement over a surface. They do not require frequent roller and pot shaft cleanings since they lack these items. They are considered superior to the older style mice. Optical mice usually come with a USB interface, although they can be adapted to use PS/2 port connectors. Figure 18.15 shows the PS/2-style motherboard connectors for the keyboard and mouse. They are color-coded (at least on the pictured computer) as follows: purple for the keyboard and green for the mouse.

18.7.3 Modems, Hubs, Switches, and Routers

Modems come in two styles: internal and external. Internal modems can be the expansion card type or integrated with the motherboard. External modems have their own case and power supply. Standard items to check when doing electrical

FIGURE 18.15 Motherboard PS/2 connectors for keyboard and mouse. The keyboard plugs into the socket closest to the motherboard surface.

troubleshooting include making sure the connecting cables are intact and correctly connected, and checking the power supply in the case of an external modem. Don't forget to check the phone line as well. Make sure the phone line is live and available when trying to use the modem. Since this text is based on electrical troubleshooting, it is beyond the scope of the text to discuss software concerns; just be aware of the need of the proper drivers, and of interrupt (IRQ) conflicts when installing or troubleshooting modems.

Many folks today use a LAN (*local area network*) at home to share files, drives, and printers. So they may be using Ethernet and a hub, switch, or router. If using a cable modem, they might be using a combination device, which consists of a modem, firewall, and router in one box. All of these devices use interconnections to the different PCs, usually in the form of Ethernet Cat-5 cables. They also use dedicated power supplies. Each PC on the network also must have a network interface card (NIC) installed. So there are many potential places to have loose or damaged cables when using a LAN.

New types of wireless routers are becoming very affordable. The speeds and distances they can handle are generally far less (often only half) of a wired system, but are generally easier to install for most people, since they do not require laying out of cables or drilling through walls and floors.

18.7.4   Digital Cameras

Digital cameras are all the rage today, and for very good reasons: they allow nearly instantaneous examinations of pictures taken, don't require using messy, toxic chemicals to develop the images, the images can be manipulated with a computer and appropriate software, and they can be quite small and inexpensive.

Images are stored on RAM which takes many forms, such as Secure Digital (SD), CompactFlash® (CF), Memory Stick® (MS), and even tiny hard drives ranging from 1.8 inches in size down to .85 inches. Storage capacity ranges up to the 10 GB range for the tiny drives and about 1 GB for the other memory forms.

To get images from the camera into the computer for processing, there are two main methods used. One is to remove the memory unit from the camera and insert it into a card reader that connects to the computer. The second method is to use a cable to attach the camera's digital output port connector to the computer. The most popular methods used to do this are Firewire, also known as IEEE-1394, and USB. Firewire offers a transfer rate of 400 Mb/s (megabits per second) for Firewire 1.0, and up to 800 Mb/s for the new Firewire 2.0 specification. The original USB 1.1 offers a transfer rate of 11 Mb/s and the newer USB 2.0 (high-speed USB) offers rates up to 480 Mb/s. Often the connector on the camera is a proprietary one, and the other end is standard USB or Firewire. Figure 18.16 shows a USB 1.1 cable designed to plug into a Nikon D100 digital camera, the one used for most of the pictures in this textbook.

FIGURE 18.16   USB 1.1 cable showing mini USB end used for a Nikon digital camera.

18.8   COOLING

The original PCs used a single fan mounted in the power supply. This one lone fan sucked the hot air generated by all the electronics out of the PC case. Air leaked in through all the small cracks between the chassis and case cover, the floppy drive disk slots, and so on. Airflow was not very good inside the PC case, but still, it was adequate for the limited amount of heat generated by the early processors, RAM, and other heat-producing components. Things have changed! This section presents some modern PC cooling techniques.

18.8.1   Case Fans

Most modern PCs use at least one fan mounted in the case, in addition to the fan in the power supply, to help with cooling. Faster processors give off vastly more heat than the early ones, and faster hard drives spinning at 5,400 or 7,200 (the standard spin rates for modern PCs), and 10,000 or even 15,000 RPM in some servers give off a large amount of heat, which must be effectively removed from inside the PC case. The much faster RAM used today also runs considerably hotter than slower RAM.

Many new PC cases include extra case fans, or prestamped holes for air and mounting screws to add one or two extra case fans. PC users who like to "overclock" their machines, by running components at higher clock speeds than originally specified, commonly add extra case fans or even more exotic cooling solutions such as water-cooling kits to their PCs. The faster components run, the more heat they produce. Extra fans mounted on the case side, directly above the main processor, or on the case top, to suck hot air out are called "blow holes" by the case modifiers (modders.) Be sure to use only ball-bearing type fans. Sleeve bearings used on the less expensive fans are prone to early failure and are not worth wasting your money on.

Careful wire dress (how the wires are installed, bundled, and secured to the chassis) is essential for maintaining good airflow through a PC case. Using wire stick-downs and tie-wraps to bundle cable, or loom to enclose multiwire cables, are good techniques to employ. Modern versions of parallel hard drive and floppy drive cables also aid in reducing air restrictions inside a case. Figure 18.17 shows two styles of round cables used to connect parallel interface EIDE hard drives to the host adapter connector.

18.8.2   Other Cooling Solutions

More radical cooling solutions include using a *Peltier* chip, a device that is essentially a semiconductor sandwich that runs on electricity and makes one side of the sandwich cool and the other side hot; it is essentially a heat-transfer system. Peltier units

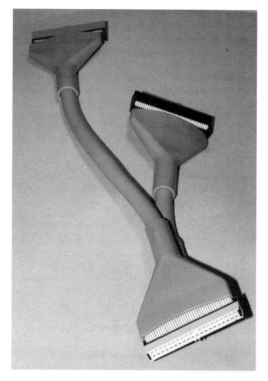

FIGURE 18.17 Round hard drive cables aid in reducing airflow restrictions inside a PC case.

are mounted in between the microprocessor and the heat sink, with the cool side toward the processor and the hot side toward the heat sink. This arrangement is purported to provide a more efficient thermal transfer. Peltier units cost around $25.

Water cooling involves a metal water block mounted to the processor through which cooling water is pumped on through a heat exchanger with a fan, similar to the cooling system used on automobiles. These systems cost about $200 and must be installed by the user. A leak in such a system will surely cause internal damage to the PC, a factor to consider seriously.

Phase-change cooling systems operate by "boiling off" a liquid and then condensing the vapor back into liquid. These systems can perform extreme cooling but are quite costly, around $1,000 or more. These are also known as "heat pipes" and are commonly used in notebook computers and some desktop models. In many notebook computers, the microprocessor is located close to the center of the motherboard, and there is insufficient space to mount a fan in that location. In these instances, a heat pipe can be used to remove the heat from the processor and move it a few inches away to a location where a sufficiently large fan can blow on the radiator portion of the cooling system.

Don't forget to keep the air inlets for any type of cooling system clean and allow sufficient air space around the inlets to ensure adequate airflow.

Some new "super quiet" desktop PCs also feature heat pipes instead of fans in order to greatly quiet the system. Some even use the case itself as a large heat sink.

18.9 UPGRADING CONSIDERATIONS

When thinking about whether to upgrade an older PC or to just replace it with a new one, there are many things to consider. Will the upgrade involve changing just a few components or many? Will it be cost-effective? Will it allow running the software intended? Many people consider a 3-year old PC to be obsolete and therefore not worth trying to upgrade. With prices of entry-level PCs around $399 and fairly capable midrange PCs with monitor, keyboard, and mouse priced under $1,000, one should carefully consider any upgrade.

18.9.1 Common Upgrades

The most commonly performed PC upgrades include adding more RAM, adding a second, larger hard drive, upgrading the video card and/or monitor, and upgrading the power supply. The cheaper PCs offer very little extra "head room" for making extensive upgrades, since their power supplies are often marginal, at best, to start with. Adding additional RAM, drives, et cetera, will tax the weak power supply and random reboots and lockups will likely result. Since a good replacement power supply might easily cost $80–100 (assuming you can even find a better supply that will fit in those minitowers), give serious thought before trying to add much to a low-end PC. If the PC is an ATX style, it might be possible to replace the motherboard and processor as a unit. This will usually also require replacing the RAM as well with the newer style required by the replacement motherboard. At this point, the cost can be in the $300–400 range, roughly the price of a brand new low-end PC. Therefore, the need for careful consideration before deciding to perform a PC upgrade is clear.

Adding USB capabilities to older motherboards that lack USB port connectors is also a worthwhile upgrade, provided your OS has proper support for it, such as Windows 98SE and later. Figure 18.18 shows two integrated USB port connectors on the I/O connector group of an ATX motherboard.

FIGURE 18.18 USB port connectors in an ATX I/O connector group.

18.10 SUMMARY

- Working with PCs involves using common sense and observing ESD safety rules.
- Treat all cards, RAM, and processors with extreme care and keep them inside anti-ESD bags when not in use.
- Don't walk around carrying unprotected PC boards, RAM, et cetera.
- As a general rule, unplug a PC before working on it, and wear an ESD wrist strap and use an ESD mat.
- If you must take voltage readings on a running PC, then do not wear an ESD wrist strap since doing so can increase the chance of shock or electrocution.
- RAM is extremely ESD sensitive. Keep RAM modules inside of their ESD bags until time for their installation. Place used RAM inside ESD bags, and store them in a cool, dry place.
- Know the difference between the connectors used on a PC. Some of the more common include: AC power, monitor power, COM port and parallel ports, USB, Firewire, and PS/2 keyboard and mouse connectors. Learn to recognize them on sight, as they are on the A+ exam.
- Motors are used to spin drives and fans. They run on 12 VDC. Some after-market fans added later by owners may be the 5 VDC varieties often available at electronics flea markets, swap meets, and surplus stores.

- Proper cooling is essential to preventing heat-related PC problems. Keep air intake and exhaust ports, as well as fan blades and air filters clean. Periodically blow dust and dirt out of the inside of the PC. Do this outside if possible.
- Know the difference between fans, Peltier chips, and heat pipes.
- Peripheral equipment includes such devices as keyboards, mice, printers, scanners, modems, hubs, switches, routers, and digital cameras. Know the interface style and capabilities of each of these.
- Know the essential components of a LAN: at least two PCs on the same workgroup, each with a NIC installed and a crossover cable for just two PCs. Three or more require a hub or switch and the appropriate type of cabling.
- The most popular form of network equipment for PCs is Ethernet. The most popular form of wiring is Cat-5 or higher.
- Wireless Ethernet is generally easier to set up than wired Ethernet as is does not require laying out of cables or drilling through walls.
- Wireless Ethernet systems usually offer slower transfer speeds and shorter distances compared with wired systems.
- Upgrading a PC is a complicated business; there are many factors to consider, including price of the upgrade, need to upgrade several mutually dependent subsystems, such as a motherboard and the style of RAM it supports, availability of power supplies that will both handle the increased load *and* fit in the case, price of brand new systems, et cetera.
- Digital cameras use one of two general interface styles to connect to a PC: Firewire or USB. There are at least two versions of each of these systems offering different transfer speeds.
- Know how to safely insert and remove different types of RAM modules.
- Know the various connectors used on a motherboard, such as those for power and for the various indicators and buttons on the PC's front panel.
- PC cooling fans with three wires provide fan speed data for a BIOS- or OS-based hardware-monitoring program. Fans with only two wires do not.
- Know the color code used for sound jacks.

18.11 KEY TERMS

Firewire	Mouse connector	GPU
USB	Video connectors	Peltier unit
COM port	Sound connectors	Heat pipe
Parallel port	Modem connectors	
Keyboard connector	Ethernet connectors	

18.12   EXERCISES

1. List the most common steps one should take to ensure that a PC will not overheat.

2. List the most common components in a PC that require periodic cleaning.

3. Describe how to properly handle ESD-sensitive PC parts.

4. Draw a picture of a serial (COM) port connector.

5. Draw a picture of a parallel port connector.

6. Draw a picture of at least two different styles of video port connectors and properly label each one.

7. What is a heat pipe and what it is used for?

8. What does GPU stand for?

9. List at least three important considerations involved with deciding to upgrade a PC.

10. Which type of cable would be the most likely one to use to connect three or more PCs connected as a LAN using Ethernet?

11. What are some of the differences in using wireless Ethernet over wired Ethernet?

12. What is the maximum specified data transfer rate of USB 1.1?

13. What is the maximum specified data transfer rate of USB 2.0?

14. What is the more common name for the IEEE-1394 specification?

15. How old do most people consider a PC must be before it is obsolete?

16. What is a common cause for a PC that randomly reboots or locks up?

17. List at least three different types of memory storage devices used in digital cameras.

18. What is the function of the third wire on a three-wire case fan?

19. List at least four parameters monitored by a hardware monitor program.

20. List at least three PC items most commonly upgraded.

21. List all the important specifications of a new PC that costs about $1,000.

22. Which connector do you plug a PS/2 keyboard into on an ATX mother-board?
    a. the one closest to the motherboard surface
    b. the one farther away from the motherboard surface

23. Identify each connector on the integrated connector I/O panel shown in Figure 18.19.

FIGURE 18.19 Integrated I/O connector panel.

24. What color jack is used to connect the rear speakers on a motherboard's sound connector panel?

Appendix

A

Electronics Careers and Hobbies

A.1 INTRODUCTION

Now that you have read this book and performed the lab exercises, you might still be wondering what a person with a background in electronics can do to make a living. What types of employment can a person with an interest and training in electronics be involved in? What kinds of recreation are based in electronics? This section will explain various examples of electronics careers, and the typical types of work those involved in the trade perform. Several popular hobbies involving electronics are presented in the last sections.

A.2 ELECTRONICS CAREERS

The next section presents the most common job types in the field of electronics.

A.2.1 Assemblers

Electronic assemblers are responsible for building the vast majority of all military, industrial, and commercial electronics equipment. Assemblers install the components on the *printed circuit boards*, (PCBs) which may consist of hundreds of *discrete* (separate) components. Typical assembly operations include insertion of small components into predrilled holes in the boards; mounting components to circuit boards or cabinets using sheet metal screws, machine screws and other hardware; soldering component wires, called *leads*, to circuit board metal traces; building-up sections of waveguides or other specialized hardware used in communication systems; mounting gauges, switches, and indicator lights on panels; installing cable harnesses; removing and replacing parts on circuit boards or chassis; desoldering and resoldering electrical components.

Assemblers must generally have good ability to understand both verbal instructions and written procedures for installing these various components. They must have excellent eye-hand coordination skills, and they must not have color-blindness. This is because many types of wires and small electronic components carry a special color code to signify the value of an important quality, called a *parameter*. Minimal knowledge of electronic theory is usually required for entry-level assemblers. Assemblers usually enjoy a very clean work environment with good lighting, a clean, uncluttered work area, and an all-around pleasant environment. Often, an assembler may work for weeks, months, or even years, putting together the same type of components or subsystems. For this reason, assembly work can tend to be boring for some people. Hourly pay is toward the low end of the electronic jobs range, but starts around the $8.00–$10.00 range, for someone with no previous assembly experience.

Depending upon the particular company's advancement policy, it is sometimes possible for an experienced assembler to move up to higher skill-level, and therefore higher-paying, jobs within the company. It is becoming increasingly rare, but not unheard of, for a person to advance all the way from an assembler position to one of "engineer." Many companies have policies requiring advanced engineering degrees in order to be originally hired as an engineer. Some electronics companies do, however, allow a person to advance beyond the position for which they were originally hired, based solely on experience and demonstrated job proficiency. Other employers require additional college-level electronics training, and some provide reimbursement for tuition and books needed for electronics classes taken outside of work hours.

Recent figures suggest that colleges and universities in the United States are graduating fewer engineers, per capita, than some of the other industrialized nations of the world, such as Japan, France, and Germany. Perhaps that is one of the reasons some companies tend to allow, and even encourage, their workers to advance into an engineering position.

In some areas of the country, assemblers are further classified as mechanical, electrical, electronic, or electromechanical assemblers, depending on the particular job specialization. All of these types of assemblers could, however, perform work on electronics equipment.

A.2.2 Technicians

Electronics *technicians* comprise the next level of experience and responsibility in the production of electronics products. A typical technician's job could range from that of an expert assembler, to one of a production tester, to that of testing a specific type of equipment, using very expensive, highly complex, test instruments. Depending on the particular requirements of the company, a technician might be required to posses all of the skills required of an assembler, including expert soldering skills, the ability to use a variety of hand and power tools, and read schematic and assembly drawings, plus the ability to accurately record data using computers. Nearly all electronics technicians must have a solid background in basic electronic circuit theory, mathematics up to the level of trigonometry, plus practical experience building and testing many different types of electronic circuits.

Many technicians gain their theoretical knowledge and "hands-on" experience, that is, circuit construction and measuring skills, by attending a two-year public or private school. There are hundreds of community colleges in the United States that offer excellent electronic technology degree programs. A person can attend day or evening classes to obtain an Associate of Science (A.S.) degree. The A.S. degree programs usually consist of general education courses as well as the required electronics and math courses. There are also many privately-operated technical schools that offer students a very concentrated program, which usually lasts less than one year. Often the cost of attending the private tech schools is many times that of attending a community college.

One of the most important skills an electronic "tech" must posses is the ability to successfully troubleshoot. *Troubleshooting* is the art and science of recognizing an existing problem, pinpointing the cause, and repairing it. A successful troubleshooter must understand what a certain system is supposed to do, in terms of many different things. The tech must understand the overall system's function, the function of each subsystem within the system, and the purpose of each discrete component within the subsystem. Knowledge of all values of things connected to the system's *front end*, called the *input*, as well as those things produced by the system on the "business end," called the *output*, is essential. Each subsystem has its own input or inputs, as well as output or outputs. You can understand, then, why a tech must undergo more advanced technical training than an assembler in order to be successful in the trade.

A.2.3   Super Techs

On the upper end of the technician job spectrum, are the engineering-aids, essentially, *supertechs*. They must be super troubleshooters, have an in-depth knowledge of specific systems, plus several years of experience calculating circuit values, and testing all sorts of components, systems, and assemblies. They must be able to work with minimal supervision. Engineering aides must communicate effectively with engineers and other coworkers. They must feel at home with highly sensitive, often delicate and expensive, test equipment. The ability to work with computers is becoming essential, and most engineering techs are required to have several years of very specialized computer training and experience in the field. Computers now do most large-scale production testing.

These techs must know how to input data to the computer, and understand the output from the computer. They must recognize problems with a computer program, and hopefully, have the knowledge to fix it. The ability to get a computer system "up and running" is of utmost importance.

The tech must be able to work independently, without the direct supervision of the engineer. Most techs must be able to produce a written document of test results. Sometimes, a tech must go before a group and orally report test results, so good communication skills are a must.

Technicians, overall, earn higher pay than assemblers. A recent graduate from a public or private technical school may find some entry-level technician positions pay less than that of an assembler who has several years of job experience. However, after a few years of job experience, the average technician earns more than the average assembler, and has a much better chance for advancement. Most techs must demonstrate the ability to work on their own a good portion of the time. Their work can be quite varied, even over the course of a single day. Some companies have demonstrated their desire to hire supertechs to do basic engineering work. The theory behind this practice is simple: the tech gains the status of good pay and interesting work, and the company saves money over what a degreed engineer would command.

The general consensus among many large electronics companies today is that they can't seem to find enough good techs, and the demand is only going to increase in the foreseeable future.

Other types of specialized technician classifications are: line maintenance tech, test tech, quality assurance tech, repair tech, and field service tech.

A.2.4 Line Maintenance Techs

Line maintenance techs are responsible for maintaining and repairing the equipment used to manufacture products on the production line. They may be required to perform repetitive testing of circuits or devices, plus assemble the particular test setup, rearrange test setups, and do minor alignment and repair of test equipment. Other times, they are responsible for just keeping all equipment used on the production line operating within acceptable limits, without actually using the equipment for production themselves.

A.2.5 Test Techs

Test technicians are concerned mainly with conducting tests on manufactured products before they are shipped to the customer. Their jobs can range from the very mundane, repetitive type, to the very complex exacting tests performed on state-of-the-art products. Often test techs become very specialized in a general area of expertise, such as analog circuits or digital circuits. They may further specialize into particular types of these broad circuit categories. A test tech may become an expert, for example, in alignment and testing of a particular type of disc drive for computers, a certain type of laser system, or a new type of communications equipment.

A.2.6   Quality Assurance Techs

Quality assurance, (often called simply QA), or quality certification (QC) techs are responsible for maintaining the highest quality of manufactured products, to company-specified limits. They test and either pass or reject items produced by the assemblers. Items rejected by QA are then either repaired and retested, or scrapped. The decision to repair and retest or to scrap a particular item depends, to a great degree, on the cost of repairing an item, called *rework*, versus the cost to produce a new item. Expensive products are usually repaired and retested. Low-cost products often would cost more to fix than to scrap. The repairs are usually done in the *rework* area by highly qualified assemblers with several years of experience, or by technicians specialized in this type of repair work.

A.2.7   Repair Techs

Repair, or service, techs do the rework in many companies, and some also perform repairs on faulty equipment returned from the customer. At other companies, repair techs strictly do only servicing of faulty equipment returned from customers. This is the type of work the TV repair technician performs. In years past, TV repair techs accounted for a large part of all electronic technicians in the United States. Today, partly because we are more of a "throw-away" society and partly because of the cost of repairing equipment compared to buying a new replacement is so close; the TV repair techs are becoming "dinosaurs" of an age gone by. However, there will always be a need for people to fix "hi-tech" products. Repair techs will have to constantly upgrade their skills in order to understand and keep up with the advancing technology. Just as compact discs have all but replaced LP vinyl records, new advances in computers, TV, radio, cellular phones, navigation systems, power generation and distribution, et cetera, force techs to stay abreast of the new technology. College and university professors report an increasing number of techs and even engineers taking classes on new devices and circuits that simply did not exist when those students first learned their trade. The field of electronics is one which demands that a person read textbooks, magazines, technical trade journals, and manufacturer's data books in order to just stay competent in a rapidly expanding technological world.

A.2.8 Help Desk Techs

Help desk techs assist in installing, operating, or upgrading software on personal computers. Other users would be customers who recently subscribed to a new service, such as an Internet, e-mail, satellite TV, or DSL service. The job requires both a good understanding of the product or service, as well as patience when answering customers' questions. Often a customer will be in a bad mood, having been frustrated with trying to deal with what the customer perceives as a problem with the service or product. Often they will not be able to clearly state what the problem is, since they are so new to using it. The help desk tech must be a good listener, and document the reported problems as well as offer support and the proper course of action to try to remedy the problem if possible. If the problem cannot be fixed with the initial phone call or e-mail, the problem may be referred to more qualified personnel, or the situation may call for the customer having to secure a *return merchandise authorization* (RMA) and then returning the item for replacement or service. Often, help desk techs work for a company other than the one that made or sells the product or service. Today, many of these positions are actually filled by people working in countries outside the USA, a process called "out-sourcing," yet they service USA companies' help desk needs.

A.3 ENGINEERS

Engineers, for the large part, are designers. They use their knowledge of existing technology to design new devices, circuits, and systems. They are the main force behind the creation of new science and technology. Radio, radar, television, stereophonic sound, compact discs, computers, car telephones, space exploration, and superconductors are just a few of the new technological areas opened-up by engineers in the last 100 years or so.

In the United States, engineers prepare for their profession by enrolling in a five-year college degree program. Their degree is called a Bachelor of Science (B.S.). The first two years of the degree program consists of taking mostly general education, science, and math classes. The last three years are spent studying advanced mathematics, physics, chemistry, and materials science. In general, college-degreed engineers decide, after the first two years of college course work, whether they intend to concentrate in civil, mechanical, or electrical engineering. From that point on, they take increasingly specialized courses. Electrical engineers, or EEs (double Es), take courses on circuit analysis and design, solid-state devices, computers, communications, plus possibly a further specialization of fiber optics, robotics, telecommunications, or solid-state integrated-circuit design.

Engineers can go on to take post-graduate degree classes, leading to a Master's degree (M.S.), or a Doctorate degree (Ph.D.). Of course, the salary an engineer can expect to earn upon graduation from college increases with the attainment of advanced degrees. After several years on the job, some engineers elect to do consulting work, outside their regular work hours, and may earn $100–$200 per hour. Others may earn their entire income from their consulting business.

Application Engineers

Application engineers help customers or "end-users" pick the correct product from all those manufactured by a certain company, in order to meet the customer's exact requirements. They must be very knowledgeable of many of the current components used for a particular application. They must work closely with both the designers and the users of these products to help the user obtain the most efficient part for the job.

Research Engineers

Engineers who perform basic research, that is, trying out new ideas to learn new scientific principles, are known as *research engineers*. They brought us such new devices as the vacuum tube, transistors, and integrated circuits (ICs).

Design Engineers

Design engineers create new device and circuit designs for specific needs. In addition to employing thousands of engineers itself, the aerospace industry has created hundreds of "spin-off" industries. These are based on new ideas developed by the research necessary to satisfy a technologically advanced nation's desire to conquer new technological areas. Airplanes and missiles require large amounts of electronics to guide and assist their flights. Thin-film circuits, surface-mount integrated circuit technology, laser and photo fiber-optic devices, and light-coupled computer "chips" are a direct result of spin-off technology developed by the design engineers working in the aerospace industry.

As the technology advances, the need for ever-increasingly specialized types of engineers will force the creation of new designations of engineers, just as it has done for technicians.

Engineers originally were those who designed, built and operated engines, as the engineer on a locomotive. Today, engineers come in many specialized varieties. Occasionally, an engineer will be required to not only design a particular device or system, but actually be responsible for the production, QA testing, and rework of the product. Electronics is indeed a field in which the "sky is the limit."

Due to the highly specialized nature of electrical (or electronics) engineering, many degreed-engineers must take courses on the latest devices to be developed.

This learning is an ongoing process, and the field of electronics is truly one in which a person must continually strive to maintain currency in the *state-of-the-art*, which is constantly growing and advancing into new areas.

The field of medicine has seen many new uses of electronics to aid in the fight to prolong life. Some recent medical uses of computers are the CT scan (computed tomography), magnetic resonance imaging (MRI) systems, biofeedback systems, and use of electricity to speed bone healing.

Computer art forms, artificial intelligence, robotics, computer-aided design (CAD), and computer integrated manufacturing (CIM) are some examples of new areas recently developed. Some applications of electronics are so new that no one has even invented a catchy acronym for them yet.

A.3.1   Field Service

Field service persons repair manufactured equipment at the customer's site. They are really traveling technicians. The field service person must travel to the customer's place of business, taking all the necessary tools and test equipment required to effect repair of the equipment. It takes a person who is not afraid to spend long hours, or days away from home, who can work with no supervision, and who has excellent troubleshooting skills. Successful field service workers can earn excellent wages and benefits. Usually, companies specifically train people to service particular types of equipment. A period of "learning the ropes" on all aspects of assembly, testing, and operation of each piece of equipment a company sells is required. It is also necessary for the field service tech to have a solid understanding of basic electronics, a good work attitude, and the ability to work well with the public.

Often the job requires speaking to people who have little or no technical background (the users) as they explain the problem with the equipment. It is sometimes necessary then, to "read-between-the-lines" while listening to the customer's description of the problem.

A.4   SALESPERSONS

For every electronic product manufactured, there exists the need for that product to be marketed. Salespersons are employed by virtually all major manufacturers of electronic equipment to accurately describe various aspects of these products to the customer, often in nonengineering terms. Customers need to ask questions in order to make intelligent purchases. The *salesperson* must be the bridge between the customer and the design engineers. Often a salesperson that is very familiar with all the pertinent qualities of a number of a company's products will be known as an applications engineer. Indeed, some devices and systems are so complex that an engineering background is required just to correctly advise customers. The technical requirements of an electronics salesperson could run the gamut from little or no technical background to that of the applications engineer.

A.4.1   Technical Writers

An offshoot of sales is the job classification of *technical writer* who produces product manuals often used to sell products. The tech writer must have excellent writing skills, of course, but also fairly good technical knowledge of the device or system to be described. Most of the specification listings (specs sheets) we read, whether they are for a new stereo system, computer, or microwave oven, were written by a technical writer. Nearly all large electronics companies employ their own tech writers, or use the services of small technical-writing service companies. Tech writers write manufacturer's data books, which describe all the important operating characteristics, voltage, and signal requirements of a certain item.

Whenever a new advertisement is created for a technical product, there is the need for a tech writer to be the bridge between the product designers and the artists and marketing specialists. We can all remember the agony of trying to assemble a child's toy at the holidays using poorly written instructions. You can understand how having clearly written documents can greatly increase the public's acceptance of a product. The more technical background the tech writer has, the easier it is to write documents that accurately describe something. The computer people have a term for this—it's called being "user friendly."

A.5   ELECTRONIC HOBBIES

Electronincs offers a wide variety of things to do as a hobby. The next section presents a few such hobbies.

A.5.1   Amateur Radio

Amateur radio, often called *ham radio,* is a fascinating hobby that combines knowledge of electronics theory with the practice of communicating with other radio operators all over the world. Ham operators must pass examinations on electronic theory, radio-operating procedures, permissible radio-operating frequencies, transmission modes, as well as proper radio operation and on-the-air conduct. They learn about radio transmitters, receivers, antennas, and all the associated equipment of a modern "ham shack." Many hams elect to build some or all of their equipment from "scratch," while others purchase commercially available products.

A perfectly workable ham shack can be assembled for several hundred dollars, if one is willing to buy used ham equipment (gear) or make things in either kit form or from scratch, using only a schematic diagram as a guide. Just as in any other hobby, such as photography or fishing, it is also possible to spend thousands of dollars on all the latest equipment. Those with a more technical background often make their own chassis and cabinets for equipment, as well as assembling the circuitry themselves.

Hams may use several different modes of transmitting information, depending on the particular class of operator license they hold. In the United States, the Federal Communications Commission (FCC) oversees the use and licensing for all radio, television, and other forms of electronic broadcast communications. The Private Radio Bureau is that section of the FCC that oversees amateur radio and CB radio. Many new hams have graduated from the ranks of CB. They grew impatient with the lack of control and poor operating habits exhibited by some CB-ers.

There are currently three different classes of amateur radio licenses in the U.S. The easiest to earn, with the least technical skill requirements, is called the Technician class. It requires a relatively easy technical exam, covering basic electronic and radio principles, FCC rules and regulations regarding the Amateur Radio Service, and no Morse code exam.

The next level is called the General Class license. Passing a tougher electronic theory test and a five-word-per-minute code exam allows the holder of this license to operate using up to 1,500 watts of radio frequency (RF) output power, and provides for greatly expanded frequency allocations. There are currently more hams in the U.S. who hold the general class license than any other.

The Amateur Extra class is the highest level of amateur radio license. Achieving this level requires passing a fairly exhaustive technical exam, demonstrating knowledge of many newer technical advances, and passing a five-word-per-minute code test. Holders of this license may operate on virtually all of the allowable ham radio frequencies and modes, at up to full legal power.

Besides the older Morse code mode of transmission, hams regularly communicate using single-sideband (SSB) and FM voice transmission, and even television and packet radio. Single-sideband is a technical improvement over the older AM method of voice transmission, and requires approximately one-third the radio spectrum space to send the same amount of information. This allows more stations to occupy a given range of frequencies, or "band." FM offers superior intelligibility under adverse receiving conditions. Just think of the improvement in listening quality of the commercial FM radio broadcasts over the older AM radio.

Sending pictures using amateur television to another amateur's station has grown in popularity over the last few years. Early slow-scan television (SSTV) was used mainly for sending still pictures, and the quality was rather poor. Modern fast-scan TV approaches the quality of a commercial broadcast.

Packet radio is a system using a computer and a unit called a terminal node controller (TNC) to change the input from an ordinary computer typewriter-style keyboard into audio tones, carried by radio waves, to another suitably-equipped ham station. There, the typed messages are displayed on the receiving station's computer monitor, exactly as sent. It is possible to carry on a "conversation" with another ham using packet radio, via computer and the ham radio link. Bulletin board systems (BBS) are able to receive and store messages sent from a station, where they may be later read by only the intended addressee of the message. This is basically the same type of technology as used in the new cellular telephones, only with text instead of voice signals.

Hams also may communicate using satellites. There have been a number of successfully launched and operated ham satellites, called *OSCARs* (Orbiting Satellite Carrying Amateur Radio). Starting in the 1960s, most of the OSCARs were funded and built entirely by hams, operating in concert, worldwide. The satellites were carried "piggyback" aboard NASA, and later some French Ariennne rockets into outer space, where they act as relay stations. Using the OSCARs, it is possible for a ham operator to communicate with another ham thousands of miles away. Even when communication by normal radio methods is not possible, hams using the satellite may communicate. The use of satellites by hams has greatly increased the knowledge and expertise of all those involved, and it has helped demonstrate to the public the expertise these hams possess.

Hams frequently provide emergency communications in times of disaster. Hurricanes, tornadoes, earthquakes, and floods may wipe out normal power and communications for days or weeks in the affected areas. Hams, using portable power generators, or operating "mobile" from their vehicles, form an emergency safety net of communications by helping the police, fire, and rescue units to relay messages for families and friends of the disaster victims. The hurricanes and earthquakes that seem to happen yearly bring the volunteer ham operators into the limelight of public disaster assistance. Their efforts provide crucial communications in place of the downed telephone lines. Recent world events have shown the absolute importance of maintaining a large reserve of persons capable of providing vital emergency communications when all other means have been destroyed, whether from Mother Nature or terrorist attacks.

Many hams enjoy "DX-ing," that is, contacting other hams in remote areas of the world. There are all sorts of awards to be earned for submitting proof of contact, known as a QSL card, with other stations, and many hams enjoy the sport of "award-chasing." It is all done in the spirit of friendship, and the FCC specifically forbids the use of ham radio for "pecuniary interest," that is, for profit. Hams enjoy being part of an international goodwill fraternity. It is possible to learn a new language, learn about geography, customs, and all sorts of other interesting topics, just by communicating with others in the world ham community. Most hams will gladly open their homes to a visiting ham and provide a tour of their "shack" (operating area). They take pride in good operating skills, and strive to improve their knowledge of technical advances in electronics and the hobby.

Figure A1.1 shows a code key, used to send CW, short for "continuous wave" transmission. CW is a misnomer, since sending using CW actually involves making and breaking a circuit with a mechanical switch, called a "key." By turning the transmitter signal on and off, letters, numbers and some punctuation symbols can be sent using Morse code. So CW is not a continuous signal, but an interrupted one.

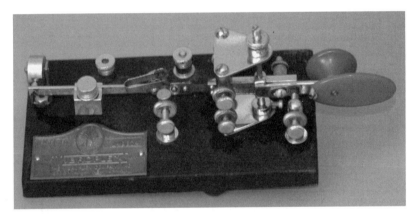

FIGURE A1.1   A morse code semi-automatic "bug."

A good place to find out how to get started in amateur radio is to contact the American Radio Relay League, 225 Main Street, Newington CT, 06111. They publish a kit, including a short booklet, and a Morse code practice tape, for about $15. Many communities have one or more ham radio clubs that offer electronic theory and code classes to the public, often for only a small fee.

It is definitely not necessary to be an electronics engineer to become a ham operator. People such as the late Senator Barry Goldwater, the past president of a mid-eastern oil-producing nation, astronauts, missionaries, carpet salesmen, in fact, anyone can be a ham operator and share in the excitement and fun.

A.5.2 Computers

In the 1960s it seemed that the only people involved with computers were the professionals working for large defense contractor companies, or the government. With the advent of home computers around 1980, it became possible for the person with an interest in computers to purchase one from a local store. Companies such as Apple®, Atari®, Commodore®, and Radio Shack® sold complete home computer systems for about a thousand dollars. These systems were quite basic, by today's standards, and had only a tape cassette unit for data storage. In these early days, it was only the well-off computer hobbyist who could afford to purchase a printer or possibly a floppy disc drive for their system. The amount of available memory was small, usually under 48K bytes (about 48,000 8-bit computer "words.")

In 1981, IBM entered the home computer market with its PC (Personal Computer), and the home computing scene hasn't been the same since. The acceptance of PCs, made by both IBM and the millions of look-alike, work-alike "clones," has brought the computer into thousands of homes and businesses. It is now possible to bring work home from the office to finish on your home PC. Many enterprising individuals have started what some have called "cottage industries." With a well-appointed home computer setup, including floppy disc drive, printer, and possibly a hard-disc drive, a person can do essentially all their office work in the privacy of their own home. Many researchers, writers, computer programmers, musicians, in fact a growing number of people with different job categories have discovered the increased freedom of working at home, thanks to the PC.

Video displays have gotten clearer, memory chips have become cheaper, and nearly all component parts of a home computer are affordable enough to allow a person to buy a package system, or assemble one from individual pieces bought from many different sources, in order to save money. There are several excellent books in print that describe how to assemble your own computer system.

Any large city is bound to have dozens of computer user groups. One may find a group consisting of people who own, or are interested in, a certain brand or style computer. They have regular meetings, at which they offer advice to those new to the PC game, and sometimes ask experts in the computer field to speak at the meetings. There are groups for every conceivable brand of computer ever made, even those brands that have not been produced for several years, such as the original Osbourne 1®, the Atari-800®, and the Sinclair®. The computer industry is constantly changing, and even the giant, IBM, has discontinued production and support of their PC Jr.

There is a virtual guarantee that a certain computer will be superseded in a few years by one that is faster, has more memory, and offers improved performance in other areas, and which will probably sell for less money. When considering the purchase of a new computer, it is necessary to weigh the desire for the best performance now, but paying a premium price for it, versus holding-off for a future model, and making do until then. Most computer advisors suggest buying the best monitor you can afford now, get a larger hard-disc drive than you think you will need, as your needs will surely grow, and the best printer you can afford. It costs more to have to upgrade in the future than it does to buy equipment that will last for many years, now.

Some of the uses people put home computers to, besides accounting or running a business are: ham operators storing records of their contacts with other hams in a permanent "log," tracking the OSCAR satellites using inexpensive programs, or operating packet radio; keeping recipes, phone numbers, and birthdays on file; controlling the turn-on and turn-off of all lights, heat, and other home appliances automatically; running telephone bulletin board services (BBS); Internet Web sites and helping your children learn their ABCs, colors, and counting using interactive learning programs specially written for children. Doing your own taxes on a PC or Mac is a good way to save paying a tax consultant to do it for you. You can still take your completed tax forms to an expert to be checked, and it will at least be in a polished, presentable form.

A.5.3   Radio-controlled Models

Radio-controlled (RC) models fascinate the old and young alike. Miniature scale-model race cars, dune buggies, boats, airplanes, and helicopters can be seen in any large park during good weather days. Nearly all the aircraft and the larger, more powerful car and truck models are gasoline-engine powered. The smaller model cars are battery-powered. The models have two basic parts—the model itself, which includes a radio receiver and decoding circuitry, plus the electrically-operated actuators that control the model, with its batteries and the radio transmitter control unit.

Most models use radio frequencies assigned by the FCC to the citizens band (CB). The control transmitters can produce a limited amount of radio power, so the models have a limited effective range, usually less than a half-mile or so. RC model enthusiasts often stage rallies and races in which all model owners can compete for prizes. The price of RC models can range from about $100 for a low-end unit, to literally thousands of dollars, with all the associated testers, extra parts, batteries and chargers, that many dedicated RC-ers eventually invest in.

A.5.4   Stereo Sound Systems

It is difficult, indeed, to find a hearing person who does not really enjoy the sound of music. The technology of sound reproduction took a giant step forward, when in the early 1960s, the technique of stereophonic sound reproduction gained wide public acceptance. Another recent development that many credit as having an equally important contribution to improving reproduced sound is the compact laser disc, or CD. The sound is so lifelike and pure; most people can't tell the sound produced by a CD played on a modern stereo system, from that of a live performance.

In the 1970s, the audio purists, so-called *audiophiles* (sound lovers) insisted that in order to gain the maximum sound reproduction quality, it was necessary to invest in separate audio components. They insisted that one should assemble a sound system consisting of separate radio tuner, preamplifier, basic (power) amplifier, and loudspeakers (speakers). As equipment quality improved, with the advance of the electronic state-of-the-art, it was no longer a necessity to invest in separates. Now, it is possible to buy a single unit that houses an AM and FM broadcast radio tuner, and pre and power amplifiers all using a common power supply. Such a unit is known as a receiver, and many offer sufficient output power to drive even the largest, most demanding speakers.

The art and science of loudspeaker manufacturing has also made great strides in quality, and the range and accuracy at which sound frequencies are reproduced is so good that the overall sound at the listener's position, the so-called *sound image,* or *sound stage,* causes the listener to almost feel present in the original recording studio or concert hall.

Some technological advances in sound reproduction never really caught on with the general public. One such advance was the idea of using two additional channels of amplification driving two additional speakers, in order to reproduce the sounds missing in normal stereophonic reproduction. The missing sounds are known as *phase-shifted,* or *ambient reflected sounds* and result from certain sounds being reflected off the side and rear walls of the recording studio or concert hall. These sounds normally arrive at the listener's ear slightly delayed, and at much lower volume level, due to the added travel time, and increased sound absorption from the walls. Normally, these sounds would be hidden by the much louder sounds coming from the two stereo speakers, normally placed in front of the listener.

The four-channel, or *quadraphonic* sound system used a special circuit, usually housed in a separate box, called a decoder, which separated out these reflected sounds, and fed only them to the rear amplifiers and speakers. The results of using this improved system were very mixed. The improvement in sound quality varied from spectacular to only a slight improvement over conventional stereo. The success of the quad system depended to a great degree on the recording technique, and the specific method of decoding used.

There were at least four different encoding and decoding schemes used, each one advanced by a different competing electronics company. Often there was no way in which the listener knew which method was used in the recording, and so no way to set the decoder to accurately recover the information. These difficulties, plus the additional cost of the decoder, two more amplifiers (or one stereo amplifier) and two additional speakers, just seemed too much for the average consumer. The idea never really died, however, and today we see the re-emergence of the idea in what is being called *surround sound*. Some of the different schemes require a total of seven separate amplifier channels and speakers. Other systems use but one additional channel and speaker, plus the two regular stereo speakers. If you have ever listened to a movie specially made to be played in a theater using *Dolby® surround sound*, you know what an improvement in listening realism this method provides. Dolby is actually the name of a sound engineer, Mr. Ray Dolby, who perfected a noise-reduction technique originally designed to reduce tape "hiss" on recorded magnetic tape.

It is entirely possible for the average person to put together a custom sound installation for their home without paying for the services of professional sound-design people. There are many consumer-oriented books and magazines on the subject, and the technical reference section of any good public library is a good place to begin.

There are a few rules that can help someone interested in assembling a stereo system today. Forget the old LP records. Unless you already have a sizeable investment in them, it would be foolish to purchase a record player. The improved sound quality, freedom from pops, clicks, warping, and record wear offered by compact discs is a giant leap ahead of records. Unless you enjoy bone-shattering bass notes, are addicted to heavy-metal rock or rap music, or want to fill a very large room with sound at high levels, it requires only about 35 to 50 watts per channel of power from the receiver. More power costs more money and you will seldom, if ever, need it. There is very little difference in quality among the better brands of receivers today, so shop competitively, and watch for sales.

Plan to spend as much, or more, for your loudspeakers, as the entire rest of the system. More than any other part of the system, the speakers contribute the most to the overall sound quality of the system. There are real differences in a certain speaker's "voice" compared to another model or brand. The absolute best way to sample speakers is to find a store that offers the customer a speaker listening room. A good example of such a listening room will have a switch panel that will allow the salesperson to instantly switch between the various speaker systems, while playing the type of music you normally listen to. If you hear one set of speakers, and then leave the store and listen to a different set at a different store, your ears will really not "remember" the first set. The near-instant comparison offered by the switch panel method will make the differences between speakers "stand out like a sore thumb." Just to convince you, ask the sales person to switch between various receivers, each tuned to the same radio station, or hooked to play the same CD or tape, and have the volume levels set the same on each receiver. Normally, it will be impossible to tell the difference between the receiver units. The speaker test, however, will sound like "night and day," especially concerning a given speaker's ability to reproduce bass notes without sounding "boxy" or "boomy."

As a rough estimate, it is possible to assemble a very good quality system, consisting of a receiver, tape player, CD player, possibly a "graphic equalizer"—basically a set of fancy tone controls—and a set of two speakers, for about a thousand dollars or less. Many stores offer closeouts on discontinued systems for far less, but stick to known name brands. Check the consumer's buying guides and magazines, as well as some of the audio hobbyist magazines for reports on overall quality and features. This technique will save you from being "pounced upon" and possibly made to feel foolish by the salesperson when you walk through the store entrance. You will be able to talk *specs* (technical specifications) with the best of them, as well as having a very good general idea of what you are looking for. It will also save you a good deal of money, as sales people often "push" the house brand, since they make a high commission on that brand. Remember, "to be forewarned is to be forearmed."

A.6 SUMMARY

This chapter was written in an attempt to give you, the reader, a brief look at just some of the possible things you might enjoy as a result of gaining knowledge of electronic technology. The field is still young, and constantly growing. There will always be new devices invented that will make our lives more fun and interesting, but also more challenging. The basics of electronics never really change, however, and a good understanding of the fundamental principles of electronics will always serve to help a person live better and smarter in an increasingly complex technological world.

Appendix

B Glossary of Terms

across Refers to a parallel connection, usually of a measuring instrument such as a voltmeter.

ambient Surrounding conditions. Usually applied to temperature of an electrical component, such as the ambient temperature of a semiconductor device, or the temperature inside a PC's case.

armature The moving part of a motor or generator that interacts with the field magnetism to produce motion in the case of a motor, or electricity from motion in the case of a generator.

artificial magnet Man-made magnet.

battery An electrical connection of two or more electric cells, in either series, parallel, or series-parallel in order to produce either a high voltage or current (or both) compared to a single cell. Common examples are "9-volt" batteries and car batteries.

branch current Current that divides at a junction of two or more components (branch), and combines again into the total circuit current after flowing through the various separate branches.

breadboard Meaning to fabricate a circuit, usually in a casual manner, in order to determine the circuit's usefulness. The breadboard is the first stage in the production of a new circuit design, after the initial engineering, but before final prototyping.

breakdown When applied to insulators, the condition describing a lowering of opposition to current flow.

carbon composition A resistor or resistive element made from carbon plus a finely ground ceramic.

cell A single unit consisting of a negative electrode and a positive electrode, that produces electricity. Examples are electrochemical cells (AAA, AA, C, and D sizes are common) and photoelectric cells (solar cells).

CEMF   Counter electromotive force. Voltage resulting from a change in current through a conductor. See *self-induction*.

chassis   A metal support structure on which all components are mounted for stability. Can also be used as a common (ground) connection.

circuit   A complete electrical pathway for current, including a source, control, and load.

color code   A method of indicating a resistor's ohmic value and tolerance by use of painted color bands around the body of the device.

conductor   A material that offers a good pathway for electricity. Conductors have only partially filled valence shells. It requires a relatively small amount of energy input to create free-electrons in a conductor.

conservation of energy   When energy changes form, its capacity for doing work remains the same; energy within a given system cannot be created or destroyed.

contactors   Large electromechanical relays used for control of high voltage or current; often used in industrial machine control applications.

control   A means of affecting the amount of voltage or current flow in a circuit. Controls can be in the form of a switch, which either allows or stops current, or in the form of a component that has a variable electrical quality such as resistance, inductance, or capacitance.

conventional current flow   Positive and negative were originally arbitrarily chosen to describe electric charges. It was assumed that electricity flowed from a positive point in a circuit to a negative one. Only years later was it discovered that electrons carry a negative charge, and it is the electrons, or the movement of negative charges in a circuit that actually takes place. Nevertheless, it had become the convention to assume electric current flowed from positive to negative. This convention is still used today in engineering applications, and is referred to as the *conventional current flow* method. The movement of electrons in a circuit from a negative point to a positive point is known as the *electron flow* method. A technician should be comfortable with either view of current flow.

conversion efficiency   Usually expressed as a percentage, describing the relative amount of energy in the desired form output from a device, divided by the total energy input to the device.

core   Material inserted inside an inductor that concentrates magnetic flux lines and increases the inductance of the winding.

current   The rate of flow of charged particles past a given point per unit of time. The unit for current is the ampere, named after Andre Ampere, a French scientist. A rate of flow of one coulomb (6.29×10^{18} electrons) per second constitutes 1 ampere of electric current. The symbol used to express amperes is A.

diamagnetic   Materials that take on a very weak magnetic field opposite in polarity to the external magnetic field to which they are exposed.

dipole field   The characteristic patten of imaginary flux lines produced by a magnet. The existence of these lines of magnetic force can be demonstrated.

domain theory   An explanation of magnetism based on the belief that small areas in a material with a similar magnetic polarity are the result of all the electrons in all the atoms in that area spinning in the same direction.

eddy current   Current that flows inside the core material in an inductor that has a fluctuating current flowing through it. The core material has resistance, so the core current causes power lost in the form of heat.

efficiency   Usually expressed as a percentage, a relative rating of how well something converts energy into some other desired form.

electrical   Concerned with the production and use of electricity.

electrochemical   Generation of voltage due to chemical action of electrolyte on dissimilar metals, called electrodes, in a cell.

electrodes   Conductor material used in a cell to generate a voltage, by the chemical action of the electrolyte on the electrodes. Also refers to any conducting part of an electrical device specifically intended to aid the transfer of electric current.

electrolyte   A chemical mixture intended to cause separation of charges within a cell by chemical action on the electrodes.

electromagnet   A magnet formed during current flow through a conductor. It is considered to be a temporary magnet.

electromagnetic   Concerned with the existence of a magnetic field during current flow, and the creation of a voltage due to the relative motion of a magnetic field and a conductor.

electron   An atomic particle that carries a negative electric charge. Much smaller then a proton, it carries an equal but opposite electric charge.

electron shell   An electron "orbits" the atomic core in a *shell* representing its energy level. A shell farther away from the core represents a higher energy level. An electron must gain or lose a discrete amount of energy to change shells.

electronics   The science that deals with the behavior and control of electrons in vacuums and gasses; use of electron tubes, transistors, photocells, et cetera.

EMF   Electro motive force. The force that drives current carriers (electrons or holes) in a circuit. Commonly called *voltage*.

energy   The capacity for doing work.

engineering notation   A value expressed as a number between 1 and 1,000 times some power of ten that is a commonly used metric prefix. In electronics work numbers are usually expressed as a decimal number between 1 and 10 multiplied by a power of ten.

ferromagnetic   Materials containing iron, which are strongly affected by magnetism, and which take on the same magnetic polarity as that of the external magnetic field to which they are exposed.

field coil   A coil of wire used to create a magnetic field in an electric motor or generator. The magnet in a motor or generator that provides the magnetic flux that cuts through the armature or moving part of the device. See also Chapter 10, "Magnetism and Electromagnetism".

film resistor   A resistor made by depositing a thin layer of carbon or metal on a ceramic tube form.

flux lines   Name given to imaginary lines of magnetic force produced by a magnet or electromagnet.

free electron   An electron that has gained sufficient energy to leave the influence of its parent atom. Free electrons are responsible for the flow of negative charges in a circuit.

frequency   The number of occurrences per unit of time.

ground   A common electrical connection between components in a circuit.

heat sink   A device made from a good conductor of heat such as aluminum, copper, or brass, that conducts excess heat away from an electrical component and allows air to circulate past itself and transfer the heat into the air. A heat sink is a form of radiator for heat-sensitive electronic components.

hole   A carrier of electric charge; part of an atom where an electron is supposed to be, but is not. Holes carry a positive charge, equal but opposite to an electron.

hysterysis   The power loss resulting from a material's tendency, when placed in a changing magnetic field to retain its degree of magnetism.

induced voltage   (1) A voltage developed in a conductor due to the interaction of a moving magnetic field, or a stationary field and a moving conductor. This is the principle upon which transformers (Chapter 13) and generators (Chapter 10) operate. (2) A voltage produced in a conductor when there is relative motion between the conductor and a magnetic field. This is the principle by which electric generators operate.

inductance   The property of an inductor to oppose a change in current by storing energy in an electromagnetic field surrounding the conductor.

inductors   Electrical devices designed to oppose a change in current flow.

input   A signal in the form of a voltage or current, applied to a circuit, in order to be acted on by the circuit.

insulator   Material offering a large opposition to current. The atoms of an insulator have near-complete or completely filled valence shells, and require a large amount of energy in order to create many free electrons.

integrated circuit   Also known as IC, a combination, on a single piece of semiconductor material, of multiple devices, usually diodes or transistors.

intermittent   The temporary, sporadic condition in a circuit of either a short or an open.

joule   The standard unit of energy or work. Originally applied to mechanical systems. When a force of 1 newton is applied over a distance of 1 meter, the amount of work done is 1 joule. A joule is equivalent in energy to 0.24 calories; that is a joule of energy will raise the temperature of 1 gram of water by 0.24 degrees C. In electrical terms, a kilowatt-hour is equivalent to 3.6 million joules.

linear   In a straight-line, as in a "back-and-forth" motion of a control such as a linear potentiometer. Also applied to a graph which shows a straight line relationship between two things, such as the resulting current when a range of voltages are applied to a resistor.

load   Any device or circuit acting as the intended end-use product connected to a voltage source. Example: the light bulb is the load for the cells in a flashlight.

magnetic compass   A direction-finding device consisting of a small, freely-pivoting magnet that reacts with the Earth's magnetic field to point north.

magnetite   A naturally-occurring magnetic iron ore.

metric prefixes   Portions of words used before a unit of measurement that indicate a power of ten. Only some metric prefixes are commonly used in electronics.

mutual inductance   The relative amount of flux lines produced by an inductor that cut through a second inductor.

NC   Normally closed. Signifies a switch that rests in a connected state. When actuated, it temporarily opens, and then returns to the open condition when released.

NO   Normally open. Signifies a switch that rests in a disconnected state. When actuated, it temporarily closes, and then returns to the disconnected state when released.

natural magnet   A magnet that is not man-made.

negative   An electrical polarity caused by the abundance or excess of electrons; the smallest unit of negative charge is carried by a single electron. Negative can also be a *relative* term, indicating a point in a circuit less positive or more negative than another point.

neutron   An atomic particle found in the core of atoms of all elements except hydrogen. Neutrons carry no electric charge and are not involved in the flow of electric current.

Ohm's law   A mathematical relationship between the potential difference (volts) applied to a device or circuit, the opposition to the current produced (ohms) and the resulting current flow (amperes). The three basic Ohm's law formulas are: $I = E/R$, $E = IR$, and $R = E/I$.

open   A discontinuity in a circuit, either intended, as the opening of a switch, or unintended, such as a broken wire.

output   A signal, in the form of a voltage or current, produced by a circuit.

paramagnetic   Materials that take on a very weak magnetic polarity the same as the external magnetic field to which they are exposed.

permanent magnet   A magnet which retains its magnetism over a very long time (years).

photocells   A class of devices that either generate a voltage, or change their resistance to current in the presence of light. Cells that generate a voltage are called *photovoltaic*, or *solar cells*.

photoelectric   Generation of a voltage directly from light, in a semiconducting material such as silicon or gallium arsenide solar cells.

PVIR table   A chart that lists power, current, voltage, and resistance values for each component in a circuit. An aid to solving for values in a circuit.

piezoelectric   Voltage produced by stress on certain crystalline or ceramic materials such as quartz, tourmaline, and Rochelle salts. Voltage applied across these materials will produce a vibration in the materials. Commonly used in thin slices mounted between two electrodes, piezoelectric crystals—commonly refereed to a simply crystals—are used to produce precise rates of vibration called *frequencies*. These frequencies are used to regulate operation of electronic circuits such as watches, computers, TVs, and radios. See also Chapters 4 and 16.

poles   **1.** Parts of a magnet that act in magnetically opposite manners. **2.** The number of discrete circuits a switch may connect-though not together—at one switch position.

positions   Similar to throws, but used to describe rotary switches.

positive   An electrical polarity caused by the deficiency of electrons in the atoms of an object. The smallest unit of positive charge in the atom is that carried by the proton. Positive can also be a *relative* term, indicating a point in a circuit less negative or more positive than another point.

potential difference   The amount two points differ in terms of quantity of charges. Also known as *electro motive force* or *voltage*. See Chapter 3 for more on voltage.

power loss   The incomplete conversion of one form of energy to another. Often some other, undesired form of energy is created, such as heat. This causes the conversion to be less than the desired 100 percent.

power   The rate at which energy is *expended or dissipated*. Power is expressed in joules per second, more often called watts (named after James Watt, the inventor of the steam engine) in electronics use. A power level of 1 watt is a *rate of expenditure* of 1 joule per second. The symbol used to express watts is W.

practical time-constant   The amount of time required for the value of current to change by 50%. It is found by multiplying the circuit time-constant by 0.7.

product-over-the-sum   Name given to the formula used to find the equivalent resistance of two resistors of different values in parallel.

proton   An atomic particle in the core that carries a positive electric charge, equal to but opposite that of an electron.

recombination   The action of an electron falling into the valence band (hole) of an atom. Energy is given off when an electron and a hole recombine.

relay   An electromechanical device used to affect one circuit with another, without any electrical connection. The transfer is made through magnetic means, opening or closing a switch.

resistance   The opposition offered to steady-state or direct current (DC) flow by an electric circuit or device. Named after George Simon Ohm, a German scientist, the units of resistance are ohms. A resistance of 1 ohm will produce a current of 1 amp in a circuit with 1 volt applied. The symbol used to express ohms is the omega, Ω.

retentivity   The capability of a material to retain magnetism over time.

right-hand rule for conductors   With the fingers of the right hand pointing in the direction of conventional current flow in a conductor, the fingers will indicate the direction of the magnetic flux lines around the conductor, from north to south.

rounding off   A method of increasing or not increasing the next most significant digit, based on examining the next least significant digit of a number. If the next least significant digit is 5 or greater, the previous more significant digit is increased by 1, or rounded up. If the next least significant digit is 4 or less, we do not round up.

saturation   The point at which an inductor core cannot accept any more magnetic flux lines.

scientific notation   A shorthand system of expressing very large or very small numbers as a decimal number between 1 and 10 times a power of 10.

self-induction   The result of a changing current flowing in a conductor producing an induced voltage in the conductor. The voltage is always opposite to the original source voltage, and will tend to cause a counter current in the conductor, which limits the rate of current change.

semiconductor   Materials that are better conductors than insulators, but worse conductors than true conductors. Semiconductor materials have atoms with four or five valence electrons. The opposition of a semiconductor material can be set during manufacture through a process called *doping*, and when installed in a circuit, by *biasing*.

sensor   A commonly used word for *transducer*.

short, short circuit   Any unwanted connection in a circuit, usually resulting from wear, damage, or moisture. The usual result of a short is unpredictable results from the circuit, often causing failure of parts due to excess current flow.

signal path   The physical pathway taken by the input to a circuit on its way to the output from the circuit.

significant digits   The number of digits in a number. The number of significant digits expressed in the answer to a problem cannot exceed the number of significant digits in any of the numbers used to arrive at the answer.

solenoid   An electromagnet in the shape of a rod, used to actuate a mechanical device, often a heavy-current switch, as in an automobile starting system.

standard form   The commonly accepted form a number is expressed in. This text uses engineering notation for most answers.

steady-state   A constant or level value of a voltage or current in a circuit. Opposite to a transient state.

temperature coefficient   The amount and direction a device's resistance changes due to a change in temperature. A positive temperature coefficient causes an increase in resistance with an increase in temperature, or a decrease in resistance with a temperature decrease. A negative temperature coefficient causes an increase in resistance with a decrease in temperature or a decrease in resistance with an increase in temperature. Usually expressed in parts per million per degree Celsius (PPM/°C).

temporary magnet   One designed to lose its magnetic force once no longer exposed to a magnetic field.

terminal   An electrical connection on a device or circuit.

thermistor   A resistor made to have a large negative temperature coefficient, used to counteract the current surge into a circuit.

thermocouple   A device consisting of two dissimilar metal wires, joined at the end, which produce a voltage across their opposite ends. Used for the detection of heat. See also *thermoelectric*.

thermoelectric   Generation of a voltage due to heating the junction of two dissimilar metal wires connected at their ends. This device is called a thermocouple, and is commonly used as a heat sensor.

throws   The number of switch positions that actually cause a connection to be made.

time-constant   The amount of time, for a given circuit, required for the current value to change by 63.2%.

total resistance   In a series circuit, the sum of the individual resistor values: $R_T = R_1 + R_2 + R_3 + R_N$

transducer   A device that converts one form of energy to another. Also called *sensor*.

transient   A state of change in a value of voltage or current. The opposite to steady-state.

triboelectric   Generation of voltage due to transfer of charges caused by friction. Not particularly useful for generation of power. A Van de Graaff generator can produce millions of volts for high-voltage research.

troubleshooter   A person having the skills needed to diagnose a problem in an electronic circuit or system, pinpoint its location, and repair it.

valence shell   The electron shell or energy level farthest from the atomic core that still controls the electron. If an electron in the valence shell absorbs sufficient energy, it can break out of the valence shell to become a free electron.

voltage   The name for units of electrical potential difference. Named in honor of Alessandro Volta, an Italian scientist who discovered an early form of producing electrical potential, called the Voltaic pile, a form of electric battery. The symbol used to express volts is E for voltage sources and V for voltage drops.

voltage divider formula   A convenient method of finding the voltage drop across one or more resistors in a series circuit if the resistor values and the applied voltage are known.

voltage drop   A voltage produced by current flowing through an opposition (impedance). See Chapter 12 for more on impedance.

voltage rise   A voltage produced by the conversion of energy due to magnetism, chemical action, light, heat, stress, or friction. See also voltage source.

voltage source   A source of electrical potential difference, created by one of six methods: electromagnetic, electrochemical, photoelectric, thermoelectric, piezoelectric, or triboelectric.

voltmeter   A measuring instrument for quantifying electric potential difference (voltage). Available in both analog (moving needle) and digital types. A voltmeter is connected in parallel (across) a device to be measured.

watt   The unit of real power in electric circuits. A power level of 1 watt represents the *expenditure* of 1 joule of energy per second. A kilowatt-hour is equivalent to 3.6 million joules.

wirewound   A resistor or resistive element made by winding wire onto a ceramic form.

with respect to ground   A means of specifying a voltage at a point in a circuit. The ground reference point must be known. An example of this designation method is V_A with respect to ground.

Appendix
C Electronics Formulas

The following are the formulas presented in the textbook.

Chapter 3

$$P_{Out} \div P_{In} \times 100 = \% \text{ Efficiency} \tag{3.1}$$

$$I = E \div R \tag{3.2}$$

$$E = IR \tag{3.3}$$

$$R = E \div I \tag{3.4}$$

$$P = IE \tag{3.5}$$

$$P = I^2 R \tag{3.6}$$

$$P = E^2 \div R \tag{3.7}$$

Chapter 5

$$R_{eq} = (R_2 R_{Meter}) \div (R_2 + R_{Meter}) \tag{5.1}$$

$$V_{Meter} = (R_{Eq} \div R_{Total}) \times E \tag{5.2}$$

Chapter 7

$$R_T = (R_1 + R_2) \tag{7.1}$$

$$I = E \div R_T \tag{7.2}$$

$$R_T = R_1 + R_2 + \ldots R_N \tag{7.3}$$

$$V_{R_1} = IR_1 \tag{7.4}$$

$$R_T = R_1 + R_2 \tag{7.5}$$

$$P = I \times E \tag{7.6}$$

$$P_T = P_{R_1} + P_{R_2} \tag{7.7}$$

$$P_T = IE \tag{7.8}$$

$$R_T = R_1 + R_2 + R_3 \tag{7.9}$$

$$I = E \div R_T \tag{7.10}$$

$$V_{R_1} = (R_1 \div R_T) \times E \tag{7.11}$$

$$R_{Drop} = (V_{Source} - V_{Load}) \div I_{Load} \tag{7.12}$$

Chapter 8

$$R_{Eq} = (R_1 R_2) \div (R_1 + R_2) \tag{8.1}$$

$$R_T = 1 \div [(1 \div R_1) + (1 \div R_2) + (1 \div R_3 +) \ldots (1 \div R_N)] \tag{8.2}$$

$$R_T = (Value\ of\ one\ resistor) \div (\#\ of\ resistors) \tag{8.3}$$

Chapter 9

$$R_T = R_1 + R_{2,3} \tag{9.1}$$

$$R_T = R_1 + R_{2,3,4} \tag{9.2}$$

$$V_{R_1} = [(R_1) \div (R_1 + R_{2,3,4})] \times E \tag{9.3}$$

$$P = V^2 \div R \tag{9.4}$$

$$I_{R_2} = [(R_3) \div (R_2 + R_3)] \times I_T \tag{9.5}$$

$$P_{R_2} + I_{R_2} = V_{R_2} \tag{9.6}$$

$$P_{R_1} = I_T R_1 \tag{9.7}$$

$$P_{R_2} = (V_{R_2})^2 \div R_2 \tag{9.8}$$

$$R_T = I_T E \tag{9.9}$$

Chapter 11

$$t = L \div R \tag{11.1}$$

$$L_T = L_1 + L_2 \tag{11.2}$$

$$L_T = L_1 + L_2 + 2(L_M) \tag{11.3}$$

$$L_T = L_1 + L_2 - 2(L_M) \tag{11.4}$$

$$C = kA \div t \tag{11.5}$$

$$t = RC \tag{11.6}$$

$$TC = R_2 C \tag{11.7}$$

$$C_T = (C_1 C_2) \div (C_1 + C_2) \tag{11.8}$$

$$C_T = C_1 + C_2 \tag{11.9}$$

Chapter 12

$$t = 1 \div f \tag{12.1}$$

$$f = 1 \div t \tag{12.2}$$

$$\lambda \cong C \div f \tag{12.3}$$

$$X_L = 2\pi f L \tag{12.4}$$

$$X_C = 1 \div (2\pi f C) \tag{12.5}$$

$$Z = \sqrt{(X_L - X_C)^2 + R^2} \tag{12.6}$$

$$V = I X_L \tag{12.7}$$

$$V = I X_C \tag{12.8}$$

$$V = I Z \tag{12.9}$$

$$I = V \div X_L \tag{12.10}$$

$$I = V \div X_C \tag{12.11}$$

$$I = V \div Z \tag{12.12}$$

$$P = I^2 X_L \tag{12.13}$$

$$P = I^2 X_C \tag{12.14}$$

$$P = I^2 Z \tag{12.15}$$

$$f_R = 1 \div (2\pi \sqrt{LC}) \tag{12.16}$$

$$Q = X_L \div R_W \tag{12.17}$$

$$f_C = R \div (2\pi L) \tag{12.18}$$

$$f_C = 1 \div (2\pi RC) \tag{12.19}$$

$$f_C = 1 \div (2\pi \sqrt{LC}) \tag{12.20}$$

Chapter 13

$$TR = N_S \div N_P \tag{13.1}$$

$$Z_R = (N_S \div N_P)^2 R_{Load} \tag{13.2}$$

Chapter 14

$$V_{Drop} = I_{Load} R_{Wire} \tag{14.1}$$

$$P_{Diss} = (V_{In} - V_{Out}) \div I_{Load} \tag{14.2}$$

Chapter 15

$$\beta = I_C \div I_B \qquad\qquad\qquad\qquad\qquad\qquad (15.1)$$

$$A_v = V_{Out} \div V_{In} \qquad\qquad\qquad\qquad\qquad (15.2)$$

Schematic Symbols, Copper Wire Table, and Capacitor Color and Markings

Resistor-Capacitor Color Codes

Color	Significant Figure	Decimal Multiplier	Tolerance (%)	Voltage Rating*
Black	0	1	-	-
Brown	1	10	1*	100
Red	2	100	2*	200
Orange	3	1,000	3*	300
Yellow	4	10,000	4*	400
Green	5	100,000	5*	500
Blue	6	1,000,000	6*	600
Violet	7	10,000,000	7*	700
Gray	8	100,000,000	8*	800
White	9	1,000,000,000	9*	900
Gold	-	0.1	5	1000
Silver	-	0.01	10	2000
No color	-	-	20	500

*Applies to capacitors only

FIGURE D.1 Resistor-capacitor color codes. Reprinted with permission from *The 2001 ARRL Handbook for Radio Amateurs.*

Copper Wire Specifications

Bare and Enamel-Coated Wire

Wire Size (AWG)	Diam (Mils)	Area (CM[1])	Enamel Wire Coating Turns / Linear inch[2]			Feet per Pound Bare	Ohms per 1000 ft 25∞ C	Current Carrying Capacity Continuous Duty[3] at 700 CM per Amp[4]			Nearest British SWG No.
			Single	Heavy	Triple				Open air	Conduit or bundles	
1	289.3	83694.49				3.948	0.1239	119.564			1
2	257.6	66357.76				4.978	0.1563	94.797			2
3	229.4	52624.36				6.277	0.1971	75.178			4
4	204.3	41738.49				7.918	0.2485	59.626			5
5	181.9	33087.61				9.98	0.3134	47.268			6
6	162.0	26244.00				12.59	0.3952	37.491			7
7	144.3	20822.49				15.87	0.4981	29.746			8
8	128.5	16512.25				20.01	0.6281	23.589			9
9	114.4	13087.36				25.24	0.7925	18.696			11
10	101.9	10383.61				31.82	0.9987	14.834			12
11	90.7	8226.49				40.16	1.2610	11.752			13
12	80.8	6528.64				50.61	1.5880	9.327			13
13	72.0	5184.00				63.73	2.0010	7.406			15
14	64.1	4108.81	15.2	14.8	14.5	80.39	2.5240	5.870	32	17	15
15	57.1	3260.41	17.0	16.6	16.2	101.32	3.1810	4.658			16
16	50.8	2580.64	19.1	18.6	18.1	128	4.0180	3.687	22	13	17
17	45.3	2052.09	21.4	20.7	20.2	161	5.0540	2.932			18
18	40.3	1624.09	23.9	23.2	22.5	203.5	6.3860	2.320	16	10	19
19	35.9	1288.81	26.8	25.9	25.1	256.4	8.0460	1.841			20
20	32.0	1024.00	29.9	28.9	27.9	322.7	10.1280	1.463	11	7.5	21
21	28.5	812.25	33.6	32.4	31.3	406.7	12.7700	1.160			22
22	25.3	640.09	37.6	36.2	34.7	516.3	16.2000	0.914		5	22
23	22.6	510.76	42.0	40.3	38.6	646.8	20.3000	0.730			24
24	20.1	404.01	46.9	45.0	42.9	817.7	25.6700	0.577			24
25	17.9	320.41	52.6	50.3	47.8	1031	32.3700	0.458			26
26	15.9	252.81	58.8	56.2	53.2	1307	41.0200	0.361			27
27	14.2	201.64	65.8	62.5	59.2	1639	51.4440	0.288			28
28	12.6	158.76	73.5	69.4	65.8	2081	65.3100	0.227			29
29	11.3	127.69	82.0	76.9	72.5	2587	81.2100	0.182			31
30	10.0	100.00	91.7	86.2	80.6	3306	103.7100	0.143			33
31	8.9	79.21	103.1	95.2		4170	130.9000	0.113			34
32	8.0	64.00	113.6	105.3		5163	162.0000	0.091			35
33	7.1	50.41	128.2	117.6		6553	205.7000	0.072			36
34	6.3	39.69	142.9	133.3		8326	261.3000	0.057			37
35	5.6	31.36	161.3	149.3		10537	330.7000	0.045			38
36	5.0	25.00	178.6	166.7		13212	414.8000	0.036			39
37	4.5	20.25	200.0	181.8		16319	512.1000	0.029			40
38	4.0	16.00	222.2	204.1		20644	648.2000	0.023			
39	3.5	12.25	256.4	232.6		26969	846.6000	0.018			
40	3.1	9.61	285.7	263.2		34364	1079.2000	0.014			
41	2.8	7.84	322.6	294.1		42123	1323.0000	0.011			
42	2.5	6.25	357.1	333.3		52854	1659.0000	0.009			
43	2.2	4.84	400.0	370.4		68259	2143.0000	0.007			
44	2.0	4.00	454.5	400.0		82645	2593.0000	0.006			
45	1.8	3.10	526.3	465.1		106600	3348.0000	0.004			
46	1.6	2.46	588.2	512.8		134000	4207.0000	0.004			

Teflon Coated, Stranded Wire

(As supplied by Belden Wire and Cable)

Size	Strands[5]	Turns per Linear inch[2] UL Style No.		
		1180	1213	1371
16	19×29	11.2		
18	19×30	12.7		
20	7×28	14.7	17.2	
20	19×32	14.7	17.2	
22	19×34	16.7	20.0	23.8
22	7×30	16.7	20.0	23.8
24	19×36	18.5	22.7	27.8
24	7×32		22.7	27.8
26	7×34		25.6	32.3
28	7×36		28.6	37.0
30	7×38		31.3	41.7
32	7×40			47.6

Notes

[1]A circular mil (CM) is a unit of area equal to that of a one-mil-diameter circle (π/4 square mils). The CM area of a wire is the square of the mil diameter.

[2]Figures given are approximate only; insulation thickness varies with manufacturer.

[3]Maximum wire temperature of 212°F (100°C) with a maximum ambient temperature of 13°F (57°C) as specified by the manufacturer. The *National Electrical Code* or local building codes may differ.

[4]700 CM per ampere is a satisfactory design figure for small transformers, but values from 500 to 1000 CM are commonly used. The *National Electrical Code* or local building codes may differ.

[5]Stranded wire construction is given as "count"×"strand size" (AWG).

FIGURE D.2   Copper wire specifications. Reprinted with permission from *The 2001 ARRL Handbook for Radio Amateurs.*

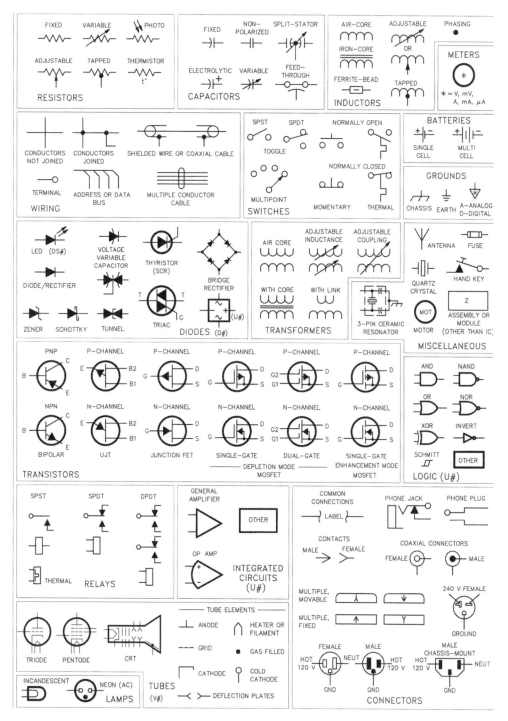

FIGURE D.3 Electronic schematic symbols. Reprinted with permission from *The 2001 ARRL Handbook for Radio Amateurs.*

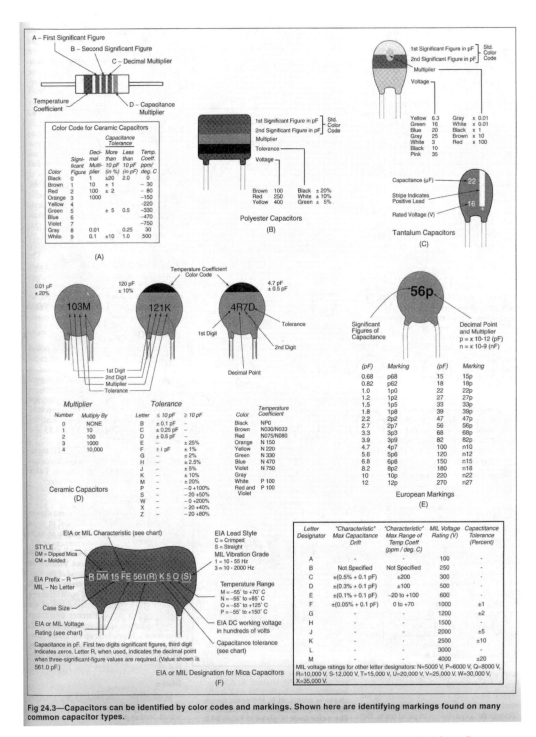

Fig 24.3—Capacitors can be identified by color codes and markings. Shown here are identifying markings found on many common capacitor types.

FIGURE D.4 Capacitor markings. Reprinted with permission from *The 2001 ARRL Handbook for Radio Amateurs.*

Appendix

E

Interpreting a PC's Motherboard Manual

E.1   INTRODUCTION

Probably the single most important document (aside from the warranty card) supplied with a PC is the motherboard manual. Always included when you purchase a separate motherboard, and only occasionally included with a new PC system, the motherboard manual provides the most important specs about the PC.

E.2   SECTIONS

Most modern motherboard manuals include the following sections:

1. Title page
2. Copyright notice
3. Table of contents
4. Introductory section
5. Hardware setup section
6. BIOS setup section
7. Companion driver disk
8. BIOS updating
9. Hardware monitoring
10. Troubleshooting section
11. Technical support section

E.2.1   Title Page

This page lists the manufacturer, model, and submodel. Usually contains a printing date which can be useful for pinpointing changes in documentation. Knowing exactly who made your motherboard and the specific model and possibly submodel is essential when you want to download the latest driver programs or flash the BIOS.

E.2.2 Copyright Notice

This tells you not to reproduce the manual without express written permission from the manufacturer.

E.2.3 Table of Contents

This is the first place to look, after the title page, to find what you're looking for.

E.2.4 Introductory Section

General specs, such as the processor(s) supported, the type(s) and amount of memory supported, general features, such as the chipset, RAID, included I/O ports, LAN, modem, sound, video, front side bus speed, etc.

E.2.5 Hardware Setup Section

This section includes detailed instructions on installing the CPU onto the motherboard, installing the heat sink/fan onto the CPU, installing RAM, and installing the motherboard into the case. Instructions are provided for connecting the various cables to the motherboard, such as to the hard drive(s), floppy drive, CD ROM drive, DVD drive, front panel switches, and indicators. This section usually includes several pictorial diagrams indicating all major features of the motherboard, including all connectors and jumpers. It usually also includes charts depicting the proper jumper settings. In general, the newer motherboards don't have nearly as many jumpers to set as the older motherboards, but there usually are still a few to set.

E.2.6 BIOS Setup Section

This section lists all the menus you will see when entering the configuration setup (BIOS) screen. There are usually around 8–12 such screens, with each screen having one or more configurable items. A general list of such screens follows:

1. **Standard Features.** This screen shows the date and time, amount of memory installed, all the drives (except for SCSI drives) and the hard drive parameters such as number of cylinders (CYL), heads (H) and sectors per track (S/T). The combination of these three major hard drive numbers is often abbreviated as CHS. SCSI hard drives are handled by the SCSI host adapter, which contains its own BIOS.

   One of the first things the author does when using this screen is to set the year, month, day, and time. One should also learn to recognize all the settings in this first screen, since if the hard drive settings are off, it can cause the PC to either fail to boot, or can cause data loss. Get used to recording the data listed on this and all the other configuration screens, as when (not if but when) the backup battery on the motherboard fails, you will have to supply all the settings to get the PC back up and running again.

2. **Advanced BIOS Features.** This screen lists much more detailed options than the first screen. One of the things listed here is the boot order. Which devices are checked, and in what order, when the PC attempts to boot up can be set in this screen, as well as a host of other options. Something of interest here is if you are running RAID. Both SCSI and ATA (EIDE) RAID are considered to be SCSI as far as the BIOS is concerned. You need to set the boot order to SCSI/RAID for either of these RAID types or your system will not boot.

As an example of the types of specific settings available in this section condsider the following:

- Quick Boot (On/Off).
- A list of possible boot devices and their order, which can be modified.
- Try Other Boot Devices (Yes/No).
- Boot Up Numlock (Enabled/Disabled).
- Floppy Drive Swap (Enabled/Disabled).
- Floppy Drive Seek (Enabled/Disabled).
- Password Check (Enabled/Disabled).
- Boot To OS/2 >64 MB (Yes/No).
- L1 Cache (Enabled/Disabled).
- L2 Cache (Enabled/Disabled).
- System BIOS Cacheable (Enabled/Disabled).
- Graphic Win Size (various sizes to choose from).
- DRAM Timing Configuration (Normal, Fast, Ultra, Safe). Do not set this yourself unless you really know what you are doing!
- SDR/DDR CAS Latency (SPD, 2T, 3T).
- SDR/DDR RAS Active Time (6T, 7T).
- SDR/DDR RAS Precharge Time (3T, 4T).
- Autodetect DIMM/PCI Clk (Enable/Disable).
- Clk Generator Spread Spectrum (Enable/Disable).
- DOS Flat Mode (Enable/Disable).
- DRAM Drive Slew Rating (Normal, Fast).
- 32k I/O Compensation (Enable/Disable).
- Memory Termination (Enable/Disable).

3. **Advanced Chipset Features.**
   Another name for Advanced BIOS Features

4. **Integrated Peripherals.**
   Typical items listed here include:

   ■ Onboard FDC (Enable/Disable). Select to use the on-board floppy disk controller.
   ■ Onboard Serial Port A (3F8h/COM1 or other combination of hex address and port number settings).
   ■ Serial Port B (2F8h/COM2-like port A, above).
   ■ Serial Port 2 Mode (Normal, IrDA, ASKIR) can be set to use a infrared sensor.
   ■ Onboard Parallel Port (378h or adjust to other addresses).
   ■ Parallel Port Mode (SPP, EPP, ECP, EPP+ECP) Select based on your printer requirements. The author uses ECP for modern printers.
   ■ Parallel Port IRQ (7, 5) 7 is standard.
   ■ Parallel Port DMA (3, 2, 1, 0).
   ■ Onboard Game Port (201h, 209h, disable) Disable this if you don't use it, so you save resources.
   ■ Onboard MIDI Port (300h, 330h, disable) again, disable if not used.
   ■ MIDI Port IRQ (5, 10, 11).
   ■ Onboard PCI/IDE (Primary, Secondary, Both, Disable) Selects which IDE channels to use or not.
   ■ Onboard AC '97 Sound (Enable/Disable).
   ■ Onboard '97 Modem (Enable/Disable).
   ■ Onboard LAN (Enable/Disable).
   ■ USB Function Support (Enable/Disable).
   ■ USB Function for DOS (Enable/Disable).
   ■ Thumb Drive Support for DOS (Enable/Disable).

5. **Power Management.**
   Some of the typical settings available in this section include the following:

   ■ ACPI Aware O/S (Yes/No)
   ■ Power Management (Enable/Disable)
   ■ Suspend Timeout (Enable/Disable)
   ■ Hard Disk Timeout (Enable/Disable)
   ■ Power On by LAN/Ring (Enable/Disable)
   ■ RTC Alarm PowerOn (Enable/Disable)
   ■ RTC Alarm DATE (Everyday, (00–31)

- RTC Alarm Hour (00–24)
- RTC Alarm Minute (00–60)
- RTC Alarm Second (00–60)
- Keyboard PowerOn Function (Enable/Disable)

6. **PnP/PCI Configuration.** Also known as PCI/Plug-and-Play Setup. Typical settings in this section include:

- Plug-and-Play Aware O/S (Yes/No)
- AGP 4X Control (Enable/Disable)
- Primary Graphics Adapter (AGP, PCI)
- Allocate IRQ to PCI VGA (Yes/No)
- PCI IDE Busmaster (Enable/Disable)

7. **PC Health Status (Hardware Monitor).**
   This section invokes a BIOS-based hardware monitor program that reports the following typical readings:

- Vcore (processor core voltage, usually 2.5 V or less, actual reading follows)
- Vcc 2.5 V (actual reading follows)
- Vcc 3.3 V (reading follows)
- Vcc 5 V (reading follows)
- +12 V (reading follows)
- SB 3 V (standby voltage, reading follows)
- SB 5 V (standby voltage, reading follows)
- -12 V (reading follows)
- V Bat (motherboard backup battery, normally about 3.5 V, reading follows)
- System Fan Speed (in RPM, 3,600 is typical of the author's PC)
- CPU Fan Speed (3,600–5,000 RPM is typical range for the author's PC)
- System Temp (30 degs C/86 degs F typical for the author's PC)
- CPU Temp (40 degs C/104 degs F is typical of the author's PC)

8. **Load Fail-Safe Defaults.** These settings *should* work, but will not be optimal, meaning the PC will boot, but run very conservatively.

9. **Load Optimized Defaults.** These are "best guess" at what settings should work well.

10. **Set Password/Change Password.**

11. **Save and Exit Setup.** This is the option to select when you have made changes that you are sure you want to save, otherwise, select the next option.

12. **Exit Without Saving.** This is what to select when you were "just looking".

E.3 COMPANION DRIVER DISK

E.3.1 BIOS Updating

This section describes how to update the BIOS firmware (the software stored on the EEPROM chip). The motherboards older than about 1996 or so required the removal and replacement of new chips programmed with the new BIOS code. The new motherboards can be upgraded by reprogramming the BIOS chip without requiring chip removal. This is known as *flashing the BIOS*. There are two general techniques for flashing the BIOS on new motherboards.

E.3.2 Making a Boot Disk

The first, and more difficult of the two, involves downloading the new BIOS program from the motherboard or system manufacturer's Web site, then creating a bootable floppy disk, transferring the downloaded BIOS code to the boot floppy disk and running a command line program. The problems involved with this method of BIOS flashing are: having to create a boot floppy and working with the command line. If you make a typing error the motherboard will not function properly. If you inadvertently type the wrong command line program, it can render your PC unbootable.

E.3.3 Running From the Desktop

The second BIOS flashing method allows the user to download the new code from the Web site and simply double-click on the program's icon on the Web site. It does not require making a boot disk, going into command line mode, and is generally a lot safer. One of the things the author always does when shopping for a new motherboard is to ascertain early on which type of BIOS flash method is used. That way, he saves a great amount of potential frustration from a failed BIOS flash session.

E.3.4   **Always Make Backups!**

Both BIOS flash methods will ask you if you want to save a copy of the old BIOS. Your answer should always be *YES!* Sometimes the new BIOS version will not work at all, or will have some problem, requiring you to revert to the old BIOS version. If you fail to save it before the new BIOS flash, it is a real pain to have to try to locate it on the web, download it, and restore it. And of course that assumes that you have a second functioning computer with which to download it. Bottom line: always make backups.

E.4   HARDWARE MONITORING

Many of the newer motherboards contain a special chip that will monitor several important PC parameters. CPU temperature, inside case temperature, RPM of several fans, and all the power-supply voltage levels are typically monitored. Some hardware monitor programs provide for alarms to be sounded if preset limits are exceeded. This is a very valuable tool, and the author insists on buying only motherboards that feature hardware monitoring. One of the major chip makers supplying hardware monitoring chips for motherboards is Winbond®. For more information go to: *http://www.winbond.com/e-winbondhtm/index.asp.*

E.5   TWO TYPES OF MONITORING PROGRAM

Just as in the case of BIOS flashing, there are two general versions of how to invoke a hardware monitoring program. The more difficult of the two requires going into the CMOS configuration setup screen and navigating to the Hardware Monitoring screen. This is a much more cumbersome method than the second style program.

The easier way to do it is to have a hardware monitoring program that will run under the operating system, such as Windows™. Then all you have to do is double-click on the monitor program's icon on the desktop to invoke it. This is a much easier method, and again, is something the author insists on when shopping for a new motherboard. Take the time when shopping for a new motherboard to investigate fully, for later you will be very pleased that you did.

E.6   TROUBLESHOOTING SECTION

This is usually very general information and not all that helpful. Advice to make sure the PC is plugged into a working outlet is pretty useless, but details on how to short a specific jumper when you need to wipe the configuration settings and start over are worth knowing. There is also usually an admonition to make sure you flash the BIOS with the proper BIOS version or you may not ever be able to boot the PC again. Don't ever stop the flash process in the middle either or you will most likely end up with an unbootable and unrepairable motherboard. The author actually knows someone who did this, and sure enough, his motherboard was toast.

E.6.1   Technical Support Section

Only somewhat useful, as you really don't want to spend hours trying to reach someone on the phone who may or may not speak English well, and whom may or may not be able or willing to help you troubleshoot your PC. Learn to avoid "phone trees" and service calls.

Appendix

F Exercise Answers

Chapter 1

1. a
3. e
5. e
7. d

9. a
11. d
13. a
15. c

Chapter 2

1. c
3. a
5. a
7. a

9. b
11. b
13. b
15. a

Chapter 3

1a. $3.45x10^{-3}$
1b. $1.45x10^{6}$
1c. $2.35x10^{4}$
1d. $8.0x10^{9}$
3a. 176 milli
3b. 793 milli
3c. 5.88
3d. 28.9 kilo
5. 166W/290W=0.57 × 100 = 57% efficiency

7. 145.390 kilo
9. 7.2×10^{4}
11. 121 Ω
13. $V=IR$
15. $R=V/I$
17. $I=9V/3.3\ k\Omega=2.73\ mA$
19. $R=5V/23\mu A=217\ k\Omega$
21. $V=3.4A*3.63\Omega=12.3V$
23. $R=(120V)^{2}\div100W=144\Omega$
25. $P_{In}=240V*15A=3600W$

Chapter 4

1. The cells are the source, the control is the on/off switch and the load is the light bulb.
3. CD Player>Amplifier>Speaker System

5. You must observe proper ESD precautions.

7. Excess current, resulting in heat and damage to components

9. Surface-mount components require special tools for removal. If the component is open, it will measure a greater than normal voltage across its terminals. The exceptions are small, non-electrolytic capacitors.

11. Ground-fault circuit interrupters (GFCIs).

13. Try to obtain proper documentation for the equipment to be serviced.

15. A positive center electrode, negative can electrode and an electrolytic paste between them.

17. Primary cell.

19. Use circuit cooling spray to cool the circuit and try to elicit the problem.

21. The heat source, the circulating fluid to transfer the heat, a turbine which the heated fluid usually steam turns, and the generator turned by the turbine.

23.

Closed (make, on)   Open (break, off)

25.   SW

Chapter 5

1. d	11. b
3. a	13. c
5. b	15. a
7. a	17. b
9. e	19. d

Chapter 6

1. Carbon composition type

3. Sliding-contact variable resistor

5. Potentiometer

7. Resistor color code

9. Change of resistance value over time, or with moisture or heat

11. Limiting current and developing a voltage drop

Chapter 7

1. $I = 100\ V \div 90\Omega = 1.11\ A$

3. $V_{R_2} = R_2 \div (R_1 + R_2)xE = 60\Omega \div (30\Omega + 60\Omega)x100V = 2 \div 3x100V = 66.7\,V$

5. $P_{R_1} = I^2 R_1 = (1.11\,A)^2 R_1 = (1.11\,A)^2\,30\Omega = 37\,W$

7. $V_{R1=}(R_1 \div R_T)V_A \qquad V_{R1}(1k\Omega \div 4.3k\Omega)V_A = (0.233)35V = 8.14\,V$

9. $V_A = 200V$

11. $V_C = (R_3 + R_4) \div R_T x E = (40\Omega + 30\Omega) \div 100\Omega x 200V = 70\Omega \div 100\Omega x 200V = 140$

Chapter 8

1. $R_T = (R_1 R_2) \div (R_1 + R_2) = (10k\Omega 22k\Omega) \div (10k\Omega + 22k\Omega) = 6.88k\Omega$

3. $I_{R_1} = E \div R_1 = 48V \div 10k\Omega = 4.8mA$

5. $P_{R_1} = E^2 \div R_1 = 48V^2 \div 10k\Omega = 230mW$

7. $P_T = E^2 \div R_T = 48V^2 \div 6.88k\Omega = 334mA$

9. $I_{R_2} = E \div R_2 = 48V \div 22k\Omega = 2.18mA$

11. First find each resistance.

$$R_{Sound} = E \div I_{Sound} = 5V \div 2A = 2.5\Omega$$

$$R_{Video} = E \div I_{Video} = 5V \div 5A = 1\Omega$$

Then use the product-over-the-sum resistance formula.

$$R_T = (R_{Sound} R_{Video}) \div (R_{Sound} + R_{Video}) = (2.5\Omega x 1\Omega) \div 2.5\Omega + 1\Omega) = 714m\Omega$$

13. Yes, since the power dissipated by each resistor does not exceed the specified resistor's power rating.

15. A parallel circuit

17. a

19. The video card looks like: $5V \div 4A = 1.25\ \Omega$
The sound card looks like: $5V \div 3A = 1.67\ \Omega$
The Ethernet card looks like: $5V \div 1.5A = 3.33\ \Omega$

21. 42.5 W (by adding up the power from each individual card)

23. To shunt to ground any noise present on the supply line before it could reach the IC

Chapter 9

1. First find the parallel resistance value of R_2, $R_3 = (50*100) \div (50 + 100) = 5000 \div 150 = 33.3\ \Omega$ then add the value of $R_1 = 20\ \Omega$ and get $20\ \Omega + 33.3\ \Omega = 53.3\Omega$

3. Their combined resistance is 33.3 ohms so $V_{R_{2,3}} = IR_{2,3} = 2A*33.3\Omega = 66.6V$

5. $I_{R_2} = V_{R_{2,3}} = 66.6V \div 50\Omega = 1.33A$

7. $P = IE = 2A*106.6V = 213.2W$

9. $P = IE = 1.33A*66.6V = 88.6W$

11.

Component(s)	R	I	V	P
R1	20 Ω	2 A	40 V	80 W
R2	500 Ω	1.33 A	66.7 V	88.4 W
R3	100 Ω	667 A	66.7 V	44.5 W
RT	53.3 Ω	2 A	106.6 V	213.2 W

13. The higher the resistor current, the higher the wattage dissipated.

Chapter 10

1. a
3. a
5. Motors, generators, electromagnetic relays
7. As a magnetic compass for direction finding
9. Like magnetic poles repel each other and unlike magnetic poles attract each other.
11. Hans Christian Oersted
13. The switch consists of two parts: a small, magnetically operated switch and a small magnet. The reed switch is mounted on the stationary part of the door or window. The magnet is mounted on the moving part of the closed door or window, immediately next to the switch. The magnetic field from the magnet penetrates the reed switch, inducing each part of the switch to act as a magnet. The two parts are drawn together due to the induced magnetic force, and the switch is closed. The closed switch completes an electric circuit to a sensor in the burglar alarm. If the door or window is opened, the magnet is moved away from the switch, removing the magnetic field from the switch, causing it to open, which trips the sensor thus sounding an alarm.
15. a
17. a
19. c

Chapter 11

1. Inductance
3. Its frequency
5. Eddy currents
7. Self-induction

9. Back voltage or CEMF
11. Time-constant
13. Magnetically saturated
15. To filter the current pulses in a power supply section and to reduce the electromagnetic radiation from cables attached to the PC.
17. b
19. Electrostatic field
21. Five
23. 5 ms (five time-constants)
25. Leakage current
27. Series voltage-dropping resistors connected in parallel with the capacitors
29. To eliminate lead inductance

Chapter 12

1. a
3. b
5. $t = 1 \div f = 1 \div 200\ Hz = 5ms$
7. $f = 1 \div t = 1 \div 50\ ms = 20Hz$
9. a
11. a
13. $X_L = 2\pi\,fL = 2\pi(2\ MHz)(35\ mH) = 440\ k\Omega$
15. $X_C = 1 \div 2\pi\,fC = 1 \div 2\pi 50\ MHz10pF) = 318\ k\Omega$
17. $f_R = 1 \div 2\pi\sqrt{LC} = 1 \div 2\pi\sqrt{1H1000\,pF} = 5k\Omega$
19. filtering the AC ripple form the output of a DC power supply
21. A band pass filter is what a radio or TV tuner is.
23. An example of a resonant circuit is a radio or TV tuner as well as the IF stage(s) in radios and TVs.

Chapter 13

1. e
3. c
5. a
7. a
9. a
11. e
13. e
15. e
17. a
19. c
21. 100 mA

23. 4.8:1 (primary to secondary) or 1:4.8 in the conventional way of describing turns ratios.
25. You risk shock as the CEMF can be very high during a connection make or break.

Chapter 14

1. d
3. e
5. b
7. b
9. a
11. a
13. c
15. d
17. c
19. e

Chapter 15

1. a
3. b
5. b
7. a
9. e
11. b
13. e
15. c

Chapter 16

1. COM port connector
3. LPT1
5. DB9, EGA, DB9 VGA, DVI XGA
7. CONFIG.SYS and AUTOEXEC.BAT
9. An NIC for each PC and a crossover cable.
11. Increased data throughput.
13. 480 Mb/s
15. by blocking, reflecting or absorbing the signals.
17. They are used on each end of cables connected to the PC to help attenuate radio frequency signals on the outside (shield) conductor.
19. The Federal Communications Commission (FCC)

Chapter 17

1. Because it is relatively easy to transform voltages up or down and lower transmission line losses
3. 240 VAC and 120 VAC
5. Most fuses are useless after blowing, while circuit breakers can be reset.
7. Because motors can generate a voltage drop when they start, and they can also generate noise, both of which can affect a PC
9. An EMI filter, which reduces electromagnetic interference (EMI)
11. To measure the limits of the AC line over which the PC will still operate properly, by monitoring its power supply output voltages
13. A UPS always runs off an internal battery. An SPS runs off the internal battery only once the commercial power fails.
15. VoltAmps (VA) or watts, and runtime.
17. A power pole-mounted power transformer
19. The ratio of real to apparent power in a circuit.

Chapter 18

1. Keep fans unblocked and clean, install extra case fans when upgrading the PC, and be sure to have one fan bringing fresh air in for every fan pushing hot air out of the case.
3. Keep boards and memory sticks inside the ESD bags until ready to install. Wear an ESD strap connected to an unplugged PC when working on that PC. Do not move or walk around while carrying unbagged boards or chips.
5.
7. It is a closed system consisting of easily vaporized liquid that boils off and takes heat from a processor to a radiator, then condenses and returns to the hot spot.
9. How long will you keep the PC? What will the PC be used for? It is cost-effective to upgrade the PC versus buying a new one?
11. Wireless is usually considered to be easier to set up, but often doesn't transfer data as fast as a wired system.
13. 480 MB/s
15. 3–5 years
17. SD. CF, and Memory Stick

19. Voltages, fan speed, temperature of the processor, and temperature in the inside of the case.

21. Which processor and speed, the amount and type of RAM, the number and capacity of hard drives, number and type of I/O connectors. What features are built-in (video, LAN, audio, modem, etc.). It is a name brand? What is the warranty?

23.

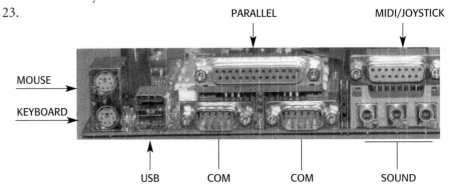

Appendix
G
About the CD-ROM

The CD-ROM included with *Essential Electronics for PC Technicians* includes all of the photos from the book as well as many drawings and tables from the textbook chapters as part of the Power Point files.

CD-ROM FOLDERS

Power Point slide presentations: Organized by chapter these slides provide an interactive learning tool for students and teachers

Figures: Contains all of the full-color photos from the book as well as the schematic diagrams and drawings, organized by chapter

OVERALL SYSTEM REQUIREMENTS

- *Windows 98SE, Me, NT, Windows 2000,* or *Windows XP Pro*
- Suggested: Windows 2000 or XP, with 256 MB RAM minimum, VGA or better video card or chipset.
- Pentium II Processor or greater
- CD-ROM drive
- Hard drive
- 128 MB of RAM, minimum; 256 recommended
- Microsoft PowerPoint 97 or higher to view PowerPoint slide presentations

Index